OCR

Address:
PONTIFICAL ACADEMY OF SCIENCES
CASINA PIO IV, 00120 VATICAN CITY

PONTIFICIAE ACADEMIAE SCIENTIARVM SCRIPTA VARIA

89

THE EMERGENCE OF COMPLEXITY IN MATHEMATICS, PHYSICS, CHEMISTRY AND BIOLOGY

PROCEEDINGS

Plenary Session of the Pontifical Academy of Sciences
27-31 October 1992

Edited by
† BERNARD PULLMAN

EX AEDIBVS ACADEMICIS IN CIVITATE VATICANA

MCMXCIV

The opinions expressed with absolute freedom during the presentation of the papers and in the subsequent discussions by the participants in the Plenary Session, although published by the Academy, represent only the points of view of the participants and not those of the Academy.

Editorial committee for the preparation of the Proceedings:
Michael Sela
Renato Dardozzi
Giuseppe Del Re

DISTRIBUTED BY PRINCETON UNIVERSITY PRESS
Princeton, New Jersey
ISBN 0 - 691 - 01238 - 5

ISBN 88 - 7761 - 055 - 7

CONTENTS

DAY FIVE - 31 OCTOBER 1992

PREFACE

"Le monde est fait pour aboutir à un livre"
Stéphane Mallarmé

La vingt-huitième réunion plénière de l'Académie Pontificale des Sciences, qui s'est tenue à la Casina Pio IV, au Vatican, du 27 au 30 Octobre 1992, a été marquée par deux événements importants dont ce Volume perpétue la mémoire.

D'abord, comme il est de tradition dans ces réunions, l'Académie a consacré une grande partie de son activité à une discussion scientifique. Le thème offert à sa réflexion par le Conseil sur proposition de son Président, le Professeur Marini-Bettolo (malheureusement absent de la réunion pour cause de maladie): "L'émergence de la Complexité en Mathématique Physique, Chimie et Biologie", a été particulièrement bien choisi à cause de son importance dans le contexte scientifique actuel, de son aspect pluridisciplinaire et aussi en raison de sa portée philosophique et des résonances religieuses que certaines de ses ouvertures ne manquent pas de susciter chez certains. Il répond par son envergure à la vocation de l'Académie Pontificale qui est de "promouvoir les progrès des sciences et l'étude des problèmes épistémologiques s'y référant ... et pouvant contribuer à l'approfondissement des questions morales, sociales et spirituelles" (articles 2 et 3 des Statuts).

Rien n'illustre mieux la richesse des débats dont le thème choisi a été l'objet que la nature des mots-clés qui apparaissent constamment sous la plume des participants et dont chacun pourrait constituer, seul ou couplé avec un terme complémentaire ou opposé judicieusement choisi, le thème d'une réunion séparée. Ordre et désordre, chaos et organisation, diversité et classification, hasard et nécessité, origine (du monde, de la vie, de la mort) et évolution, entropie et anthropie, le plein et le vide (le plein du vide et le vide du plein), la téléonomie et la téléologie, le réductionnisme et le holisme, sont quelques-uns des concepts qui constituent les multiples facettes du problème de l'émergence de la complexité, autour desquelles se sont organisées des discussions hardies et animées. Les familiers de l'histoire et de la philosophie des sciences reconnaîtront dans le choix de ces concepts l'écho des angoisses spirituelles qui accompagnent la réflexion des hommes sur la nature et le sens de l'Univers depuis la plus lointaine Antiquité.

Qui dit émergence de complexité présuppose un instant de départ où celle-ci fut absente. Il est aujourd'hui quasiment admis que tel était la situation aux origines du monde — au Big Bang, à la naissance de l'espace, de la matière, de l'énergie et du temps — et que l'univers a commencé dans l'état le plus simple d'équilibre thermodynamique. C'est en particulier la thèse défendue, entre autres, par notre éminent confrère Stephen Hawking. La complexification du monde, l'apparition et le développement des structures diverses et composées et des forces multiples est une caractéristique de son évolution et de son ... refroidissement. L'amplification de la complexité paraît avoir été progressive — confirmant ainsi la réalité d'une "flèche de temps" cosmologique —, les êtres vivants doués de conscience représentant son apogée momentanée.

Les mémoires contenus dans ce Volume traitent en grande partie de la matérialité de cette évolution dont ils soumettent le(s) mécanisme(s) à une analyse scientifique et philosophique. En effet, contrairement à ce que croyait Diderot lorsqu'il écrivait que *l'esprit doit être plus étonné de la durée hypothétique du chaos que de la naissance de l'Univers*", ce schéma évolutionnaire, qui implique la génération de l'ordre à partir du désordre, se heurte, à première vue, au 2e principe de la thermodynamique qui nous enseigne juste le contraire, a savoir que l'univers doit évoluer toujours dans le sens du désordre croissant. Il paraît établi aujourd'hui que ce conflit n'est qu'apparent et que cette loi ne s'oppose pas à la création de l'ordre (organisation des structures et leur complexification) en certain endroits de l'Univers à condition que se produise en d'autres lieux un désordre compensatoire plus grand, de sorte que le *bilan net* soit un désordre croissant. L'expansion de l'Univers et la création continue du désordre par les étoiles qui convertissent leurs atomes (d'hydrogène) en énergie (lumière et chaleur) garantissent sinon la pérennité du moins la longévité de cette compensation. De fameux développements récents auxquels sont attachés les noms d'Ilya Prigogine et de notre confrère Manfred Eigen ont montré que les systèmes physiques déportés loin de l'équilibre thermodynamique deviennent instables et assument spontanément, par l'établissement de corrélations de longue portée, des structures organisées ("structures dissipatives"). Ces phénomènes s'observent aujourd'hui facilement dans le monde inanimé. De là à imaginer qu'ils ont pu se produire à une échelle plus complexe et être à l'origine de la formation des structures primitives pouvant conduire éventuellement à l'apparition des organismes vivants, il n'y a qu'un pas, grand il est vrai, que certains n'hésitent toutefois pas à franchir. Paul Valéry, lui, exprime cette réussite de la matière par un des aphorismes, quelque peu paradoxaux, dont il a le secret: *"La Vie est un désordre qui fonctionne"*.

La structuration et la complexification ainsi permises par les lois de la physique, une question cruciale qui se pose est de savoir s'il est possible d'expliciter, dans le même cadre, les caractéristiques des "propriétés émergentes", propriétés nouvelles qui apparaissent lors de la structuration d'un ensemble plus ou moins complexe à partir de constituants plus simples. C'est le vieux problème du tout qui est plus que la somme des parties, de l'information englobée dans le tout dépassant la somme des informations contenues dans les parties. Dans la terminologie moderne il met en jeu les approches dites holiste et réductionniste de la vision des choses.

Le problème apparaît naturellement déjà au niveau de l'émergence des complexités les plus élémentaires. Toutefois, dans le monde matériel auquel nous avons aujourd'hui affaire, tout au moins en chimie, physique et biologie, il se pose d'une façon particulièrement évidente au niveau de la combinaison de ces briques fondamentales, sinon élémentales, dont sont construits tous les objets de l'Univers, à savoir au niveau de l'association des atomes(*). Rappelons que l'objection principale contre la doctrine démocritienne, qui affirmait que tous les objets étaient formés par "l'agglomération", "l'agglutination" des atomes indivisibles et impénétrables, était que cette représentation ne permettait pas d'expliquer et encore moins de prévoir l'émergence et la nature des propriétés nouvelles dont étaient doués les objets composés (nous dirions aujourd'hui, en premier lieu, les molécules). L'objection, formulée déjà par Aristote et reprise au cours des siècles, en fait des millénaires, par tous les antiatomistes de l'histoire — et Dieu sait s'il en eut et ceci jusqu'au début du siècle actuel —, n'a été mise en défaut que durant les dernières décennies, en fait depuis l'avènement de la

_____

(*) Que l'on ne s'imagine pas pour autant que le deuxième "principe" constitutif de l'Univers selon Démocrite — à côté des atomes — à savoir le vide est lui le prototype de simplicité. Les théories actuelles nous montrent un vide aussi complexe, sinon plus, que le plein, comme l'ont souligné notre Président Nicola Cabbibo et Carlo Rubia.

mécanique quantique dont les méthodes permettent de comprendre le mécanisme de la formation des associations d'atomes (atomes "modernes", il est vrai, qui ne sont plus aussi simples que les atomes "antiques") et de prévoir la nature des propriétés émergentes. A ce point de vue, l'étude quantique de la molècule d'hydrogène, en 1924, par Heitler et London représente une étape cruciale dans le traitement explicite des propriétés émergentes des structures composées (à distinguer de l'émergence des propriétes statistiques, résultant d'une action combinée d'une grande quantité d'entités indèpendantes, telles les propriétés thermodynamiques, dans l'étude desquelles se sont illustrés au siècle dernier James Clerk Maxwell et Ludwig Boltzmann).

Il est long, bien sûr, le chemin qui conduit de la molécule d'hydrogène aux grandes macromolecules biologiques et encore plus long est celui qui aboutit aux êtres vivants. Si nul ne saurait objecter que, quelle que soit la dimension et la complexité d'un arrangement polyatomique, cet arrangement est déterminant pour la nature des propriétés émergentes, les avis deviennent plus nuancés quant à la portée exhaustive de ce facteur au fur et à mesure de l'accroissement de la complexité. De plus, si personne aujourd'hui ne défend la cause "d'atomes vivants, sensibles et intelligents" (comme le firent, au Siècle des Lumières, Maupertuis et Diderot), l'étude des caractéristiques de la complexité dans les structures vivantes ne saurait se passer d'une prise en compte du rôle de l'environnement: une bactérie, en fait n'importe quel être vivant, a la même structure atomique dans la seconde qui a précédé sa mort et dans la seconde qui l'a suivi. Ce sont ses modalités d'interaction avec l'environnement qui font la différence. De même, il convient de ne pas oublier que la mécanique quantique, doctrine holistique par excellence, souligne la nécessité incontournable d'inclure dans cette vision l'observateur et ses moyens d'investigation et de mesure.

Le problème que pose alors certains est de savoir s'il convient d'attribuer à de tels systèmes très complexes des *lois holistiques* qui se surajouteraient aux lois qui définissent les propriétés des unités constituantes. Greffé sur ces réflexions est encore l'interrogation, non moins énigmatique, que pose le principe anthropique, avec son parfum finaliste d'un univers réglé de façon à conduire nécessairement, à une époque de son existence, à l'émergence des êtres conscients.

Tous ces problèmes furent évoqués et discutés au cours de notre réunion. Heureusement, le devoir de réserve auquel est tenu le présentateur d'un Symposium polyvalent et multifacial m'autorise, m'invite même, à ne pas prendre position sur ces questions délicates et qui suscitent beaucoup de passions, comme en témoignent certaines des présentations et des interventions reproduites dans ce Volume. Je me bornerai simplement à rappeler ici la phrase célèbre de Blaise Pascal qui, je l'espère, satisfera, elle, tout le monde: *"Toutes choses étant causées et causantes, aidées et aidantes, médiates et immédiates, et toutes s'entretenant par un lien naturel et insensible qui lie les plus éloignées et les plus différentes, je tiens impossible de connaître les parties sans connaître le tout, non plus que de connaître le tout sans connaître particulièrement les parties"*. Nul n'a mieux exprimé la complémentarité des visions réductionniste et holiste du monde. Dans l'exploration de l'Univers, elles correspondent aux deux vues que l'on obtient par les deux bouts d'une même lorgnette.

Si ce premier événement, que j'ai qualifié d'important, de la 28e Session Plénière de l'Académie Pontificale s'est déroulé souvent sur un terrain mouvant et semé d'embûches, le deuxième frappe, au contraire, par son allure de grandiose solidité. Il s'agit de l'Audience solennelle que Sa Sainteté le pape Jean-Paul II a bien voulu accorder à tous les participants à la Session Plénière et à laquelle furent aussi conviés les dignitaires de l'Eglise et les membres du corps diplomatique accrédités au Vatican. Solennelle déjà par le cadre dans lequel elle s'est tenue (Sala Regia) et la composition de l'auditoire, elle l'a été surtout par ce qui en fut l'épisode historiquement le plus significatif et le plus émouvant et qu'il est conventionnel d'appeler la "réhabilitation" de Galilée. "L'affaire Galilée", qui fut depuis plus de 3 siècles et demi une pomme de discorde entre la Science et l'Eglise a trouvé ce jour-là (31 Octobre 1992) un dénouement heureux. Ce fut l'aboutissement d'une initiative prise par Jean-Paul II, peu de temps après son élévation au Pontificat, désirant voir le cas Galilée réexaminé dans un esprit d'équité par un ensemble de personnalités polyvalentes de stature incontestable. Dans ce but, le Pape a institué le 3 Juin 1981, une "Commission Pontificale pour l'étude de la controverse ptoléméo-copernicienne aux XVIe et XVIIe siècles", une manière élégante de considérer le cas Galilée dans une optique plus générale et de le replacer dans l'atmosphère de l'époque, conditions jugées indispensables pour un jugement éclairé et équilibré. Les conclusions auxquelles ont abouti les efforts de cette Commission ont été exposées à l'Académie Pontificale par son Président, Paul Cardinal Poupard, Président du Conseil Pontifical pour la Culture. Le texte de ce rapport figure dans l'annexe de ce Volume et chacun peut donc se rendre compte de l'important travail d'exégèse historique et de réflexion qui fut accompli par les membres de la Commission. On ne peut aussi qu'admirer la franchise intellectuelle et la noblesse de termes qui caractérisent ce texte, rédigé dans le but évident de mettre un terme définitivement aux querelles du passé. La reponse du Saint-Père, également reproduite *in extenso* dans l'annexe de ce livre, non seulement endosse les conclusions de la Commission et reconnaît les torts dont Galilée a eu à souffrir de la part de l'Eglise mais, decidément tournée vers l'avenir, réaffirme l'orientation franchement positive que ce Pape a imprimée, dès le debut de son Pontificat, à l'attitude de l'Eglise envers la recherche scientifique et qui repose sur le respect mutuel de l'indépendance de la démarche scientifique et de la démarche religieuse, dont les méthodologies propres, estime-t-il, *"permettent de mettre en évidence des aspects differents de la réalité"*.

De par ces événements importants et réconfortants qui ont jalonné la 28e Session Plenière, cette réunion occupera une place particulièrement significative dans l'histoire de l'Académie Pontificale.

Je ne saurais terminer cette Préface sans exprimer, au nom de tous les participants, notre profonde reconnaissance au Professeur Cabibbo, Président de l'Académie Pontificale et à Monseigneur Dardozzi, Chancelier, pour l'excellente organisation intellectuelle et matérielle de cette réunion.

BERNARD PULLMAN
Académicien Pontifical

PONTIFICAL ACADEMICIANS
PRESENT
AT THE 1992 PLENARY SESSION[*]

A. BLANC-LAPIERRE	*(France)*
G. COYNE	*(Vatican City)*
H. CROXATTO	*(Chile)*
N. DALLAPORTA	*(Italy)*
J. DÖBEREINER	*(Brazil)*
J. ECCLES	*(Switzerland)*
P. GARNHAM	*(United Kingdom)*
P. GERMAIN	*(France)*
M. HELLER	*(Poland)*
S. JAKI	*(U.S.A.)*
T. LAMBO	*(Switzerland)*
J. LEJEUNE	*(France)*
A. LICHNEROWICZ	*(France)*
J. LIONS	*(France)*
S. ŁOJASIEWICZ	*(Poland)*
J. McCONNELL	*(U.S.A.)*
K. MALU	*(Zaire)*
J. METZLER	*(Vatican City)*
M. MOSHINSKY	*(Mexico)*
R. MÖSSBAUER	*(Germany)*
C. PAVAN	*(Brazil)*
B. PULLMAN	*(France)*
G. PUPPI	*(Italy)*
C. RAO	*(India)*
P. RAVEN	*(U.S.A.)*
M. REES	*(United Kingdom)*
C. RUBBIA	*(Italy)*
K. RUNCORN	*(United Kingdom)*
M. SELA	*(Israel)*
M. SINGER	*(U.S.A.)*
J. SZENTÁGOTHAI	*(Hungary)*
W. THIRRING	*(Austria)*
C. TOWNES	*(U.S.A.)*
H. TUPPY	*(Austria)*

[*] Several outside experts contributed to the scientific part of the Session. They are listed in the opening remarks.

DAY ONE
27 OCTOBER 1992

OPENING REMARKS

G.V. COYNE

Pontifical Academician

Fellow colleagues, Ladies and Gentlemen,

It is my pleasant duty to formally welcome you all to this Plenary Session. I should let you know why I am doing this and why Professor Dallaporta is here with me to welcome you in the name of our President G.B. Marini-Bettòlo who, as most of you are well aware, after a serious sickness but fine recovery is unable to be with us at this meeting.

Our first point of business is the message that we would like to send to the Holy Father. The text in Italian is:

Il corpo accademico, riunito in Sessione Plenaria per trattare insieme agli esperti del campo il tema della Complessità e per esaminare il ruolo dell'Accademia Pontificia delle Science di fronte ai problemi della modernità, invia un devoto pensiero alla Santità vostra e attende la Vostra Illuminata parola in attesa dell'Udienza di sabato 31 Ottobre 1992.

Con devozione.

The English translation is:

The Academic body of the Academy gathered for the Plenary Session to deal with the theme of Complexity and also to re-examine the role of the Academy in modern society sends its most devoted thoughts to Your Holiness and we are waiting for your words to us on the occasion of the solemn Papal Audience to be held on this coming Saturday 31st October.

With great devotion.

We also propose the following message to the President.

Dear Professor Marini-Bettòlo, as we convene our meeting this morning our first moments were spent in thinking of you. We know that you are with us in spirit and we were happy to hear from the Council of the Academy on their visit to you yesterday of your splendid recovery and we send you our finest best wishes that this recovery will continue.

I now ask Professor Dallaporta to continue the introduction.

N. DALLAPORTA
Pontifical Academician

Excellencies, Ladies and Gentlemen,

As an introduction to the 1992 Plenary Session of the Pontifical Academy of Sciences I would like, as Professor Coyne has already done, to express our best wishes for the rapid and complete recovery of our president Professor Marini-Bettòlo whom Professor Coyne and myself have been invited to represent on this occasion.

I should like to give a brief outline of the general structure of our meeting. First of all the Council of the Academy, as already mentioned yesterday evening by Monsignor Dardozzi, has chosen *Complexity* as the theme for this year's discussions.

The reason for this choice is already known to most participants but perhaps I can briefly remind you of it. We should see that our Plenary Session is dedicated to dicussing some important interdisciplinary questions of great topical and cultural relevance.

Now, Complexity certainly falls into this category and is perhaps on the scientific horizon the theme that appears to be the newest, the most fundamental and the richest as regards its cultural implications. Therefore this choice was almost obligatory.

As this is a theme which has been in existence for only a few years and yet at the same time is a highly extensive one, it was considered advisable that speakers belonging to the Academy should be supplemented by a number of outside experts whose help was likely to enable a sufficiently complete survey of the topic to be made.

In respect to the programme as presented a small change has to be announced Professor Ennio De Giorgi will be unable to participate in this session but if his contribution is received in time it will be read by Professor Lichnerowicz.

The remaining contributions as announced are those of the following Academicians: W. Thirring, C. N. Rao, M. Moshinsky, M. Heller, J. Lions, P. Raven, J. Eccles.

The following Academicians will be chairmen of the various sessions: A. Lichnerowicz, G. P. Puppi, G. V. Coyne, B. Pullman, J. Szentâgothai, J. McConnell, R. Mössbauer.

The invited experts whom I have the pleasant task of introducing to the members of the Academy are, according to the order of their contributions as set out in the programme. Professors:

Enrico Berti	Philosophy	University of Padua
René Thom	Mathematics	Institute des Hautes Études Scientifiques, Paris
Lázló Lovász	Informatics	University of Budapest
Adi Shamir	Informatics	Weizmann Institute of Science, Israel
Peter Richter	Theoretical Physics	University of Bremen
Tito Arecchi	Theoretical Physics	Istituto Nazionale di Ottica, Florence
Giuseppe Del Re	Physical Chemistry	University of Naples "Federico II"
Giuseppe Geraci	Molecular Biology	University of Naples "Federico II"
Rémi Chauvin	Biology	Sorbonne, Paris
Josef Seifert	Philosophy	International Academy for Philosophy, Lichtenstein
Basarab Nicolescu	Theoretical Physics	Université de Paris VI

On behalf of the President and all the Academy I warmly thank them for their collaboration. Special thanks are also due to Professor Giuseppe Del Re.

As has been emphasized by the Holy Father on several occasions, our Academy does not consider itself as being simply a body of high scientific competence, but as a centre for the promotion of science as an integral part of our general culture. It therefore seemed right that our meeting, apart from the closed session which is to deal with internal matters on Thursday 29th October, should be open to a small number of invited guests, who will thus be directly informed of our work in person and prior to the publication of the proceedings. On behalf of the President and all of us I welcome them most heartily to our Academy.

COMMEMORATION OF RECENTLY DECEASED PONTIFICAL ACADEMICIANS

The first act performed by the 1992 Plenary Session was to honour the memory of those Members who had died in the preceding year: Otto Creutzfeldt and George Speri Sperti. The commemorations, by J. C. Eccles and G. V. Coyne, are published in the order in which they were presented

OTTO CREUTZFELDT
B. April 1, 1927. Professor of Neurology at the Max-Planck Insitute for Biophysical Chemistry in Göttingen, Germany. Member of the Pontifical Academy of Sciences since October 1990. Died January 23, 1992.

Mr. Chairman, Academicians

It is a great honour and a privilege for me to be chosen to present a short statement on the life of my dear friend Otto Creutzfeldt, a Pontifical Academician from 1990 till his death on 23rd January 1992. He was one of the leading brain scientists in the world with a wonderful performance as a scientist and teacher and with his entrancing spirit. He had been for over 20 years Director of the Department of Neuro-Biology that he set up at the invitation of Manfred Eigen at the Max Planck Institute for Biophysical Chemistry in Göttingen. Before that, he had been for nearly 10 years at the Max Planck Institute for Psychiatry in München. As I contemplate the greatness of Otto as a scientist and scholar I come to his great teacher Richard Jung of Freiburg, who had a miraculous success for German Neuroscience, attracting wonderful students who were enthused by Richard's great insights in neurology and in humanity, and by one of the best electrophysiological laboratories in the world. Richard Jung was the first to appreciate my intracellular recordings for cortical neurons in 1951-52. And Richard Jung's institute was always visited when I came to Europe. So, I met there Otto as a young man in his early thirties and was enthused by his success in intracellular recording of neurons of the cerebral cortex, which was a great advance. Soon after my election to the Pontifical Academy I was invited by the President Lemaître, the great cosmologist, to organise a meeting of the Academy on *Brain and Consciousness* and I chose Otto as one of my participants. That was in 1964. He spoke on trasmission in the visual system. I think he was the youngest participant, 37, in this room here. He illustrated the story of the visual system with intracellular recordings of the neurons to the visual cortex responding to specific patterns of light on the retina. It can be seen in *The Brain and Conscious Experience*, a publication of the Academy.

Otto's next association with the Pontifical Academy was in 1988 when he participated in a conference I organized at the invitation of the President Chagas. It was entitled *The principles of design and operation of the brain* which Otto and I co-edited for Scripta Varia with the international publication in *Experimental Brain Research*, Series 21.

At that meeting he gave a talk with G. Ojemann of Seattle. It was the last of Otto's papers, entitled (and here is a very advanced subject), *Neuronal Responses in the Human Lateral Temporal Lobe to Speech*. This was done remarkably on conscious human subjects and recordings were of neurons in their brain. This study of Otto's can be regarded as the beginning of Otto's final goal in the study of the brain with his lifetime interest in the mind/brain problem and the philosophy of the person. But he did not live to go on with this. We will sadly miss his great wisdom. Had he lived he could have led us on to an understanding of the self transcending the dominant materialism of today with its banal value system. The Academy suffers a great loss after only two years of his fellowship, but today we are electing Professor Wolf Singer, a great friend and kindred spirit. The greatness of Otto is not only as a neuroscientist and philosopher but also as a human person radiating love particularly to his wife Mary and his children. May I ask that we all rise for a moment of silence in memory of Otto Creutzfeld.

J. C. Eccles

GEORGE SPERI SPERTI

B. in Covington, KY (USA) on January 17, 1900. Biochemist. Professor at St Thomas Institute for Advanced Studies, Cincinnati, Ohio (USA). Member of the Pontifical Academy of Sciences since October 28, 1936. Died on April 24, 1991.

I would like with these few brief words to commemorate George Speri Sperti, who was an important figure in what I think can be defined as the second wave of the establishment of Catholic intellectual life in the United

States of America, the wave which saw the wide diffusion of Catholic centres of learning in the U.S.

He was among the most prominent scientists in the St. Thomas Institute for Advanced Studies in Cincinnati, Ohio. Born on January 17th 1900 in Covington, Kentucky, he pursued all of his studies in that geographical area and received his doctorate in Physics in Duchaine University, Pittsburg, Pennsylvania. He interestingly combined both an academic and a business career. He was for instance responsible for the contact labs of the Union Gas and Electric Company, Cincinnati and he directed the basic science research laboratory of the University of Cincinnati. He was a professor at the University of Dayton. He was issued 160 patents for his inventions, among these the Sperti Solar Lamp, and I am personally, if he would permit me today a personal note, grateful to him that he was one of the first to investigate low pressure lamps for commercial use. This has resulted today in the production of low pressure sodium lamps which have helped save our precious skies for astrophysical investigation. At the beginning I noted that George Speri Sperti was a prominent figure in the movement in the United States for the development of higher learning. He indeed did this mostly through his research in applied physics. However, an interesting final note is the following, that his research was conducted in order to gain financial support for basic research which would not be adequately supported by the University of Cincinnati or by the Catholic Church of Cincinnati. I believe his interesting combination of a business and an academic life devoted to the service of the development of intellectual centres in the United States is a fine commemoration to him. I would ask you to rise for a brief moment in commemoration of George Speri Sperti.

G. V. Coyne

PRESENTATION OF THE NEW MEMBERS

The next act of the Plenary Session consisted in the welcome to the new members, who answered with brief speeches of thanks.
Academician G. V. Coyne said:

Life in general, and the life of our Academy, has to deal with suffering and death. But there are also many positive events. We have the great joy of seeing our society revitalized with new members. Against this background I would like to welcome to our society the newly elected members, nominated on the 18th of September 1992.

Bernardo Maria Colombo of Italy,
Minuro Oda of Japan,
Wolf J. Singer of Germany,
Richard Southwood of Great Britain.

The solemn ceremony will take place before the Holy Father on Saturday, and you will receive the insignia of your office. It is my pleasure, however, to present to you the brief which nominates you as a Papal Academician.

During the Plenary Session, the nomination (on the 28th of October 1992) of Georges M. M. Cottier OP, of Switzerland, as an honorary member of the Pontifical Academy was communicated to the Academicians by Mrg. Renato Dardozzi, Director of the Academy. Father Cottier's acknowledgement speech, although actually delivered during the scientific meeting, is printed with the others in this part of the Proceedings.

BERNARDO MARIA COLOMBO

B. in Olgiate (Como), Italy, in 1919. Professor of Demography at the University of Padua. Member of the Italian National Commission for the guarantee in statistical information.

Ladies and Gentlemen,

I am sorry for my limited command of the English language, which does not allow me to express in the right words how deeply I am touched by the nomination as a member of this Academy. I fully appreciate the honour paid to my person and what the nomination means in terms of confidence in my possibilities for collaboration. I am sure you will kindly forgive me and I wish to express my best thanks.

To overcome the embarrassment in introducing myself, I shall start by speaking of another person, Marcello Boldrini, a Pontifical Academician. I was fortunate to be a pupil of Professor Boldrini's when at 19 I started attending the Statistical Institute he directed in Milan while preparing to graduate in Economics. There, I soon learned what it means to do research, with rigor, tenacity, open-mindness, and last but not least, with intellectual honesty. War events then stole 5 years of my youth. But, in a concentration camp, I luckily had at my disposal the manual on Statistics that Boldrini had published in 1942. In more than 1,000 pages, making use of his deep knowledge in many fields, he was able to present statistical tools through a host of examples, showing the power of these procedures. I think I have been a little infected by this insatiable curiosity for everything new and intriguing.

After the war, I started as an Assistant of Statistics in Venice. My first interests were methodological ones and concerned mainly themes of foundations for statistical inference, at that time under discussion in my country. It was proposed that I should give a look at the strange phenomenon of birth rate recovery during the war in countries involved in the conflict or near the area of operations. I worked hard on this and I was able to discard some explanations which had been advanced but found it difficult to propose solutions that could be adequately supported by evidence. The book I wrote on the subject opened several doors for me. One of these was as a Rockefeller fellow for Social Sciences at the Office of Population Research at Princeton University. There, I enjoyed the opportunity of being in contact with several research workers who then became prominent figures in the world of demography. For quite a while, this remained my milieu, for research and various engagements, until problems of health — later solved — compelled me to restrict my activity.

Among my interests, there have always been biometrical ones. On this point, for instance, I gave attention to the sex ratio in man. Another book came out, in which I succeeded in drawing seemingly plausible conclusions about the level of the primary sex ratio. But I ended up stating, as I did in a paper read at a Symposium in Cold Spring Harbor, that in the interest of both science and life the problem of the determination of the human sex ratio remained unsolved.

When I returned to Italy from Princeton, I enjoyed the opportunity of being called to give lectures in Padua. There were many more departments at

Padua than at Venice. In its Statistical Institute, I started being contacted by many research workers from various fields asking for advice on statistical techniques. From that experience several things came out: papers of my own, both in statistical methods and substantive matters, co-authorships, acknowledgements. Among the colleagues seeking advice, was a young Botany student. She became my wife.

At that time I obtained a Chair of Statistics in Venice, but later on I switched to Padua and Demography, where I had the possibility of pushing for the creation of a new statistical Faculty. Soon other engagements fell on my shoulders, I was nominated as a member of the Parliamentary Commission for evaluation and proposals on the school system in my country: 110 meetings in Rome in eight months and a book on higher education were the consequence. Experts in educational sciences did not care much for it and continued to discuss the question among themselves.

I was also called as an external *peritus* in the preparation of what became a pastoral constitution of Vatican Council II. It was embarrassing to vote raising hands at the same level as the chairing Cardinal.

One subject which has occupied my mind since Princeton is that of demographic policies. At several international gatherings, I had also the opportunity of seeing in action the politics of demographic policies, but my personal interests were mainly scientific ones. A big challenge came when I was invited to give a lecture on human rights, ideology and population policies at the closing plenary session of a Conference of the International Union for the Scientific Studies of Population. But I have continued thinking about these problems, and have had other occasions to go further. The road for solutions is steep, and full of traps.

Later on, another avenue was opened up in my research, involving official statistics. This is a very rich and largely unknown field. Several papers on the subject and a series of administrative undertakings which I do not mention, were the result.

I wish to jump to my main current activity in research. I have come back to biometry, this time on the menstrual cycle and its relation to fertility regulation. It is a fascinating field. Much work, with collaborators, has been done on some first class documentation we have collected and more will come with an international multicenter prospective study.

May I conclude my talk with a prayer, taken from the Vespers of a day last summer. It is good for me, but perhaps also for the Academy. "To those who search for the Truth, may You grant the joy of finding it, and, having found it, the desire to seek for it again."

GEORGES M. M. COTTIER

B. in Geneva, Switzerland, in 1922 Professor at the Diocesan Seminar, Geneva,
Theologian of the Pontifical House.

Je tiens à vous remercier vivement du grand honneur que vous me faites
en m'accueillant parmi vous.

Je me suis demandé quelle signification pouvait bien avoir cette
nomination, qui est pour moi une surprise. Si je pose la question, c'est parce
que la présence d'un philosophe dans votre Académie n'est peut-être pas
évidente pour tous. C'est aussi parce que je dois m'interroger sur ce que
pourra être ma propre contribution.

J'ai donc enseigné la philosophie jusqu'à ce que, il y a deux ans, un
nouveau travail m'appela ici, a Rome.

Je me suis beaucoup occupé de Karl Marx. Pour un homme de ma
génération, il n'y a là rien d'extraordinaire. Marx m'a conduit a Hegel, qui est
un penseur d'une bien autre envergure. Or l'un et l'autre, Hegel et Marx,
étaient habités par une même ambition, d'enfermer dans une connaissance
exhaustive la totalité de la réalité qui, pour eux, est avant tout la réalité de
l'histoire.

Karl Marx, sans cesser de dépendre de lui, critique Hegel, en qui il voit
l'achèvement et la perfection de la philosophie spéculative. Il annonce la
mort de la philosophie, à laquelle devrait succéder une nouvelle discipline,
qu'il appelle le *matérialisme historique*. Ce qui fait la nouveauté et
l'originalité de celui-ci c'est qu'il est présenté come une *science de l'histoire*,
de ses structures, de la necessité de son cours et de son avenir. Le
matérialisme historique prétend aussi nous apporter une connaissance
prédictive, de nature scientifique, de l'avenir.

De là une première série de problèmes auxquels je n'ai cessé de réfléchir.
Abstraction faite de la question, à vrai dire essentielle, de la verité ou de la
fausseté du contenu du matérialisme historique, que peut bien signifier ici le
terme de science, si on le compare à son usage dans les sciences de la
nature? La question posée à propos de l'histoire des historiens et de la
philosophie de l'histoire, doit l'être également pour des disciplines come
l'économie, la sociologie, ou encore la psychologie et la psychanalyse, — en
un mot, pour l'ensemble des sciences humaines. Il apparait aussitôt que la
parole science est polysémique et que son usage est analogique.

Hegel — qui est à l'horizon de Marx — pose une question qui, a mon
avis, est beaucoup plus radicale. Cette question, elle ressort de deux

affirmations du philosophe. La première est: *tout ce qui est rationnel est réel, tout ce qui est réel est rationnel.* Réel signifie ici le cours de l'histoire humaine et le sens de cette histoire. La seconde est: la philosophie de l'histoire est la *véritable théodicée,* c'est-à-dire la justification de Dieu, que Leibniz avait tenté de construire sur des bases métaphysiques.

Jamais sans doute, dans l'histoire de la philosophie, la raison n'avait affiché une telle ambition. Mais jamais sans doute n'aura-t-elle fait une démonstration aussi patente que la raison humaine, quand elle perd le sens de la mesure et le sens de ses limites, est livrée à l'égarement.

Hegel, en effet, ne réussit à mener à terme son projet qu'en limitant à l'extrême la part du mal dans le cours des choses humaines, et ceci à un tel point que ce qui devait, dans son intention, constituer une théodicée, est devenu une justification du mal. De même la contingence se voit rejetée, comme un facteur négligeable, dans les nuages du système.

Pour les hommes du 20ème siècle que nous sommes, aprés tant d'expériences terribles, il n'est plus permis de banaliser ainsi le mal; qui le tenterait encore recevrait des événements les plus immédiats un cinglant démenti. Ainsi, ce que nous devons considerer comme l'échec de la philosophie hégélienne de l'histoire nous oblige à regarder en face les questions de l'irrationnel et de la contingence.

Plus radicalement, puisque une raison qui prétend tout expliquer aboutit ainsi à des impasses majeurs, nous sommes conduits à nous interroger sur la raison elle-même et sur la rationalité, — quelles sont leur signification, leur autorité, et aussi leurs limites?

Ces problèmes philosophiques, sous un certain aspect, interessent aussi l'epistémologie des disciplines scientifiques. Quand ces dernières réflechissent sur elles-mêmes, il ne leur est peut-être pas inutile de se référer à un point de comparaison extrinsèque. En ce sens, le philosophe peut se sentir d'une certaine utilité vis-à-vis des ses collègues scientifiques.

Si tel n'était le cas, il lui resterait, en ce qui le concerne, la conviction socratique que le premier pas de la philosophie est de savoir qu'on ne sait pas. Telle est bien ma première réaction devant des exposés scientifiques d'une haute complexité qui recourent de surcroît à un technicité poussée.

C'est dans cette disposition d'esprit que je vous remercie chaudement de m'accueillir parmi vous.

MINORU ODA

B. in Sapporo, Japan, in 1923. Professor of Astrophysics at the University of Tokyo and at the Institute of Space Science and Astronautics. President of the Japanese Institute of Physics and Chemistry (RIKEN). Member of the Academy of Japan. Received two Japanese and two international awards for work in Astrophysics and Astronautics.

Distinguished Chairman and Academicians;

First of all, I can't find words to express how deeply grateful I am to the Pontifical Academy for giving me the honour of being a member, and specifically to thank those who initially introduced me to the Academy.

As a self introduction, I have been basically an experimental physicist.

After working in fields of microwave physics, solar radio-astronomy and cosmic ray physics, then in high energy physics, during the period when my country was recovering from the destruction caused by the war, I joined a Massachusetts Institute of Technology (MIT) group under Prof. B. Rossi who was originally from Padua.

Since then, I have defined myself as a pupil of Rossi, though he calls me a young friend of his.

Since the early 1960's, when he and his colleagues initiated X-ray astronomy which is one of the most intriguing fields in Astrophysics, I have worked with him at MIT and also in Japan.

Now, my colleagues in Japan form firm components of the internationally collaborative community in the field of Astrophysics.

For four years I have directed the RIKEN Institute which consists of a complex of about 50 autonomous laboratories covering nuclear science, a big facility of synchrotron radiation, laser science, fundamental technology, a variety of chemical sciences, neuro-science etc.

Finally, I should add how impressed I was when I heard that the Academy unlike some other honourary bodies immediately took up the current key theme of complexity for the Plenary Session.

Complexity was what I had learnt, as a young student, to avoid in research.

But now some of my friends half-jokingly say that, thus far we have considered quantum mechanics and general relativity as the two major achievements of this century in Physics, but towards the end of the century the third major achievement in Physics, complexity and deterministic chaos, might appear.

If it is appropriate to mention here, I would like to thank my wife and children who have formed an ideal secretariat for my academic life. Thank you.

WOLF JOACHIM SINGER

B. in Munich, Germany, in 1943. Professor of Physiology at the Technische Hochschule of Munich. Director of the Max-Planck Institut für Hirnforschung in Frankfurt a.M. Member of the European Academy. Received the IPSEN Foundation award 1991 (with U. Bellugi and T. Wiesel).

Mr. President, Director Dardozzi,

Excellencies,

Sir John's touching commemoration of Otto Creutzfeldt cannot but add to my emotional confusion and to my deep feelings of gratitude for having been accepted in your distinguished community. Prof. Creutzfeldt was my first academic teacher, he paved my path into the world of science and as time passed on, Otto became one of my best friends. One of the last and happiest moments we had together was, when he returned from Rome, two years ago, landing in Frankfurt, staying overnight with us, the day after you had nominated him member of the Pontifical Academy of Sciences. His excitement, his happiness, his gratitude and his sincere feelings that he did not deserve this honour are now mine. I feel like preserving them for him and experience this as a deep satisfaction and at the same time I am sorrowful, wishing him to be here rather than me.

Getting to know Otto Creutzfeldt and entering science were one step. It occurred in 1965, in Munich. I was a medical student and Creutzfeldt had offered a lecture series on the neurobiological correlates of higher cognitive functions, organized together with Paul Matussek, a very distinguished psychoanalyst, one of the orthodox Freudians who, probably more than his master, believed in the power of words rather than that of physico-chemical interactions in tangible brains. One of the lectures, I remember, was devoted to the joint effort to relate the dissociated experience of reality which characterizes certain schizophrenics to the dissociation of perceptual phenomena that can be found in split brain patients, then studied extensively by Roger Sperry. Here then seemed to be a field of research, wide enough not to require immediate decisions from the young student who still oscillated between the conviction that the only key to truth and ultimate understanding were the natural sciences, physics above all, - and the equally compelling intuition that what really matters were mental phenomena, cultural achievements, poetry and music, in brief, the subjects of the humanities. - It took me quite some time to understand that also the models and theories in the basic sciences are nothing but constructs within well delineated, often isolated description systems, that they are creations ultimately based on primary experience whereby the criteria for "truth" are not too different from those applied to pieces of art: aesthetics and consistency.

Anyway, the student felt that getting involved in the neurosciences would allow him to move freely back and forth or up and down across the different levels into which the world seemed to be subdivided - as Nicolai Hartmann had told him - and which - in some mysterious way - appeared to be interdependent. Otto Creutzfeldt accepted me as student and so I found myself in the laboratory, studying the exchange of electrical signals between the two cerebral hemispheres of animals, looking for the substrate that synchronizes their activity and assures that they wake up at the same time. In parallel, the medical curriculum required me to work in hospitals, to see patients, to act and to do something whose relevance was directly accessible. For many years I experienced the medical and the scientific approaches to the solution of problems as equally fascinating but they seemed difficult to reconcile. One required rapid assertive decisions, and pragmatic action, even if the database was poor - the other, by contrast, necessitated doubt and hesitation, even if the data seemed excellent. Finally, wanting to know more won over wanting to apply what we know and my attempts to explore the world became more and more indirect. Increasingly sophisticated equipment became intercalated between my senses and the world that I wanted to understand and rather soon, the phenomena which attracted most attention were those which could only be made to exist by using electron microscope amplifiers and chemical reactions, and the significance or meaning of the invisible and intangible phenomena most of the time escaped primary intuition it is only within the framework of theories that we manage to assign a particular meaning to the phenomena that we believe we can isolate with our methods - and very often we even ignore how and why we were driven to base our interpretation on that particular theory and not on another.

Of course, reality fell short of my expectations. We have not succeeded in moving freely back and forth between the description systems for mental and physical events, even if, in unreflected lab talks, we say things like "information about this event is only attended to and reaches consciousness if the responses of the cells signalling the event are sufficiently coherent". What we probably mean is that phenomena such as attention, consciousness and intentionality are emergent properties of highly complex aggregates of matter such as our brains, that these phenomena come into existence only through interactions between such highly organized brains, when these observe one another and invent descriptions for the observed phenomenology. What we probably mean is that the biophysical processes that we define in individual brains and in the framework of description systems of the natural sciences, are somehow and obviously causally related to the mental phenomena that we define in the description systems of the humanities - but until now we seem to be unable to design even intuitively graspable concepts or metatheories which would allow us to unify the different systems of description, and for many who do not adhere to monistic concepts of emergentism this is a relief rather than a worry.

I avoided recapitulating those parts of my scientific vita which must have been made available to you when you had to decide on me, but feel I owe you a few words on our current interests. Above all we want to understand how the brain constructs representations of its environment, how it codes and represents perceptual objects, how it stores and how it recalls this information. To this end we study developmental learning processes which are known to lead to drastic and experience-dependent changes in the architecture of neuronal connections. We consider this early learning process as an interesting transition between embryonic, purely genetically determined development and adult learning. In addition, we investigate the functional organization of the cerebral cortex, in particular its dynamic properties and temporal codes, because it is generally assumed that cognitive representations reside in the neocortex. We pursue these questions primarily in experiments with animals when we examine system properties and in *in vitro* preparations of brain tissue when we analyse cellular and molecular mechanisms, but we attempt, whenever possible, to relate our findings to brain processes in humans. We do this by investigating the development of cognitive functions in babies and by testing predictions on cognitive performance in psychophysical experiments with adult subjects.

To me the most fascinating and at the same time puzzling issue of this research is the growing evidence, that there is no single site in the brain where all information converges and where a *homonculus* could be seated which decodes the results of the many parallel processes, interprets them and reaches decisions - as Descartes and most before and after his time had postulated. Rather, we face an immensely complex and distributed system which lacks a coordinating metastructure and nevertheless, through the action of local rules, selforganizes towards ordered and coherent states. I believe that our social and economical systems show some of these features and the recent and dramatic collapse of those systems which had the most hierarchical and centralized organization seems to indicate, that coordinating metastructures may not be a good solution for the management of systems once they have reached a critical level of complexity. Thus, studying the brain may not only assist us in improving the therapy of its diseases but it may help us in the search for principles of organization that are required to stabilize highly complex systems. To identify such principles and to translate them into political and economical structures appears to me as the most important challenge to which we are at present exposed. I assume that it is your expectancy that what I have learnt, which is in no way special, can make a small contribution to the interdisciplinary endeavours of this distinguished academy and that it can, in combination with the wealth of scholarship assembled here, make a modest contribution to the better understanding of the *conditio humana* and the world in which it evolved, and that this will help in turn to improve the management of the future of our biotope. I shall try my best to meet these expectancies and want to join this promise with the expression of my deep gratitude for your confidence. Thank you for accepting me.

RICHARD SOUTHWOOD

B. in Northfleet (Kent), Great Britain, in 1931. Linacre Professor of Zoology and Rector of the University of Oxford, England. Member of the Royal Society, of the U.S. National Academy of Science, of the American Academy of Arts and Science, of the Norwegian Academy of Arts and Science. Baronet since 1984 and "Cavaliere Ufficiale dell'Ordine al Merito" of the Republic of Italy since 1991.

Ladies and Gentlemen

I am a biologist, whose current interests lie particularly in the field of Environmental Studies. From my childhood I have had a keen interest in Natural History and at the age of 17 had the good fortune to go to Rothamsted Experimental Station (a large agricultural research institute) and work in the Entomology Department as a vacation worker. From the work I undertook there and some private studies, I was able to publish a few short scientific papers whilst still an undergraduate. For one of these papers I needed to read another study which had been published in Italian and it seems appropriate to recall here that the translation was kindly undertaken for me by a Catholic priest. I studied for my first degree at Imperial College London, graduating in 1952. I was then given a scholarship by the Agricultural Research Council to return to the Entomology Department at Rothamsted (which was associated with London University) and work towards a Ph.D. The department was much concerned with the movements and migrations of insects and one part of my own studies was concerned with the ecology of field-margins: the interface between the natural habitat and what we would now call the agroecosystem. Those two threads have been an important part of my subsequent interests.

Returning to Imperial College as a teacher of Applied Entomology in the Department of Zoology, I continued my research into animal ecology, being more particularly concerned with insects. Why do some insects fly around a great deal whilst others remain fairly stationary? I was able to relate this in a semi-quantitative way to the permanence of the habitats that the insects occupied and that led me on to consider the role of the habitat in shaping the ecological strategies of species. One can view the biological characteristics of species, their size, shape and life cycle as exhibiting the characters which have been forged by the forces of evolution on the anvil of their particular environments. Those organisms that live in short-term situations, such as carrion, breed rapidly, move around a great deal and produce large numbers of offspring. Those that live in more stable situations, such as an insect living on an oak tree, will be more sedentary, will produce smaller numbers of offspring and will have defensive mechanisms to increase the probabilities of their survival. These sorts of patterns of nature have always fascinated me, and another variable which I noticed as a young student and which has now been thoroughly explored is the number of different species of insects associated with different varieties of tree: some such as the oak in Britain have many hundreds of insect species while others, such as the cedar in Britain, have very few. Studies have shown how this phenomenon may be related to many different factors including the length of time the tree has been present in the region, the relative abundance of the tree species and the level of its chemical defences.

Studies made on the edges of agricultural land have shown the striking impact of agricultural activities on these habitats. Why were the populations of partridges falling all over Western Europe? The answer proved to be the effect of herbicides in eliminating weeds from the cereal fields, which had provided shelter and food for the insects which were a vital food for young partridge chicks in the first few weeks of life. One can thus easily see how one particular action can have repercussions which extend both in time and space. Such repercussions may be quite detrimental compared with the original procedure and so ecologists have met economists in trying jointly to address a cost benefit analysis of environmental change.

After a period as Professor of Zoology and Dean of Science at Imperial College, I came to Oxford in 1979 as Linacre Professor of Zoology and for the last four years have held office as Vice-Chancellor, the University's executive head. However, through work with a number of Government bodies, I have been concerned for the last twenty years with the environmental impact of various human activities ranging from the problems presented by the addition of lead to petrol, by the pollution caused by the excessive use of pesticides, by radiation from nuclear power and other sources to the impact of power station emissions on lakes and waterways, particularly in Scandinavia, through the phenomenon of 'acid' rain. A contribution to the resolution of these problems is the focus of my current activity.

THE PIUS XI GOLD MEDAL AWARD

After the speeches of the new members, Academician V. G. Coyne announced that the PIUS XI Gold medal had been awarded to Dr. Adi Shamir, of the Weizmann Institute of Science, Israel. Dr. Shamir's presentation of his work is published here. A more scientific contribution is to be found in the proceedings of the sessions on complexity.

ADI SHAMIR

Dr. A. Shamir was born July 6, 1952 in Tel-Aviv, Israel. Ph.D. in Computer Science at the Weizmann Institute, Israel, 1977. Worked in research at Warwick University and MIT in the years 1976-1980. Associate Professor (1980-84) and then Professor (1984-) at the Dept. of Appl. Maths., Weizmann Institute of Science.

After getting my Ph.D. at the Weizmann Institute of Science in Israel in 1977, I was invited to join the department of Mathematics at MIT. Shortly after my arrival Prof. Ron Rivest drew my attention to the paper "New Directions in Cryptography" which had just been published by two Stanford researchers, Diffie and Hellman. The paper described a revolutionary type of cryptosystem called "public key cryptography" which enables any pair of users to protect the privacy and authenticity of their electronic messages without the prior exchange of secret keys. The paper described the benefits of such schemes and conjectured that they exist, but described only partial constructions. I started to discuss this new concept with my MIT colleagues Rivest and Adleman, and after several months and many false starts we developed the first practical public key cryptosystem, known as RSA (our initials). The scheme is based on the difficulty of factoring large (200 digit) numbers, a problem which had been investigated for hundreds of years by some of the greatest mathematicians. Today, this cryptosystem is still the best known and most trusted public key cryptosystem, and has numerous practical implementations, ranging from smart cards and special purpose chips to operating systems and electronic mail packages.

Shortly after the introduction of the RSA scheme, Merkle and Hellman developed a different type of public key cryptosystem, based on the combinatorial "knapsack problem". Its security became a major open problem in the field, and after 5 years of on-and-off analysis I managed to find a new cryptanalystic attack, which broke the basic version of this scheme. Within months, my technique was improved and extended to pratically all the knapsack-based cryptosystems.

In the early 1980's, a more rigorous approach to the question of cryptographic security was developed by Yao, Goldwasser, Micali, and others. They formalized the tricky concepts of pseudo randomness, bit security, cryptographic knowledge, etc. I contributed to this new wave of research by publishing several papers, among them the first contribution of a provably secure pseudo random number generator, and the first strong result on the bit security of the RSA function (in a joint paper with Ben-Or and Chor).

In 1985 Goldwasser Micali and Rackoff published their seminal paper on "zero knowledge interactive proofs". This is the seemingly contradictory concept of a surprising mathematical proof which yields no new information to its verifier. At first the concept looked like a philosophical curiosity, but in 1986 Goldreich Micali and Wigderson proved that it is applicable to a large class of computational problems known as NP (as well as to some problems believed to be just outside NP), and I published (with my student Amos Fiat) a very efficient identification and signature scheme based on this concept. In fact, zero knowledge proofs turn out to be ideally suited to identification protocols, in which users to establish their identity by proving their knowledge of some secret information without enabling the verifier to measure the protocol by repeating the proof to someone else.

In the next few years, research on zero knowledge interactive proofs mushroomed, and literally hundreds of papers were published on the topic. One of the major open problems was to characterize exactly which class problems had such proofs. Most researchers believed that this class was a slight extension of NP. However, in a rapid sequence of papers (aided by quick dissemination of draft by electronic mail), Lund Fortnow, Karloff and Nisan proved that the class extended well beyond NP, and a few days later I published the final characterization of this class in a paper titled "IP-PSPACE".

In the last couple of years I started (together with my student Eli Biham) a detailed study of the best known and most widely used (non public-key) cryptosystem, known as the DES. It is a national standard adopted by the US government in 1977, and is used extensively in civilian applications such as banking, telephony, and pay TV. We developed a new type of cryptanalytic attack which we called "differential cryptanalysis", and applied it successfully to a wide variety of DES-like cryptosystems. It is currently the only known technique which is capable of breaking the full DES cryptosystem in less than the 2^{56} complexity of the obvious attack of trying out all the possible keys.

ORDRE ET DÉSORDRE DES GRECS À GALILÉE ET DE GALILÉE AUX TEMPS MODERNES

ENRICO BERTI

Università di Padova, Italia

Avant tout il nous faut définir les notions d'ordre et de désordre. Par "ordre" nous entendons la disposition des éléments d'un ensemble selon un ou plusieurs critères reconnaissables: par exemple, si ces éléments sont des nombres, un ordre possible sera la série des nombres naturels, une progression arithmétique, une progression géometrique, ou encore une succession plus compliquée, résultant de plusieurs critères de succession (addition et soustraction, etc.). En tout cas, l'ordre implique la possibilité de réduire une multiplicité illimitée à l'unité ou à une multiplicité limitée, ou bien la possibilité de réduire l'indéterminé au déterminé. Le "désordre", par conséquent, n'est que l'absence d'ordre, c'est-à-dire de critère, de détermination.

Dans l'histoire de la science et de la philosophie — qui à l'origine étaient confondues ou étroitement liées — on rencontre deux types fondamentaux d'ordre reconnus par l'homme dans la realité naturelle, c'est-à-dire dans l'univers. Le premier type est l'ordre "simple", fondé sur un critère unique: il pourrait aussi être appelé ordre "mécanique", non par référence à la science mécanique que nous connaissons aujourd'hui, mais parce que ses partisans assument comme modèle pour l'interprétation de la réalité la machine (en grec *mechanè*), réduisant tout changement au mouvement mécanique, c'est-à-dire au déplacement de parties matérielles dans l'espace. Le deuxième type est l'ordre "complexe", fondé sur plusieurs critères: il pourrait aussi être appelé "ordre biomorphique", parce que ses partisans présupposent comme modèle l'organisme vivant, où l'on observe plusieurs changements, qui ne sont pas réductibles au simple mouvement mécanique, et cependant sont tous orientés vers le même but, la conservation de la vie[1].

Dans la première période envisagée par notre exposé, c'est-à-dire des Grecs à Galilée, on peut constater une évolution progressive des visions du monde, de la prépondérance de l'ordre complexe, de type biomorphique (antiquité et moyen âge), à la prépondérance de l'ordre simple (Galilée), tandis que dans la deuxième période, c'est-à-dire de Galilée à nos jours, on peut constater le passage inverse, de la prépondérance de l'ordre simple, de type mécanique (XVIIe et XVIIIe siècles), à un ordre complexe qui est encore, d'une certaine façon, biomorphique (XIXe-XXe siècle). Naturellement dans les deux périodes, et même chez les auteurs qui soutiennent l'un ou l'autre type d'ordre, il y a des exceptions à la tendance dominante, mais on peut parler également d'une prépondérance de l'un ou de l'autre type d'ordre. Je chercherai à illustrer avec plus de détails ces deux passages.

I. Des Grecs à Galilée

1. La naissance du concept d'ordre et sa double signification

On a l'habitude de faire coïncider la découverte du concept d'ordre avec la naissance de la philosophie (ou de la science, qui dans l'antiquité coïncide avec la philosophie), c'est-à-dire avec la pensée des philosophes de Milet (Thalès, Anaximandre, Anaximène, VIIe et VIe siècles avant J.C.). Dans les oeuvres des poètes qui leur sont antérieurs, en effet, surtout chez Hésiode, on rencontre seulement la notion de désordre, indiquée par le mot *chaos*, faisant allusion à un espace immense et vide, à une sorte d'abîme, qui serait à l'origine de toutes les choses. Chez les premiers philosophes au contraire, en particulier chez Anaximandre, on rencontre la notion d'ordre, indiquée par le mot *taxis*[2] ou, plus généralement, par *kosmos*, qui désignait à l'origine seulement l'ornement (d'où la notion de "cosmétique"), mais a pris ensuite la signification d'"ordre" (peut être avant tout l'ordre de la cité et ensuite l'ordre

---

[1] J'adopte l'expression "ordre biomorphique" au lieu de "ordre biologique" pour éviter toute réference à la biologie, telle qu'elle est pratiquée aujourd'hui, tenant compte des remarques qu'à ce propos m'ont été adressées par les professeurs B. Pullman, N. R. Rao e M. Sela.

[2] Anaximandre, frg. 1 Diels-Kranz, selon lequel les êtres s'engendrent d'un principe indéterminé et se corrompent en faisant retour dans ce dernier selon l'"ordre" *(taxis)* du temps.

du monde entier)[3] et plus simplement de "monde" (d'où les mots "cosmique", "cosmologie", etc.).

Chez les philosophes de Milet, et même chez Héraclite, l'ordre du monde est essentiellement de type biomorphique, parce qu'ils réduisent toutes les choses à un principe qu'ils appellent *physis*, c'est-à-dire principe de la génération, ou de la vie, d'où les choses prennent leur naissance et auquel elles font retour d'une manière qui rassemble à la mort. Cela n'empêche pas l'existence, dans l'univers, d'un rythme cyclique (surtout chez Empédocle) où l'ordre alterne avec le désordre, ni l'existence de processus de type mécanique à la base de la génération et la corruption, comme la condensation et la raréfaction (Anaximène), ou l'agrégation et la désagrégation des quatre éléments (Empédocle). Le philosophe chez lequel cette conception de l'ordre atteint le maximum de complexité est Anaxagore, qui considère l'ordre cosmique comme le resultat de l'action ordonnatrice accomplie sur le mélange initial, qui était une forme de désordre, par une intelligence transcendante (*noûs*).

Si cette conception "complexe" de l'ordre peut être considérée comme la principale durant cette période, elle n'est pas la seule professée par les premiers philosophes grecs, puisque dans le même temps les Pythagoriciens ont élaboré, probablement sous l'influence de l'astronomie chaldéenne, une explication des mouvements des astres qui est essentiellement de type mécanique, parce qu'elle consiste à admettre des mouvements circulaires du soleil, de la lune, de la terre et des autres planètes autour d'un "feu" qui serait au centre de l'univers. Cet ordre est beaucoup plus régulier, c'est-à-dire plus simple, que l'ordre admis par les Milésiens et surtout il est fondé, comme par ailleurs, selon ces philosophes, l'est toute la realité, y compris les phénomènes terrestres, sur des rapports mathématiques, c'est-à-dire sur les nombres.

Une conception mécanique de l'ordre cosmique, qui admet un ordre beaucoup moins régulier, mais paradoxalement encore plus simple, que celui des Pythagoriciens, est representée par l'atomisme de Leucippe et Démocrite (Ve siècle), selon lesquels, comme il est bien connu, toutes les choses dérivent du mouvement mécanique des atomes dû au simple hasard. Il faut dire, cependant, que cette conception n'a jamais eu beaucoup d'audience en Grèce: elle a été reprise, en effet, par Epicure (IIIe siècle) et par Lucrèce (I[e] siècle), mais n'est jamais devenue la vision du monde dominante dans la culture grecque. En tout cas, il faut remarquer que, dès son origine,

---

[3] W. JAEGER, *Paideia. Die Formung der griechischen Menschen*, Berlin 1933, vol. I, ch. IX.

le concept d'ordre s'est présenté dans la double signification que nous avons signalée, celle d'ordre biomorphique et celle d'ordre mécanique.

2. Le développement des deux concepts d'ordre

Chez les deux philosophes grecs les plus célèbres, ceux qui ont exercé l'influence la plus grande dans toute l'histoire de la pensée, c'est-à-dire Platon et Aristote, nous retrouvons les deux concepts d'ordre déjà mention- nés, combinés entre eux de sorte que l'un s'applique au ciel et l'autre à la terre, mais avec, dans l'ensemble, une suprématie de l'ordre biomorphique. Pour ce qui concerne les phénomènes célestes, Platon accepta le modèle introduit par les Pythagoriciens, le modifiant simplement sur un point, c'est- à-dire en substituant la terre au feu central et en faisant tourner tous les autres corps autour d'elle. De cette manière il créa ce qu'on a appelé le "modèle à deux sphères", selon lequel l'univers est constitué par deux sphè- res concentriques: la terre immobile au centre et le ciel qui tourne autour d'elle[4]. Il s'agit d'un modèle mécanique, parce que le phénomène principal auquel tous les autres sont réduits est le mouvement circulaire; cependant ce mouvement est produit par une âme immanente au monde, ce qui ressort du modèle biomorphique, parce que l'âme, chez les anciens, était essentielle- ment le principe de la vie.

Il semble, en outre, que Platon ait posé à ses disciples le problème sui- vant: comment expliquer, c'est-à-dire réduire à une règle, les mouvements apparemment irréguliers des planètes, en employant la formule devenue fameuse de "sauver les apparences" (sózein ta phainomena)[5]. Cela donna l'oc- casion à Eudoxe de Cnide de formuler sa célèbre théorie des sphères concen- triques, selon laquelle le mouvement apparemment irrégulier de chaque planète serait le résultat de la composition des mouvement réguliers, c'est-à- dire circulaires, de trois ou quatre sphères ayant comme centre la terre, mais avec des axes différemment inclinées et avec les pôles de l'une fixés sur la surface de l'autre. Cette théorie, un peu modifiée quant au nombre des sphè- res par Callippe, disciple d'Eudoxe, fut acceptée par Aristote, qui par consé- quent n'est ni l'inventeur ni le principal représentant du système géocentri- que.

---

[4] TH. KUHN, The Copernican Revolution. Planetary Astronomy in the Development of Western Thought, Cambridge, Mass., 1957, ch. I.

[5] Simpl. in de Caelo, p. 488 Heiberg = Eudoxos, frg. 121 Lasserre.

Un autre disciple de Platon, Héraclide de Pontus, donna au même problème une solution différente, en faisant tourner les planètes autour du soleil et le soleil autour de la terre, ce qui ressemble beaucoup au fameux système proposé au XVIe siècle par Tycho Brahe. Ensuite, Aristarque de Samos (IIIe siècle avant J.C.) résoudra le problème posé par Platon en admettant le mouvement des planètes et de la terre autour du soleil, ce qui ressemble au système héliocentrique de Copernic. Hipparchus de Nicée (IIe siècle avant J.C.) reprendra l'hypothèse géocentrique de Platon substituant aux sphères d'Eudoxe des mouvements circulaires excentriques et des épicycles (c'est-à-dire des mouvements circulaires ayant comme centre un point qui tourne lui aussi autour de la terre). C'est à cette dernière théorie que se rattachera Ptolémée d'Alexandrie (IIe siècle après J.C.), le fameux astronome auteur du système dominant jusqu'à Galilée, qui lui ajoutera seulement quelques complications ultérieures. Dans tous ces cas nous sommes en présence de modèles d'ordre de type mécanique, qui réduisent tous les phénomènes célestes à des mouvements circulaires.

Il faut ajouter, cependant, que Platon, et probablement ses disciples aussi (sauf Aristote), considéraient ces théories comme simplement vraisemblables, et non comme nécessairement vraies, parce que pour Platon il n'y a pas de science des phénomènes sensibles, de sorte qu'on peut faire à propos d'eux seulement un "discours vraisemblable" (*eikôs logos*), étant donné que la véritable science a pour objet seulement les réalités transcendantes. Un discours encore plus éloigné de ce qui est vrai concerne, selon Platon, les phénomènes terrestres, où l'on retrouve des mouvements complètement irréguliers. Cela n'empêche pas, toutefois, qu'aussi bien le ciel que la terre, pour Platon, soient ordonnés selon des rapports mathématiques, qui s'expriment dans "les formes et les nombres", rapports dont l'auteur est une intelligence démiurgique[6]. On peut conclure, par conséquent, que chez Platon nous avons une suprématie de l'ordre mécanique sur l'ordre biomorphique.

Aristote, comme cela est bien connu, conserve pour les phénomènes célestes l'explication donnée par Platon, Eudoxe et Callippe, qui était de type mécanique, attribuant aux astres une matière incorruptible et inaltérable (l'éther) et admettant comme cause de leurs mouvements des moteurs immobiles, qui agiraient en exerçant une sorte d'attraction amoureuse sur les âmes des astres (explication de type biomorphique). Pour Aristote aussi les mouve-

---

[6] PLATON, *Timée*, 27 C-29 D; 52 D-53 B.

ments circulaires des corps célestes obéissaient à des régles mathématiques, étant mesurés par le temps.

Un ordre complètement différent est admis par Aristote en ce qui concerne les phénomènes terrestres, qui sont caractérisés par des changements irréductibles au mouvement circulaire, dont le plus important est la génération et la corruption des substances, par exemple la naissance et la mort des êtres vivants. Tous les corps "naturels", c'est-à-dire non artificiels, selon Aristote ont en soi un principe de mouvement et de repos, leur «nature», qui se manifeste de la manière la plus complète dans les êtres vivants, dont il forme l'âme. Ce principe agit toujours en vue d'une finalité, qui pour les êtres vivants est l'accroissement jusqu'à la perfection complète et la reproduction. Nous sommes donc en présence, pour ce qui concerne la terre, d'un ordre de type biomorphique, qui coexiste avec l'ordre mécanique des cieux. Aristote compare l'"ordre" (*taxis*) de l'universe, considéré par lui comme un "bien", à l'ordre d'une armée commandée par le général ou à l'ordre d'une maison, comprenant femme, enfants et esclaves, commandée par le chef de famille[7], de sorte qu'on peut penser que pour lui aussi l'auteur de cet ordre est l'intelligence suprême, constituée par le premier des moteurs immobiles, c'est-à-dire le moteur du premier ciel.

A la différence de Platon, Aristote ne fait aucun recours à la mathématique pour expliquer les phénomènes terrestres, qu'il considère comme étant trop complexes pour être réduits à des mouvements mesurables; et à la différence de Platon, Aristote considère la connaissance de la nature comme une véritable science, à laquelle il donne pour la première fois le nom de "physique", qui a pour objet des régularités (par exemple le fait que "l'homme engendre un homme"), même si celle-ci admettent des exceptions (par exemple les monstres). La physique en effet, pour Aristote, est la science de ce qui arrive "toujours ou dans la plupart des cas". Tous deux, Platon et Aristote, admettent qu'au dessus de l'ordre du monde il y a un principe transcendant, dont l'ordre cosmique dépend et qui est l'objet d'une science différente de la physique, appelée par la suite "métaphysique".

Dans le moyen âge, tant les philosophes arabes et juifs que les philosophes chrétiens ont conservé les deux formes d'ordre développées par Platon et Aristote, attribuant l'ordre mécanique aux phénomènes célestes et l'ordre biomorphique aux phénomènes terrestres, en leur ajoutant, comme fondement, la notion biblique de création. Pour ce qui concerne, en particu-

---

[7] ARISTOTE, *Métaphysique*, XII 10, 1075 a 11-25.

lier, l'explication des mouvements des planètes donnée par Ptolémée, tant Averroès que saint Thomas d'Aquin lui ont attribué explicitement un caractère seulement hypothétique, parce qu'à leur avis les épicycles auraient été en contraste avec la physique d'Aristote, qui considérait aussi les sphères célestes comme composées d'éther[8].

A la Renaissance, les premiers astronomes modernes, c'est-à-dire Copernic et Képler, ont fait explicitement appel aux Pythagoriciens pour leurs théories héliocentriques et, surtout le dernier, pour leur conception mathématique des mouvements célestes. Ils ont donc gardé, sauf pour l'aspect géocentrique, la conception platonicienne de l'ordre mécanique de l'univers, tout en admettant, dans le cas de Képler, une âme (résidu de l'ordre biomorphique) comme cause des mouvements célestes. Les aristotéliciens, au contraire, ont attribué à la physique, du point de vue de la méthode, le caractère de nécessité absolue qui appartenait aux démonstrations mathématiques (Zabarella), négligeant presque complètement la métaphysique.

3. Galilée

Chez Galilée nous assistons au triomphe de l'ordre mécanique sur l'ordre biomorphique, triomphe qui se poursuivra pendant les trois premiers siècles des temps modernes. Galilée applique, en effet, également le modèle mathématique, qui jusqu'à son époque était appliqué de préférence aux phénomènes célestes, aux phénomènes terrestres, considérant seulement les aspects mesurables de ceux-ci et négligeant complètement le problème de leur "essence", c'est-à-dire de leur nature intrinsèque, de leur origine et de leur signification. On connaît ses affirmations sur l'univers conçu comme un livre écrit en langue mathématique, dont les caractères seraient des triangles, des cercles et d'autres figures geométriques[9], et sur l'impossibilité de connaître l'essence des choses[10]. De cette manière non seulement Galilée assimile la physique à la mathématique, mais il exclut la possibilité de la métaphysique, comme il résulte de son affirmation, selon laquelle dans le domaine des mathématiques l'intelligence humaine, au moins pour ce qui est de l'intensité de sa connaissance, est égale à l'intelligence divine, ce qui veut dire que la mathématique est un savoir absolu et constitue par conséquent le savoir suprême[11]. Il faut

---

[8] AVERROES, *De Caelo*, II, comm. 35; THOM. AQ., *In Aristot. De Caelo*, II, 17, 451.

[9] G. GALILEI, *Il Saggiatore*, dans *Opere*, Firenze 1890-1909, vol. VI, p. 232.

[10] GALILEI, *Istoria e dimostrazioni intorno alle macchie solari*, ibid., vol. V, pp. 187-188.

[11] GALILEI, *Dialogo sopra i due massimi sistemi del mondo*, ibid., vol. VII, pp. 128-129.

dire, d'ailleurs, que cette dévaluation de la métaphysique était partagée, à l'é-
poque, par les adversaires aristotéliciens de Galilée eux mêmes.

Lorsque Galilée découvre, grâce à sa lunette d'approche, que les corps
célestes ont les mêmes propriétés que les corps terrestres, par exemple que la
lune a des aspérités et que le soleil a des taches, réfutant de cette manière la
croyance d'Aristote en l'inaltérabilité des cieux, il semble assimiler les premiers au second, mais en réalité il trouve dans cette découverte la confirma-
tion de l'homogéneité du ciel et de la terre, et par conséquent de la validité de
l'application du modèle mécanique et mathématique à la terre aussi bien qu'en
ciel. De cette façon, comme il l'admet lui-même, Galilée ne reduit pas le ciel à
la terre, mais il porte la terre au ciel[12]. Là nous voyons le triomphe de l'ordre
mécanique, très simple, sur l'ordre biomorphique, bien plus complexe, c'est-à-
dire l'extension de l'ordre mécanique à l'univers tout entier.

Cela induit Galilée à croire que le mouvement circulaire, propre aux
corps célestes, appartient à tous les corps, c'est-à-dire qu'il est un état d'inertie, de même que le repos[13], ce qui est notoirement faux, même si cette thèse
mènera à la découverte du principe d'inertie, formulé par Descartes, selon
lequel le mouvement rectiligne et uniforme d'un corps n'a besoin d'aucune
cause, mais se maintient dans son être si rien ne vient l'empêcher[14]. Elle semble être manifestement en contraste avec la philosophie d'Aristote, selon
laquelle tout changement requiert une cause, mais sa validité est vérifiable
seulement dans des conditions idéales, c'est-à-dire pour des mouvements
sans frottement, qui dans l'univers physique, n'existent pas. Encore une fois,
donc, la mathématique l'emporte sur la physique dans le sens que la première absorbe la seconde.

Fort de cette assimilation, Galilée croit trouver la démonstration physique de la théorie copernicienne dans le phénomène des marées — qui au
contraire, comme l'on sait, est produit par l'attraction lunaire —, et par conséquent il donne à cette théorie non seulement la valeur hypothétique qui lui
était reconnue par le cardinal Bellarmin (par analogie à la valeur hypothétique reconnue à la théorie ptolémaïque par les commentateurs mediévaux
d'Aristote), mais il lui donne la valeur d'une description fidèle de la réalité
physique, c'est-à-dire la valeur d'une vérité ontologique. C'est pour cette raison, et par le fait qu'il donne la même valeur à toutes les théories physiques,
que Galilée demanda d'être nommé par le Grand Duc de Toscane non seule-

---

[12] Ibid., p. 62.
[13] GALILEI, ibid, p. 53.
[14] Voir A. KOYRÉ, *Etudes galiléennes*, Paris: Hermann, 1966, ch. III.

ment "mathématicien", mais aussi "philosophe" (naturel), c'est-à-dire connaisseur de la véritable nature des choses (naturelles).

Mais, comme il assimile la physique à la mathématique, Galilée doit donner aux démonstrations physiques la même valeur qu'il donne aux démonstrations mathématiques, c'est-à-dire la valeur de démonstrations nécessaires, qui non seulement montrent comment les choses se comportent, mais démontrent qu'elles ne peuvent pas se comporter autrement. Cela signifie relier les causes aux effets par un lien de nécessité, tel que les effets decoulent nécessairement de leurs causes. Du reste, la convertibilité des propositions, et donc la convertibilité entre les prémisses et les conclusions, c'est-à-dire entre les causes et les effets, est un caractère déjà reconnu comme propre à l'analyse mathématique par Aristote[15].

Pour cette raison le pape Urbain VIII a cru que l'argument des marées violait la toute-puissance divine, contraignant Dieu à créer le monde d'une certaine manière plutôt que d'une autre, et il se mit en colère lorsqu'il constata que dans son *Dialogue* Galilée non seulement reléguait l'affirmation du caractère hypothétique de la théorie copernicienne dans une prémisse imprimée en caractères différents du reste de l'ouvrage, mais mettait l'objection, selon laquelle l'argument des marées serait incompatible avec la toute-puissance divine, dans la bouche de Simplicius, c'est-à-dire du personnage ridiculisé par l'ouvrage tout entier[16]. C'est pour cette raison, probablement, que Galilée a été condamné par le Saint-Office[17].

Il n'y a pas de doute que le Saint-Office a eu tort de condamner Galilée, mais cela ne signifie pas que celui-ci avait raison: Galilée en effet n'a jamais fourni la démonstration scientifique que Bellarmin lui avait demandée pour pouvoir changer l'interprétation traditionnelle de la Bible, et que Bellarmin lui-même, il faut l'ajouter, tenait pour impossible, parce que lui aussi croyait qu'elle devait être une démonstration impliquant une nécessité. Le Saint Office a eu tort parce que les membres de cette institution, et aussi Urbain VIII, n'avaient pas compris l'absence totale de signification des théories physiques, quel que soit le type de démonstration qu'on attend d'elles, pour les vérités de la foi, et donc l'incompétence de l'Eglise dans les théories scientifiques.

Celles-ci sont d'une autre nature que la foi, mais non pas pour la raison

---

[15] ARISTOTE, *Seconds Analytiques*, I 12, 78 a 6-13.

[16] GALILEI, *Dialogo* cit., dans *Opere*, vol. VII, p. 488.

[17] Pour la documentation sur ces affirmations je dois renvoyer à mon étude *Implicazioni filosofiche della condanna di Galilei*, "Giornale di Metafisica", n. s., 5, 1983, pp. 239-261.

avancée par Galilée, c'est-à-dire parce que la science nous enseigne comment "le ciel va" tandis que la foi nous enseigne comment "on va au ciel". Il y a des véritées de foi, en effet, qui sont des vérités de fait, indépendamment de la valeur sotériologique qu'elles ont pour ceux qui croient en elles, et qui semblent clairement en contraste avec la science, par exemple la résurrection de Jesus Christ, qui est le fondement de toute la foi chrétienne. Les théories scientifiques sont étrangères à la foi parce qu'elles n'atteignent pas le niveau où se trouve le principe qui rend possible toute révélation et donc tout contenu de la foi, c'est-à-dire le Dieu transcendant et tout-puissant, qui peut faire exception aux lois de la nature parce qu'il les a créées lui-même. Il s'agit du niveau métaphysique, c'est-à-dire d'un niveau qui peut être atteint seulement par un discours qui n'est ni science ni foi, et qui ouvre entre la science et la foi l'espace nécessaire pour rendre possibles (non pour démontrer) les vérités de la foi.

Ni Galilée, ni les aristotéliciens de son époque, ni le pape Urbain VIII et ses ministres, n'avaient une idée claire de cette métaphysique. Les premiers, c'est-à-dire Galilée et les aristotéliciens, prétendaient que la science, physique et mathématique, par ses théories copernicienne ou ptolémaïque, était la seule connaissance possible de la réalité, et donnaient à la foi une signification simplement sotériologique ou bien pratique. Les autres, c'est-à-dire le pape et ses ministres, niaient à cette science la valeur de vérité, craignant qu'elle puisse menacer les vérité de la foi, et de cette manière eux aussi sous-entendaient que la science embrassait toute la réalité, excluant toute autre forme possible de connaissance.

II. De Galilée à nous jours

1. Développement de l'ordre mécanique

Après Galilée, l'ordre mécanique triompha presque complètement sur l'ordre biomorphique grâce à l'oeuvre des principaux philosophes et savants des temps modernes. Descartes réduisit l'univers matériel simplement à l'extension et au mouvement (mécanique), formulant pour la première fois le principe d'inertie, selon lequel le mouvement rectiligne et uniforme peut se maintenir indéfiniment dans son être, s'il n'y a pas d'obstacles, et donc n'a besoin d'aucune cause. L'univers tout entier, y compris les êtres vivants et même le corps de l'homme, est une grande machine, créée par Dieu, mais capable de fonctionner indépendamment de lui, privée de toute finalité. Pour sauveguarder les vérités de la foi chrétienne, c'est-à-dire l'immortalité de l'âme humaine et la transcendance de Dieu, Descartes conçoit l'âme comme une substance indépendante du corps, c'est-à-dire comme un pur esprit, et considère les lois mathématiques et physiques qui gouvernent l'univers comme l'expression d'un choix fait par Dieu d'une manière complètement arbitraire.

Spinoza n'aura plus ces scrupules et, proclamant le mécanisme absolu de la réalité, c'est-à-dire la nécessité mathématique de sa structure, y inclura même l'âme humaine et Dieu, considérant la première comme un simple aspect, de même que le corps, de la substance unique qui forme l'univers et le second comme identique à cette substance tout entière. Dans sa philosophie une physique mécaniste et déterministe conduit à une métaphysique de l'immanence de Dieu dans l'univers, c'est-à-dire au panthéisme. Une réaction à cette métaphysique sera tentée par Leibniz, qui réaffirmera la transcendance de Dieu et la spiritualité de l'âme humaine, mais d'une manière plus dogmatique que critique.

La version proprement scientifique de l'ordre mécanique de l'univers fut rendue possible par Newton. Celui-ci réussit, en effet, à réduire tous les phénomènes, tant célestes que terrestres, à une même loi, la loi de la gravitation universelle, qui s'exprime uniquement en termes de masses, forces motrices et mouvements mécaniques. On a pu concevoir, de cette manière, l'existence d'un ordre unique, de type très simple, auquel se ramènent les phénomènes les plus complexes. La caractéristique principale de cet ordre, comme il a été remarqué récemment[18], est la "réversibilité", c'est-à-dire la possibilité, pour les phénomènes mécaniques, de renverser la direction du temps dans lequel ils se produisent, sans qu'ils changent aucunement. Le modèle parfait de ces phénomènes est représenté par le mouvement d'un pendule sans frottement.

Probablement Newton lui même ne partageait pas complètement cette vision de l'ordre cosmique, parce qu'il considérait la force de gravitation comme capable d'expliquer seulement les phénomènes macroscopiques et non les microscopiques (c'est-à-dire les mouvements des corpuscules, que cependant il admettait), et comme demandant, à la fois, une explication ultérieure. En outre il était convaincu que cette vision du monde était valable seulement pour la réalité physique, c'est-à-dire pour la nature entendue comme différente de l'homme, et il admettait même que l'univers dans son ensemble était sujet à un processus irréversible[19]. Mais les philosophes qui

---

[18] I. PRIGOGINE - I. STENGERS, *La nouvelle alliance. Métamorphoses de la science*, Paris 1979.

[19] J'accepte, à ce propos, les précisions qui ont été faites, au cours de la discussion, par les professeurs S. L. Jaki et M. Pullman.

s'inspirèrent de cette vision du monde, en particulier les "lumières", n'hésitè-rent pas à appliquer le modèle mécanique à la réalité toute entière, y compris l'homme, faisant de celui-ci une simple machine, comme il résulte du célèbre traité, *L'homme machine*, de La Mettrie. Newton en outre, en tant que croyant, était convaincu que l'univers mécanique, décrit et expliqué par ses théories, était créé par Dieu. Laplace, le philosophe et physicien qui reprit la théorie kantienne de la nébuleuse primitive comme origine de l'univers, con-clut que Dieu était désormais devenu une hypothèse inutile[20].

Il faut ajouter que les philosophes et les savants modernes (par exemple Galilée, Gassendi, Newton lui-même) reprirent aussi la théorie atomiste, éla-borée par quelques philosophes grecs, qui nous présente un modèle d'ordre sans doute moins facilement réductible à la regularité et à la réversibilité des mouvements des astres. Mais il s'agit également d'un ordre de type mécani-que, parce que pour la théorie atomiste tous les phénomènes sont clairement réduits au déplacement des masses dans l'espace.

2. La crise de l'ordre mécanique

Le modèle mécanique, qui pendant les XVIIe et XVIIIe siècles s'était affirmé en tant que le type d'ordre dominant, subit une crise au cours du XIXe siècle à cause des développements de la thermodynamique, d'un côté, et de la biologie évolutionniste de l'autre. L'univers de Newton était fondé sur la réductibilité de toute forme d'énergie, et de la matière elle-même, à l'é-nergie mécanique, c'est-à-dire à une force capable de produire des mouve-ments de masses dans l'espace. Aux début du XIXe siècle le physicien anglais James P. Joule démontra la transformabilité complète de l'énergie mécani-que en énergie thermique, découvrant de cette manière le premier principe de la thermodynamique, c'est-à-dire que la quantité totale de l'énergie exi-stant dans l'univers reste immuable.

Mais le physicien français Sadi Carnot et le physicien allemand Rudolph J.E. Clausius demontrèrent l'impossibilité de la transformation inverse de celle admise par Joule, c'est-à-dire l'impossibilité de transformer complète-ment l'énergie thermique en énergie mécanique: chaque fois, en effet, qu'on tente de réaliser cette transformation, par exemple par la machine à vapeur,

---

[20] Evidemment, du point de vue scientifique, Laplace avait raison, comme l'a observé le prof. M. Moshinski, parce que l'existence de Dieu relève du domain de la philosophie, pas de celui de la science. Mais Laplace n'admettait autre domain de connaissance que celui de la science, ce qui n'est pas un point de vue scientifique, mais philosophique.

une quantité d'énergie thermique se perd (par exemple à cause du frottement) et la quantité d'énergie mécanique qu'on obtient est inférieure à la
quantité d'énergie thermique dont on disposait au départ.

Cette découverte venait démentir la croyance en la réductibilité de tous
les phénomènes physiques à des phénomènes mécaniques, qui était à la base
de la physique de Newton, et par conséquent en la réversibilité de tous les
changements physiques (caractéristique propre aux mouvements mécaniques), montrant l'existence dans la nature de changements irréversibles, qui
en plus se produisent vers un état d'équilibre caractérisé par une transformabilité mineure et donc doué d'une moindre utilisabilité de la part de l'homme.
Le physicien anglais Lord Kelvin nomma cette tendance "entropie",
employant un terme qui à la lettre signifie "involution", c'est-à-dire le contraire d'une évolution entendue comme progrès, et donc une forme de regression.
Bien plus, Clausius formula le deuxième principe de la thermodynamique en
terme d'entropie, affirmant que "l'énergie de l'univers est constante, mais l'entropie de l'univers augmente constamment vers une valeur maximale".

De cette manière la thermodynamique montra que l'ordre de l'univers est
plus complexe que l'ordre simplement mécanique et ses resultats furent
interprêtés comme attestant l'existence d'un certain désordre à l'intérieur de
l'ordre. Si, en effet, l'univers tend vers un maximum d'entropie, cela signifie
qu'il tend vers un état d'équilibre complet caractérisé par le désordre. Du
reste, on a la même impression lorsqu'on agite un vase contenant une certaine quantité d'objets disposés initialement dans un certain ordre: ce qu'on
obtient à la fin, c'est-à-dire quand ils arrivent à un état d'équilibre spontané,
est un mélange qu'on appelle naturellement désordre. On a l'impression, en
somme, que l'ordre mécanique est un ordre qui tend au désordre.

L'autre découverte qui produisit la crise de l'ordre mécanique fut la théorie de l'évolution des espèces biologiques proposée par Darwin. Comme il est
bien connu, selon cette théorie, en effet, à la suite d'une série de transformations dues à la sélection naturelle — c'est-à-dire à la survivance des individus
les plus doués pour faire face à la lutte pour la vie, et à la transmission de
leur caractères par hérédité —, les espèces vivantes se transforment, assu-
mant une structure qui est toujours meilleure, c'est-à-dire plus apte à affronter la lutte pour la vie, elles se transforment donc dans des espèces meilleu-
res, plus fortes, plus "évoluées". Darwin, probablement sous l'influence de
Spencer, a nommé ce processus "évolution", donnant justement à ce terme la
signification de progrès[21].

---

[21] E. GILSON, *D'Aristote à Darwin et retour*, Paris 1971.

Dans le cas de l'évolution des espèces vivantes, comme dans celui de la transformation de l'énergie mécanique en énergie thermique, nous sommes en présence d'un changement irréversible. Il n'est pas possible, en effet, qu'une espèce subisse une "involution", c'est-à-dire une transformation dans une espèce pire: elle doit ou bien évoluer ou bien périr. Mais dans ce cas, à la différence de ce qui arrive dans la thermodynamique, la direction des changements n'est pas de l'ordre au désordre, mais au contraire du désordre à l'ordre, ou au moins d'un état moins ordonné, dans le sens de moins résistant, moins riche, vers un état plus ordonné, en somme supérieur. Il arrive quelque chose de semblable à ce que nous pourrions observer si, agitant un vase contenant une quantité d'objets mêlés entre eux, et donc en désordre, nous obtenions, grâce à l'introduction d'un filtre, une séparation de certains objets des autres objets, et donc une forme d'ordre. Dans l'évolution biologique, le filtre est constitué par la sélection naturelle.

Cette différence de comportement entre les espèces vivantes et les réalitées non vivantes ouvre, dans la visione moderne de l'univers, un abîme entre deux mondes, le monde de la matière inerte et le monde le la vie, qui reproduit la distincion ancienne entre l'ordre mécanique, plus simple, et l'ordre biomorphique, plus complexe. D'où la necessité de chercher un nouveau modèle d'ordre, qui soit capable de réunir, ou bien de réconcilier, ces deux mondes et de constituer l'ordre total de l'univers. Ceci sera l'oeuvre du XXe siècle, en particulier de la tendance, ou du phénomène culturel, appelé "émergence de complexité", qui, comme nous allons le voir à l'instant, replacera même l'ordre mécanique à l'intérieur d'un ordre plus général, qui ressemble plus au type biomorphique qu'au type mécanique, et donc est plus complexe que simple.

3. L'émergence de la complexité

Par "émergence de la complexité" on entend aujourd'ui une série de recherches, qui vont du domaine de la sociologie (Edgar Morin) à celui de la psychologie (Jean Piaget), du domaine de la cybernétique (par exemple, "l'intelligence artificielle") à celui de la mathématique et de la physique (Réné Thom, Ilya Prigogine): toutes ces recherches aboutissent à une extraordinaire complexité de la connaissance et de la réalité[22]. Celles qui nous intéressent

---

[22] Voir G. BOCCHI e M. CERUTI (edd.), *La sfida della complessità*, Milano 1985. Comme il a été observé au cours de la discussion par le prof. P. Germain, il ne faut pas parler d'une "théorie de

à présent sont les recherches de Prigogine, parce qu'elles sont liées directement aux concepts traditionnels d'ordre et de désordre employés par les
visions philosophique et scientifique du monde[23].

Dans leur livre récent, *Exploring Complexity*, Nicolis et Prigogine affir-
ment que notre univers physique n'a plus comme symbole le mouvement
régulier et périodique des planètes, qui est à la base de la mécanique classique. C'est au contraire un univers d'instabilité et de fluctuation, qui sont à
l'origine de l'incroyable variété et richesse de formes et de structures que
nous voyons dans le monde. Dans le monde qui nous entoure nous trouvons
aussi bien des régularités inattendues que des fluctuations, également inattendues, d'une échelle très ample (par exemple dans le climat). A côté de
changements réversibles comme le mouvement d'un pendule sans frottement
où le passé et le futur sont interchangeables, il y a des changements irréversibles comme la diffusion et les réactions chimiques où il y a une direction pri-
vilégiée du temps. En outre nous sommes forcés d'admettre l'existence de
processus "stochastiques", c'est-à-dire gouvernés non par la nécessité déterministe, mais par une probabilité statistique[24].

Selon la vision classique de la science (siècles XVII-XIX), il y avait une
distinction très nette entre les systèmes simples, étudiés par la physique et la
chimie, et les systèmes complexes, étudiés par la biologie et les sciences
humaines. Or, cette différence est aujourd'hui en train de disparaître et dans
le même temps s'affirme une vision pluraliste du monde physique, où différents types de phénomènes coexistent l'un à côté de l'autre lorsqu'on change
les conditions auxquelles le système est soumis. Nicolis et Prigogine donnent
plusieurs exemples de comportements complexes: l'auto-organisation des
systèmes physico-chimiques, comme la formation d'un flocon de neige, cristal de forme dendritique, à partir d'un cc. d'eau lorsque la température arrive à 0 degrés; la convection thermique, les phénomènes induits par la ten-
sion superficielle et étudiés par la science des matériaux; les changements de
climat et leur variabilité; l'histoire de la terre à travers l'alternance de glacia-
tions et de diminutions des glaciers; l'histoire elle-même de notre univers à

---

la complexité", parce qu'il ne s'agit pas d'une théorie scientifique dans le sens courant du terme,
mais d'un ensemble de recherches et d'interprétations de la réalité.

[23] Malgré les réserves soulevées par beaucoup d'interventions à propos des théories de
Prigogine, j'estime celles-ci comme une référence très utile pour une discussion philosophique
de la complexité, à cause des considérations très générales qu'elles contiennent et qui se prêtent
facilement à des évaluations interdisciplinaires.

[24] G. NICOLIS - I. PRIGOGINE, *Exploring Complexity. An Introduction*, Munich 1987, Prologue.

partir du *Big Bang* vers une expansion progressive; le développement d'un embryon; les stratégies d'adaptation des insectes sociaux; l'auto-organisation des systèmes humains[25]. On pourrait ajouter les changements qui se produisent, au niveau de la neurophysiologie des états de conscience, à la suite d'une "information"[26].

Nicolis et Prigogine introduisent la distinction entre systèmes conservatifs, constitués pas des changements réversibles et qui tendent à conserver inaltérée leur propre structure, par exemple l'univers de Newton, et systèmes dissipatifs, constitués par des changements irréversibles et qui augmentent de plus en plus leur complexité, par exemple — ce sont nos auteurs qui l'écrivent — la région sublunaire de l'univers comme elle a été décrite par Aristote. Or, les systèmes dissipatifs forment une classe vaste et importante des systèmes naturels. L'analyse des systèmes physiques, par conséquent, ne peut pas être réduite à un jeu de mathématique: les systèmes physiques sous beaucoup d'aspects doivent être regardés comme hautement atypiques, ou non-génériques, du point de vue mathématique[27].

Parmi les systèmes dissipatifs il y en a d'ouverts, qui échangent continuellement de l'énergie avec leur entourage, provoquant des changements même dans celui-ci: par exemple tous les êtres vivants sont des systèmes dissipatifs ouverts qui soustraient énergie à leur milieu, produisant dans celui-ci une augmentation d'entropie, et la transforment en vue d'une organisation toujours plus complexe. Dans ce cas, les systèmes vivants et les systèmes non-vivants forment un seul ensemble, où l'entropie et l'évolution sont complémentaires[28]. Dans beaucoup de changements, affirme Prigogine, le passage vers la complexité est étroitement lié à la "bifurcation" de nouvelle branches de solution, qui naissent de l'instabilité d'un état de référence causée par la non-linéarité et par des liens qui agissent sur un système ouvert.

En conclusion, nous vivons dans un monde pluraliste, où des phénomènes simples, comme le mouvement périodique de la terre autour du soleil, nous apparaissent comme réversibles dans le temps et nécessaires, mais où

---

[25] Ibid., ch. 1.

[26] Voir G. DEL RE - E. MARIANI (edd).), *Il rapporto di Napoli sul problema mente-corpo*, Napoli 1991.

[27] NICOLIS-PRIGOGINE, *op. cit.*, ch. 2.

[28] Sur ce sujet on peut voir aussi le fascicule *Simplicité et complexité*, edité par M. Ceruti et E. Morin, dans "50 rue de Varenne", supplément italo-français de "Nuovi argomenti", n. 25, de l'Istituto Italiano di Cultura di Paris (mars 1988), et encore R. Morchio, *Sistemi ordinati e disordinati nell'evoluzione dell'universo*, "Giornata di astronomia", XV, 1989, nn. 3-4; *Quale portata ha l'evoluzione biologica in cosmologia?*, ibid., XVII, 1991, nn. 1-2.

nous trouvons aussi des changements irréversibles comme les réactions chimiques, et des changements stochastiques comme le choix entre des états de bifurcation. Selon la conception classique, l'irréversibilité et le hasard à l'échelle macroscopique étaient seulement des apparences dues à la complexité des comportements collectifs d'objets intrinsèquement simples. La croyance dans la simplicité du niveau fondamental était l'une des convictions dominantes de la science classique des trois derniers siècles. Aujourd'hui il faut reconsidérer la situation: le niveau fondamental n'est pas simple. Par conséquent il faut élaborer un formalisme plus ample capable de comprendre, à côté des systèmes les plus simples, le caractère stochastique et l'irréversibilité des systèmes les plus complexes[29].

L'"émergence de la complexité", que nous avons très brièvement décrite, rappelle de très près la vision de l'univers élaborée pas les philosophes anciens, en particulier par Aristote, c'est-à-dire un univers où l'ordre mécanique et l'ordre biomorphique cohabitent et forment un ordre global très complexe. Dans cet ordre, il y a de la place aussi pour ce qui peut être interprété comme une forme de désordre, c'est-à-dire qu'il y a des marges d'indétermination, de probabilité, de hasard, mais ceux-ci font partie eux aussi de l'ordre complexe.

Il ne faut pas, cependant, interpréter ces résultats comme une revanche de l'ancienne métaphysique par rapport à la science moderne, ni d'Aristote par rapport à Galilée. La théorie de la complexité, en effet, est une théorie simplement scientifique, qui n'a rien à voir ni avec la métaphysique ni avec la religion, et qui peut être interprétée aussi bien comme compatible avec una métaphysique de la transcendance, c'est-à-dire théiste, qu'avec une métaphysique de l'immanence, c'est-à-dire athée. Les théories scientifiques, en effet, sont complètement différentes par nature aussi bien de la métaphysique, parce qu'elles concernent l'ordre de l'univers et non son principe dernier, que de la foi, dont la possibilité suppose l'existence d'un principe trascendant par rapport à l'ordre du monde.

Il ne faut pas répéter, par conséquent, l'erreur commise par le Saint-Office aux temps de Galilée, c'est-à-dire surévaluer la signification des théories scientifiques et se prononcer sur leur valeur par rapport aux vérités de foi. Peut-être, la théorie de la complexité rend-elle plus facile l'acceptation d'une métaphysique de la transcendance, parce qu'elle nous présente un ordre cosmique qui, à cause de sa complexité, semble beaucoup plus récla-

---

[29] NICOLIS-PRIGOGINE, *op. cit.*, ch. 5.

mer un principe trascendant et intelligent, que ne le faisait l'ordre simple du modèle mécanique. Mais elle ne peut pas être un argument suffisant pour fonder cette métaphysique.

Les scientifiques aussi, par conséquent, ne doivent pas répéter l'erreur, commise par Galilée, de surévaluer les théories scientifiques, les considérant comme la seule forme possible de connaissance authentique, au delà de laquelle il n'y a de place que pour la foi. Si, en effet, la science était la seule connaissance possible, elle devrait embrasser toute la réalité qui est à la portée de la raison humaine, y compris le principe dernier de l'ordre cosmique, et inévitablement elle réduirait ce dernier à une partie de cet ordre, se transformant de cette manière en une métaphysique de l'immanence, incompatible avec la foi. C'est ce qui est arrivé, nous l'avons vu, aux temps modernes. Il faut espérer que les développements de la science contemporaine, en particulier les recherches sur la complexité, qui vont dans la direction opposée à celle qui a caractérisé le développement de la science dans les temps modernes, nous aideront à éviter que cette expérience se répète[30].

---

[30] Malheureusement, à cause du temps trés limité dont je disposais, je n'ai pas pu tenir compte des théories épistémologiques de L. Boltzmann et de J.-H. Poincaré, qui m'ont été justement signalées respectivement par les professeurs A. Lichnerowicz et G. Puppi. Je remercie aussi les professeur N. Cabibbo, Th. A. Lambo, W.K. Malu et B. Nicolescu, qui sont intervenus dans la discussion, apportant des précisions très utiles.

L'ÉMERGENCE DES STRUCTURES

RÉNÉ THOM

I.H.E.S., 91440 Bures-sur-Yvette, France

1. Introduction: A propos du vocabulaire

Ce titre qui m'a été proposé n'est pas sans comporter une certaine ambiguité. Les deux substantifs qui le forment, en effet, sont affectés d'une redoutable polysémie. L'émergence, en principe, désigne l'entrée à l'existence d'une entité nouvelle. Mais l'émergence n'est pas la naissance: le neuf est immergé dans l'ancien, qui subsiste - au moins un certain temps - en concurrence avec lui. De plus, il n'est pas évident que ce phénomène d'émergence soit fondamentalement *temporel*, au sens qu'il aurait nécessairement lieu dans le temps historique ou physique. L'émergence peut avoir la signification abstraite d'un processus continu au cours duquel, en fonction d'un paramètre (s) paramétrisant cette continuité, l'état final $s(1)$ du système considéré diffère de l'état initial $s(0)$ par l'addition d'un nouvel objet (S) qui sera précisément la structure émergeante. Qu'on pense à la *procession* du Néo-Platonisme et de la théologie comme exemple possible d'un processus intemporel d'émergence.

Quant au mot "structure", on connait la vaste multiplicité de ses significations. Disons, pour simplifier, que ces sens sont distribués le long d'un spectre unidimensionnel (un segment) (Figure 1), qui, à gauche, désignerait les structures abstraites (le paradigme de la structure abstraite étant la structure formelle, logique ou mathématique), et à droite les structures concrètes, comme la structure d'un pont, d'un bâtiment ou d'un outil.

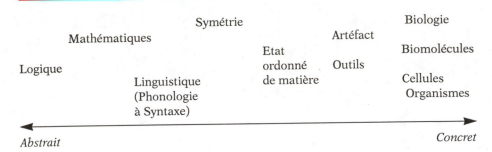

Figure 1: Axe des structures

En intermédiaire, on trouvera à gauche les structures linguistiques (par exemple, celles de la grammaire formelle, de la phonologie), les structures mentales ou culturelles (celles de Lévi-Strauss), les structures de la biologie (anatomie ou physiologie). Le problème de l'émergence d'une "structure" n'est alors qu'un cas particulier du problème général du changement des objets de pensée, dont certains, dits scientifiques ou artéfactuels (techniques), sont considérés comme représentant aussi fidèlement que possible la réalité objective, celle du monde sensible, ou agissant dans celle-ci en tant qu'instruments.

Il faudrait ici résoudre une question préjudicielle: un objet concret peut-il être une structure? Dans mon sentiment de francophone, je crois qu'il faut répondre par la négative. Certes, on désigne tel ou tel échafaudage, tel ou tel bâtiment comme une structure. Mais il n'empêche que le mot structure réfère toujours à une classe d'équivalence entre objets concrets, c'est un "universel". De plus le mot contient implicitement l'idée d'une certaine cohérence entre Ies parties, cohérence qui se manifeste par des propriétés internes globales d'opérativité, de régulation, de stabilité, voire de régénération après une lésion accidentelle de l'objet (ainsi qu'il arrive pour les êtres vivants). Très souvent ces propriétés de stabilité intrinsèque de la structure sont liées à des capacités fonctionnelles. Mais il est difficile de se replier sur une définition abstraite et générale, sauf à pratiquer l'impérialisme intellectuel de Bourbaki... Dans une optique plus philosophique, il est bon d'en revenir au vieux schéma hylémorphique d'Aristote: une structure concrète est une matière subordonnée à une forme, où la forme est la structure, le matériel sous-jacent pouvant varier en sa nature (mais non arbitrairement!). Par exemple le mot "pont" réfère à une structure concrète dont l'organisation (la forme) est parfaitement définie (un "tablier" reposant sur deux culées) - et dont la fonction est d'assurer un passage à travers l'obstacle sis entre les deux culées. En ce qui concerne l'émergence d'une structure concrète, cela se dit essentiellement d'un changement de forme où apparait une nouvelle structure (avec, en principe, conservation du substrat matériel). On parle beaucoup, actuellement, d'état *chaotique*. Je suppose qu'on admettra qu'un système chaotique - par antithèse - ne peut présenter de structure. Car d'une structure, on peut dire quelque chose; elle peut être décrite et, dans les meilleurs cas, définie, alors que du chaos, on ne peut rien dire (sauf - parfois - dans le cas d'un système chaotique à la Boltzmann) un milieu qui conduit par moyennisation à *l'apparence homogène* du substrat, c'est-à-dire qu'il n'existe aucun détail stable dans l'objet, ni aucune opérativité globale (symétrie) agissant dans cet espace.

Par là, évidemment, la théorie des structures se rattache à une théorie du changement de formes des systèmes naturels. Là le paramètre d'évolution est le temps physique (ou, en sciences humaines, historique). Si l'on disposait d'une théorie autonome des structures et de leurs connexions intrinsèques en tant que structures, (l'émergence étant considérée comme une connexion particulière), on pourrait essayer d'appliquer cette théorie à l'évolution *temporelle* des systèmes naturels.

L'émergence est un phenomène qui, psychologiquement, se rattache à la fascination qu'exerce sur tout esprit *l'innovation* en matière de production mentale.... Nous allons d'abord considérer l'émergence des structures au pôle abstrait du champ continu défini plus haut (Fig. 1). Ce sont les mathématiques qui nous en fourniront l'exemple.

2. *Les structures mathématiques.*

Contrairement à l'opinion standard (celle de Bourbaki), j'estime que ce qui importe en mathématiques, ce ne sont pas les structures, mais bien plutôt les *objets*. Ce qui - dans le point de vue moderne - justifie la prépondérance des structures sur les objets, c'est qu'un objet paradigmatique comporte une structure algébrique qui lui est canoniquement attachée, à savoir le groupe de ses automorphismes, et ce groupe peut agir sur d'autres objets. Mais d'un point de vue génétique - non structural - ce qui fonde le développement des mathématiques, c'est la construction, *l'émergence* de ses objets. La première illusion à détruire, c'est que l'émergence des structures en mathématique est une déduction logique. Voici le premier contre-exemple. L'objet

premier en mathématique, c'est l'ensemble N des entiers naturels (les cardinaux, qui dénombrent les ensembles finis). Si on lui applique le schème hylè-
morphique, on pourrait dire que la *matière* de N c'est la totalité des entiers
naturels (*n*), sa *forme* sera alors l'opération $n \rightarrow n + 1$, qui, appliquée à
l'entier initial 1, engendre tout l'objet. Vient ensuite, comme second objet, le
groupe Z des entiers relatifs. Pour l'obtenir il faut ajouter à N l'ensemble
défini par 0 et les entiers négatifs (*-m*). On voit de suite que cette opération
n'est pas une déduction logique, car elle comporte *un engagement ontologique,* celui qui consiste à admettre qu'un symbole tel que 2 - 3 a un sens. On
crée ainsi une nouvelle matière - à adjoindre à celle de N -, les entiers négatifs (y compris zéro), et la forme acquiert un nouvel opérateur $n \rightarrow n - 1$ qui
complète la forme de N de manière à former un groupe (le groupe abélien
des entiers relatifs Z). Il serait aisé de voir que tout grand progrès de la
mathématique a été lié à la construction d'un nouvel objet. Le continu (R)
serait alors le troisième objet (défini par exemple selon Dedekind à l'aide des
coupures de l'ensemble φ des nombres rationnels). En postulant que l'équation algébrique $x^2 + 1 = 0$ a deux racines *i* et *-i* (nouvel engagement ontologique), on crée une nouvelle "matière", scientifiquement fort importante: celle
qui constitue les objets analytiques. Une fonction analytique présente cette
propriété fondamentale qu'elle est déterminée par sa valeur au voisinage
d'un point. Tout comme Cuvier - à ce qu'on dit - reconstituait le squelette
complet d'un fossile à partir d'un seul os, le mathématicien peut reconstituer
une fonction analytique à partir de ses valeurs au voisinage d'un seul point
(son "germe" en ce point) - par le processus dit du *prolongement analytique.*
En un certain sens, toute fonction analytique a la propriété structurale de
pouvoir se régénérer à partir d'un fragment. C'est ce fait, duement exploité
en Physique, qui donne toute sa valeur à l'adage Galiléen: le livre du monde
est écrit en langage mathématique, et le fameux miracle *de l'efficacité déraisonnable des mathématiques* à décrire le monde (E.P. Wigner)[1] repose sur ce
principe.

Mais nous savons que ce miracle a ses limites; il ne fonctionne bien que
dans le domaine où jouent les symétries globales de l'univers physique, donc
en fait grâce à des hypothèses cosmiques. Pour les phénomènes locaux qui
intéressent notre monde sublunaire - donc en particulier pour les structures
locales qui peuvent apparaître ou disparaître du monde à notre échelle

---

[1] E.P. Wigner, The unreasonable effectiveness of mathematics in natural Sciences,
Communications, Pure and Applied Math., 13, n° 1, February 1960.

humaine, l'application du formalisme analytique est, nous le verrons, fort suspecte. Mais ne peut-on croire que néanmoins le prolongement analytique y garde une certaine validité locale et approchée? Pratiquement, ce vaste domaine qu'on appelle les *mathématiques appliquées* repose sur cette confiance; admettons-la: car toute extrapolation d'une fonction numérique repose - en dernière analyse - sur le prolongement analytique, et c'est sur cet algorithme que repose tout l'efficace du formalisme mathématique en science. De plus, chose essentielle, le prolongement analytique est un *principe d'individuation* des phénomènes physiques: le mouvement de la pierre qu'on lance vers le haut et qui retombe consistait pour Aristote de deux mouvements: un mouvement *forcé (biaios)* vers le haut, suivi après le sommet de la trajectoire d'un mouvement *naturel (phusikos)* de chute vers le centre de la terre. Galilée demontre experimentalement - à ce qu'on dit - que ces deux mouvements ont même équation ($z = z^0 -1/2\ gt^2$), établissant ainsi *l'unicité* du phénomène de la gravitation sur Terre.

Tout ceci ne permet pas d'affirmer que les êtres mathématiques non analytiques sont sans intérêt. En effet on sait qu'un objet non analytique peut être dans certains cas rendu analytique par un changement de variable approprié - au moins localement -. L'exemple type de ce phénomène est donné par le *théorème classique des fonctions implicites:* si une fonction numérique lisse F (pourvue de dérivées partielles continues $\partial F/\partial x_i$ autour de l'origine O, $x_i = 0$) admet une dérivée première non nulle en O, on peut, par changement de variables locales $x_i \rightarrow u_j$ transformer F en une coordonnée u_k (donc en une fonction linéaire). Bien entendu ce changement de coordonnées ne sera pas analytique (en général). Si on prend pour objet l'ensemble défini par l'équation $F = 0$, ce théorème montre que cet objet peut être transformé en $u_k = 0$, un "hyperplan" local. Un défaut dans la définition de l'objet peut être compensé par une transformation du milieu (un homéomorphisme). Par une telle transformation, on renforce la structure intrinsèque de l'objet - on le *linéarise* - au détriment des propriétés structurantes, intrinsèques, du milieu ambiant: la structure différentielle analytique de l'espace euclidien ambiant doit être perturbée - on doit donc y renoncer - pour régulariser l'objet. Ce type de transformation est utile pour une classification des formes fondée sur leur équivalence *qualitative,* phénoménologique comme celle que nous proposerons à propos des formes biologiques.

On a vu que l'édification des structures usuelles en mathématique ne procède pas d'une déduction logique (où il y aurait nécessairement perte informationnelle), mais au contraire *d'engagements ontologiques* successifs et superposés, qui créent une hiérarchie de "matières" (hylè) au sens de la *materia signata* de l'aristotélisme. Notre manière de voir, évidemment, n'est pas sans contredire le point de vue aristotélicien sur les mathématiques. En mon sens, on devrait dire, en Mathématique, que *la matière est engendrée par la forme*, et plus la forme, en tant que processus génératif, est subtile, plus est subtile la matière engendrée. Ceci est déjà bien visible sur l'exemple de la hiérarchie entiers naturels → entiers relatifs. La matière constituée par zéro et les nombres négatifs a évidemment un caractère beaucoup plus irréel que celle des entiers naturels. Ce phénomène explique les difficultés de l'enseignement des mathématiques, où l'étudiant - confronté aux premiers rudiments de l'analyse -, ne peut s'empêcher de penser: où *diable est-on allé chercher cela?* (Par exemple: Pourquoi s'intéresser à une expression telle que $\lim_{n = +\infty} (1 + 1/n)^n$?) On pourrait penser que l'émergence d'objets nouveaux au sein d'une morphologie empirique procède d'un mécanisme analogue. Si, *Galileo dixit*, la nature calcule, pourquoi ne pourrait-elle pas inventer? Et si, dans l'esprit de l'analogie Microcosme - Macrocosme, elle inventait ses objets comme le cerveau du mathématicien les invente? La spéculation peut paraître audacieuse .... C'est néanmoins celle que nous allons explorer.

3. *Structure et Morphogénèse: L'émergence des structures concrètes*

Ici, je crois qu'il nous faut quelque peu changer notre recherche. Nous ne sommes pas parvenus à définir ce qu'est une structure. A tout le moins, dans le cas d'un objet matériel, on sera en mesure de définir sa forme spatiale (et l'évolution de sa forme) en tant qu'objet plongé dans l'espace-temps. On sera donc amené à considérer la *morphogénèse* des entités matérielles. Certes, beaucoup d'objets du monde ne sont pas des structures ni, a fortiori, des "organisations". Mais au moins, dans la mesure où on peut localiser et décrire les objets, on saura ce dont on parle .... (ce qui, hélas, dans le discours de certains férus d'épistémologie, n'est pas toujours le cas des structures ...). Considérons donc une forme (F) comme un ensemble fermé plongé dans l'espace (ou l'espace-temps). Entre points de F on définit une relation d'équivalence (f), l'équivalence phénoménale. Deux points x, y de F sont (f) équivalents, s'il existe des voisinages U de x, V de y, et un homéomorphisme local $h_{xy}(U) \to V$ tel que si z et w sont des points de U, resp V, avec $h_{xy}(z) = w$, alors les points z et w ont des voisinages "phénoménalement homéomorphes", i.e. ont *qualitativement* la même apparence. Cette relation (f) est une relation d'équivalence et elle partitionne l'espace en sous-ensembles fermés, les "strates". A condition de ne pas regarder trop fin (cf. en Géologie la notion de "facies" d'un minéral laquelle précisément, sur une coupe de terrain, définit une strate de la géologie), il pourra se faire qu'on n'ait qu'un nombre fini de classes d'équivalences pour l'apparence ponctuelle du milieu en un point. C'est la situation des ensembles algébriques (et analytiques) dans l'espace euclidien (conduisant à la notion récente d'ensemble "stratifié"). Evidemment il se peut que certaines propriétés qualitatives d'un milieu puissent varier continuement (par exemple la couleur), constituant en ce cas ce que classiquement, on appelle un genre (γένος). Il faut alors décomposer ce "genre" continu en espèces et procéder comme fit Aristote pour définir l'organisation biologique grâce aux notions d'homéomère et d'anhoméomère[2]. On obtient ainsi une définition rigoureuse des différentes *parties* d'un objet spatialement défini, mathématisant la vague notion de *"situs partium"* par laquelle les Anciens définissaient l'organisation biologique. Cette systématisation, si imprécise et grossière qu'elle paraisse au réductionnisme moderne, n'en est pas moins un preliminaire nécessaire à la classification *linguistique* des êtres et des choses. Mais avant d'aborder les structures biologiques, on évoquera, à titre d'intermediaire, les instruments, les outils, les machines de l'homme.

4. L'émergence des structures artéfactuelles

Par structure artéfactuelle, j'entends les structures liées à la forme et au fonctionnement des outils. A cela on pourra aussi ajouter les instruments à caractère ludique, comme les instruments de musique et les jouets. Quand on considère l'apparition historique des techniques et des outils, on ne peut manquer d'être frappé par un fait: Très frequemment, un appareil ou un outil est apparu bien avant qu'une théorie cohérente, scientifiquement fondée, en ait été formulee et en justifie le fonctionnement. L'invention de l'horloge à foliot, qui réalisait la première forme de l'échappement en horlogerie, date du XIIIᵉ siècle, époque où aucune théorie de la dynamique n'existait (sauf celle, erronée comme chacun sait, d'Aristote). De même, on a taillé des verres de lunette dans la même époque, bien avant Snell et Descartes. Et

---

[2] Sur ce sujet, voir R. THOM, *Homéomères at Anhoméomères en Théorie Biologique d'Aristote à aujourd'hui* in *Biologie, Logique et Métaphysique chez Aristote.* Editeur P. Pellegrin. Paris: Editions du C.N.R.S. 1990.

en Anthropologie, il existe de nombreuses découvertes de techniques très importantes (comme l'agriculture, la domestication du bétail ... etc) apparues indépendamment et pratiquement de manière simultanée en des points trop éloignés du globe pour qu'on puisse raisonnablement invoquer une propagation geographique. Je ne crois pas non plus qu'on puisse uniquement faire appel à une expérimentation individuelle issue d'un *bricolage*[3] couronné de succès pour justifier ces faits. On ne peut guère — me semble-t-il — éviter une théorie du style de la réminiscence platonicienne (le Menon) que j'esquisserais comme suit: la dynamique de notre organisme est régie localement par des objets mathématiques dotées de matière et de forme. (N.B: la matière est ici à concevoir sur Ie mode des êtres mathématiques définis au §1, et elle ne s'identifie que partiellement avec la matière physico-chimique.) Lors de l'activité motrice de nos membres, il y a inscription dans l'espace-temps de ces dynamiques locales qui se réalisent alors comme champs moteurs dans les activités usuelles de la vie. Certaines de ces activités, les plus simples, apparaissent nécessairement dans le développement canalisé de l'ontogenèse. Le jeu apparait alors comme l'apparition de ces dynamiques, dans un cadre biologiquement non motivé, et s'appliquant sur une matière non vivante extraite de l'environnement. Par exemple, on a vu, au §1, que la mathématique partait de l'opération $n \to n+1$. Réalisée géométriquement sur la demi-droite R^+ , cette transformation a pour espace-quotient le cercle (S^1) (positivement orienté). Par localisation au bout d'un doigt, par exemple, une telle structure dynamique — une rotation — se projette éventuellement matériellement dans un matériau plastique. D'où la possibilité de fabriquer des objets qui réalisent (par leur bord) ces structures cinétiques. On pensera ainsi à *!a roue*, que les Mayas connaissaient dans leurs jouets, mais qu'ils n'eurent pas l'idée d'appliquer à la construction d'un chariot. (La roue a d'ailleurs une réalisation organique dans la rotule du genou). La bicyclette peut être regardée comme une simplification (la section par le plan de symetrie) du schema tétrapodal du vertébré - où chaque roue réalise l'effet d'une paire de membres. Finalement, la morphologie de l'objet artéfactuel est le fruit d'un compromis entre une structure abstraite, de nature algébrico-dynamique - le schéma dynamique intrinsèque de la fonction opératoire -, et une dynamique concrète issue des déplacements du corps humains, ou visant à satisfaire imaginairement des besoins humains (cf. à ce sujet les idées de G.

---

[3] Le *bricolage* a été suggéré par Lévi-Strauss comme responsable de l'origine des mythes. Voir à ce propos La Pensée Sauvage, Paris: Plon 1962, pp 26-27. Le biologiste F. Jacob a repris la même explication pour l'origine des morphologies vivantes.

Simondon[4]). Autre exemple: la mesure du temps pourra se faire d'abord par un phénomène physique de vitesse approximativement constante (c'est le principe du sablier, de la clepsydre antique). Dans l'horloge à échappement, mue par un poids, on pourrait théoriquement mesurer Ie temps écoulé par la chute du poids; mais on préfère freiner cette chute par liaison avec un pendule réalisé matériellement dans les oscillations du balancier; le nombre des oscillations s'inscrit alors comme autant d'arcs sur le cadran, trajectoire de la pointe d'une aiguille mue par la chute du poids. La structure algébrique sous-jacente est en fait une rationnelle de la forme a/b, où a et b sont non des nombres, mais des cercles, le cercle a étant un revêtement du cercle b. La raison d'être du cadran (a) est alors de fournir un modèle de la périodicité nycthémérale, alors que le cercle (b) symbolise la période - beaucoup plus petite - du pendule.

Un des premiers outils de l'humanité a été l'instrument tranchant, réalisé par le *silex biface* qui a connu plus tard - en changeant de substrat matériel - une grande postérité (le couteau, la hache, l'épée...). Là la structure dynamique théorique génératrice est le "cusp dual" de la théorie des catastrophes[5]. Le fait qu'il en existe de nombreuses réalisations organiques (par exemple, parmi les dents, les incisives) témoigne du rôle fondamental de ces structures dans la morphogenèse des structures artéfactuelles et biologiques. C'est là, où évidemment, le problème de l'émergence des structures s'est le plus clairement manifesté.

5. L'émergence des structures biologiques

Dans le discours actuel de la Biologie contemporaine, on considère que le problème de la morphogenèse est résolu par la Génétique: *tout sort de l'ADN du génome,* lequel se transmet pratiquement identiquement de génération en génération. Nous observerons simplement qu'une telle affirmation présuppose résolu le problème de l'organisation d'un système de molécules données "a priori" dans une enceinte dans des conditions physiques fixées.

---

[4] G. SIMONDON, *Du mode d'existence des objets techniques,* Paris: Aubier 1969, p. 58: "Le dynamisme de la pensée est le même que celui des objets techniques; les schèmes mentaux réagissent les uns sur les autres pendant l'invention comme les divers dynamismes de l'objet technique réagiront les uns sur les autres dans le fonctionnement matériel."

[5] Sur le cusp dual, voir E.C. ZEEMAN, *Catastrophe Theory,* Reading, Mass.: Addison-Wesley 1977, p. 60.

L'évolution d'un tel système est-elle régie par un déterminisme rigoureux?
Peut-on prévoir l'état final du système après un temps fini connaissant l'état
initial et la variation temporelle imposée des paramètres physiques de contrôle (température, etc)? Je suppose qu'on m'accordera que hormis quelques
cas purs obtenus par idéalisation (cristal parfait, gaz dilué), ce problème
général est non résolu. Pour des raisons principielles l'existence même d'un
déterminisme pragmatiquement contrôlable est incompatible avec un phénomène comme la *métastabilité* en Thermodynamique. Comme par ailleurs
le milieu général du monde vivant (le cytoplasme) est hautement polyphasi-
que (un mélange de structures descriptibles - organelles -, et de fluides à
organisation variable ou nulle), on doit considérer que l'affirmation d'un
strict déterminisme génétique en Biologie ressort de l'acte de foi. Ce qui est
en jeu, ici, c'est décider comment deux organismes sont "les mêmes": ce qui
peut se définir de manière conceptuellement acceptable si l'on s'en tient à
une équivalence phénoménologique grossière (l'équivalence "stratifiée" définie plus haut); même en ce cas, le fait que deux jumeaux homozygotes peuvent souvent être distingués par examen direct grossier montre les limites du
déterminisme génétique.

La variété quasi-infinie des formes vivantes est bien connue. Mais si l'on
considère, non plus l'anatomie, mais la physiologie, on observera une très
grande régularité dans la définition des grandes fonctions physiologiques.
C'est sur cette régularité qu'est fondée la distinction entre Animaux et
Végétaux. Les premiers tirent leur énergie par prédation, les seconds par
saprophytisme ou photosynthése[6]. De même, en Anatomie animale, si l'on
s'en tient à une équivalence "topologique" modulo la direction des grands
gradients externes comme la gravitation, il n'y a qu'un nombre restreint de
plans généraux d'organisation (une dizaine au plus). En sorte que le problème de la genèse évolutive de ces plans peut fort bien être considéré comme
un problème "in abstracto". C'est ce point de vue que j'ai développé dans
mon livre[7]. Considerant l'embryologie comme une succession de bifurcations
d'une dynamique initialement ponctuelle (l'ovule quiescent avant féconda-
tion), je me suis efforcé de déterminer le *modèle minimal* de succession de
bifurcations pouvant conduire à une dynamique assez riche pour engendrer
la totalité des grandes différenciations cellulaires indispensables à l'orga-

---

[6] La distinction Animal-Végétal peut être embigüe chez certains Unicellulaires (Euglena
Gracilis).
[7] Pour la définition du modèle de la Blastula Physiologique, voir: *Esquisse d'une
Sémiophysique*, (Paris: Interéditions, 1988), chap. 4, §D.

nisme animal (Modèle dit de la *Blastula Physiologique*, escalier composé de quatre cellules de feed-back). Je crois qu'on peut considérer ce modèle comme une illustration de cette idée générale d'une évolution biologique construisant ses structures comme le mathématicien construit ses theories. Je suis trop grand admirateur d'Aristote pour ne pas déplorer que, dans une telle construction, j'ai quelque peu négligé le "substrat" (hypokeimenon). Mais ce qui importe dans un substrat, ce sont les contraintes qu'il apporte aux bifurcations virtuelles de la dynamique qu'il supporte. L'oscillation continuelle entre un point de vue "gradient" donc dissipatif, et un point de vue purement hamiltonien y est génerale (cf. la transformation de Van der Pol d'un cercle "hamiltonien" en un "feed-back loop", cf. [7]). C'est celle qui s'impose dans la construction de deux "feed-back loops" fonctionnellement associés (*la coexistence des coplis*[8] dans la concaténation de deux catastrophes cusps), dans une articulation, entre deux os contigus et les tendons mobiles qui les lient. Il faut en Biologie accepter une finalité de fait, dont on s'efforcera de trouver un modèle dynamique qui minimise le caractère arbitraire de la description et qu'on pourra s'exprimer comme un "changement de phase" (en un sens généralisé et *local*) du milieu. Il me semble qu'on retrouve ici sous une forme simple toute la complexité efficiente que montre souvent l'interaction à effet enzymatique d'une protéine et de son ligand. Ici, je ne peux m'empêcher de penser à la mystérieuse théorie du *Nous* aristotélicien: une entité impersonnelle qui investit un être local qu'elle rend "intelligent": une sorte d'énergie dirigée permettant une bifurcation originale et stable de la dynamique ambiante à l'objet.

6. *Le problème philosophique de l'innovation.*

J'aimerais terminer par une réflexion générale sur la question de l'émergence - ou plus généralement - de l'innovation. Si l'on considère que tout savoir scientifique doit être transmissible (d'homme à homme), un tel savoir est nécessairement exprimé par un texte d'une langue reçue. (Il n'y a plus de science en tradition orale ...).

De ce fait, le savoir se formalise et s'enregistre en un texte comportant uniquement des symboles admis dans la collectivité sociale et *en nombre fini*. De là vient ce que Borgès a évoqué dans son essai: *La grande bibliothèque de*

---

[8] La coexistence des coplis, marque de finalité, est explicitée au Chap. 3, § H du livre cité supra [7].

Babylone[9]. Le texte innovateur existe déjà virtuellement dans le corpus des textes formels lisibles par un homme au cours de son existence: il suffit d'aller l'y lire... Evidemment, la grande biblothèque est très grande — et ceci explique qu'on puisse, parmi tous les textes existants, n'en retenir qu'un petit nombre comme réellement innovateurs et significatifs. Il n'en demeure pas moins que sur le plan de la production symbolique, l'innovation radicale n'existe pas, elle peut être mécanisée ...; on bute seulement sur le temps immense qu'il faudrait consacrer à cette recherche.

Tout le génie revient à savoir discerner dans l'ensemble des textes ainsi construits ceux qui présentent des idées innovantes particulièrement promet-teuses. Au XIX siècle, l'Anglais George-Henry Lewes avait proposé le terme d'*émergence* pour caractériser l'apparition de propriétés nouvelles dans un composé, inexistantes dans chacun des composants pris isolément (Observation faite par le P. Jaki dans son rapport à cette Rencontre). "Au XX siècle le zoologue allemand Konrad Lorenz[10] avait qualifié de *fulguration* l'apparition brutale d'une structure présentant des propriétés radicalement nouvelles par rapport au milieu (physique ou culturel) existant. (Ainsi l'appa-rition de la vie par rapport à la physico-chimie; cf. à ce sujet l'article de

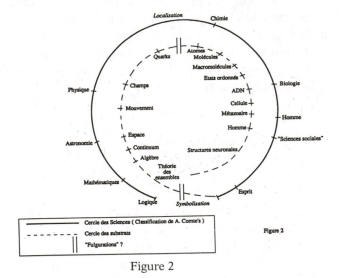

Figure 2

---

[9] JORGE LUIS BORGES, *La Bibliothèque de Babel*, in *Ficciones*, Buenos Aires: Emece Editores S.A. 1956.
[10] K. LORENZ, *Die Rückseite des Spiegels, Versuch einer Naturgeschichte Menschlichen Erkennens*, München: R. Piper 1973.

Klaus Ruthenberg[11].) Cette problématique peut être symbolisée par la figure 2, où on a superposé un cercle des théories (des sciences), reprenant la hiérarchie classique d'Auguste Comte - à un cercle de leur substrats.

Contrairement à la vision habituelle, ce schéma suggère qu'on devrait remplacer la fulguration classique - apparition de la matière vivante émergeant de la physico-chimie - par une double émergence qu'on a representée sur le cercle de la figure 2 par deux doubles barres verticales. L'une, en haut, exprime la localisation de la matière dans le noyau atomique (la matière se séparant du plasma des gluons, en fait l'ancrage matériel de la géométrie euclidienne), l'autre, en bas, précipitant la matière vivante dans la pensée symbolique.

L'impossibilité théorique de l'innovation scientifique radicale exprime un paradoxe:

Il y a une contradiction de principe entre l'induction par voie experimentale (la description et l'enregistrement des régularités phénoménales) et le progrès scientifique par innovation.

Car si l'innovation est effectivement un phénomène concret, ce phénomène, en tant que processus descriptible, doit être plus qu'une conséquence déductible du savoir reconnu: il ne peut être une conséquence linguistiquement ou formellement dérivable des connaissances reçues antérieurement - auquel cas il n'y aurait pas innovation (cf. les deux exemples classiques: la supernova de Tycho-Brahé, la découverte de la radioactivité par Becquerel).

Comme d'autre part chacun accepte que la science progresse, on ne peut échapper au paradoxe qu'en remarquant le fait: dès qu'un phénomène est linguistiquement exprimé, il s'insère dans un ensemble théorique de propositions que véhicule avec lui le langage (et ceci, aussi bien dans le langage usuel que dans la formulation proprement scientifique du phénomène). C'est quand l'expérience *infirme* une de ces conséquences qu'a lieu la découverte innovante. En ce sens l'activité de la Science est *fondamentalement critique*, elle ne progresse qu'en niant ses progrès. Le plus souvent, au prix d'une complication du formalisme existant, on va pouvoir rendre compte du phénomène innovant. Il faut alors apprécier le caractère *ad hoc* du nouveau formalisme en liaison avec l'ampleur et les perspectives offertes par l'extension du langage. De là vient que le progrès de la Science provient plus d'une amélioration *des capacités déductives du langage descriptif* que de l'extension proprement dite du savoir empirique, une situation que le développement moderne

---

[11] KLAUS RUTHENBERG, *Is there a philosophy of Chemistry?* 9th International Congress of Logic, Methodology and Philosophy of Science, Abstracts Vol. III, Section 11, Harper Torchbook 1990, p. 377.

des techniques expérimentales pris comme un but en soi tend à occulter dans l'esprit des scientifiques d'aujourd'hui.

Par une technique qualitative comme la Théorie des Calastrophes (fondée sur une structure fibrée où la Dynamique-fibre est rapide, la Dynamique-base est lente, le rapport de ces vitesses tendant vers l'infini), on n'obtient pas des conséquences vérifiables *stricto sensu* comme dans les modèles classiques de la Physique; les prédictions ne sont que qualitatives, et le modèle proposé n'en tire qu'une validité hypothétique. Mais là où, comme c'est souvent le cas en Embryologie, il n'y a aucune intelligibilité du processus, la seule possibilité d'une interpretation intelligible devrait, raisonnablement, être considérée comme un gain.

Partons du fait que tout savoir scientifique, étant verbalement exprimé, participe nécessairement de la "métaphysique implicite" que véhicule tout langage. Il ne faudrait surtout pas croire que le langage de la science est une langue formelle pure. Une des grandes difficultés (théorique - cela va sans dire -) de la Mécanique Quantique réside dans la jonction entre le schéma mathématique mis en oeuvre (Espaces de Hilbert, opérateurs, représentations des groupes, etc), et l'usage inevitable de concepts "classiques" dans l'interprétation expérimentale des previsions théoriques. (Détecteurs, interface du monde quantique et du monde classique, mesurement ...). Le langage usuel est vecteur de la causalité, et toute expérimentation vise à vérifier une hypothétique causalité. Il y a deux grandes structures théoriques du langage usuel: la prédication, qui apparait dans la copule être: *Le ciel est bleu*, et la phrase transitive: *Pierre bat Paul*, ces deux types de phrase représentent (génériquement) les structures minimales porteuses d'une signification autonome. On peut penser (et beaucoup de linguistes semblent actuellement l'accepter) que ces deux structures apparaissent comme des universaux linguistiques, présents - *mutatis mutandis* - dans toutes les langues du monde. La description d'un savoir scientifique experimental met en jeu ces deux structures, bien que ces formes de causalité ne soient pas toujours prises en compte dans le formalisme mathématique. Par ailleurs, notre problème de l'émergence est lié à une question récurrente depuis la Relativité d'Einstein: le temps n'est-il que la quatrième coordonnée d'un espace-temps minkowskien, ou est-il porteur d'un caractère spécifique qui le sépare des coordonnées d'espace? Le temps a-t-il des vertus créatrices de phénomènes et d'événements, ce que n'aurait pas l'espace, entité purement statique? Un philosophe comme A.N. Whitehead[12] identifiait Dieu à l'action continuellement

---

[12] A.N. WHITEHEAD, *Apart from the intervention of God, there could be nothing new in the*

innovante du temps (d'où, aussi, l'irréversibililé foncière du temps). On se rappellera, dans cet esprit, la célèbre discussion de 1922 entre Bergson et Einstein, à Paris[13] (laquelle ne fut guère qu'un dialogue de sourds).

Dans la perspective einsteinienne, on ne peut guère croire à une spécificité du temps qui rende possible l'innovation radicale. Un tel point de vue met en question la validité des "épistémologies évolutives" qui, depuis Darwin, ont cru expliquer le changement en faisant appel à la pérennité d'un substrat. Si on explique la stabilité des choses par une propriété formelle des configurations locales des phénomènes en tant que formes géométriques, on ne voit pas pourquoi le même type d'explication ne serait pas valable dans l'espace-temps, en dimension quatre. Je crois qu'on n'a pas assez réfléchi aux conditions mêmes, *d'origine linguistique*, qui sont à l'origine du phénomène (au sens étymologique du Grec *phainomenon:* paraissant), vu comme transfert irréversible d'une émanation du réel dans le psychisme de l'observateur. L'arc-en-ciel est un phénomène, mais il n'a pas de substrat ...

La nécessité de prendre en compte le caractère virtuel du phénomène (un contour apparent) oblige à introduire le conflit de deux dynamiques, une lente (de type substrat, support de la substance qui engendre l'étendue de la spatialité), une rapide, qui engendre la qualité, le prédicat. Nous avons dit, au début, que l'émergence n'était pas la naissance. La naissance est (sous sa forme originelle) la scission de l'être maternel représentée graphiquement par le symbole de la fourche; les deux entités émergeantes sont qualitativement de même nature.

Or il existe une forme typique d'émergence, où l'entité émergeante est de nature fondamentalement différente des entités qui l'engendrent: c'est le cas de la lumière issue d'un corps émissif, et qui permet au mouvement des planètes d'être une phénoménologie. C'est, dans un domaine différent, le cas du sens, de la signification, qui émane du symbole lorsque dans l'activité symbolique le symbole est compris. Une fois l'intellection faite, le symbole perd pratiquemcnt toute importance, il est anéanti. On peut représenter le processus par la fourche inverse.

Contrairement au symbolisme de l'antique *sumbolon,* dans la phrase autonome minimale ("le ciel est bleu"), les deux composantes (substance prédicat) ne sont pas ontologiquement équivalentes, elles reflètent l'opposition: "temps lent - temps rapide" évoquée plus haut (ou encore l'opposition sail-

---

world and no order in the world. In: *Process and Reality,* New York: Harper & Row, Publishers, Harper Torchbook 1960, p. 377.
    [13] BERGSON, EINSTEIN, Bulletin de la Société Française de Philosophie, Paris, 1922.

lance - prégnance définie en [7]).

Seule cette dualité substance - prédicat peut engendrer le sens, la signifi-cation, la copule "est" ne jouant le rôle que d'un auxiliaire grammatical indi-spensable à l'autonomie de la signification. Cette binarité du symbole paraît essentielle pour la constitution du sens. Il faut laisser aux métaphysiciens et théologiens l'exclusivité de l'emploi *absolu* du verbe être.

DISCUSSION
(A. LICHNEROWITZ, chairman)

LICHNÉROWICZ: Je remercie René Thom de cette contribution extrèmement riche et qui va certainement animer et conduire à une discussion. À propos de la notion de fulguration, il y a des gens qui s'intéressent à cette notion. Peut-être pourrait-on en discuter.

THOM: J'ai essayé de traiter ici ensemble d'une part les sciences existantes et d'autre part ce qu'on pourrait appeler *les matières* associées à ces sciences, les "matières" en un sens aristotélicien. On pense d'ordinaire que la fulguration essentielle est celle qui sépare la vie de la physico-chimie, comme je l'ai dit. Ici je serais tenté de dire que cette "fulguration" se dédouble, manifestant deux discontinuités: localisation et symbolisation. Il y a un élément tout à fait neuf dans la localisation spatiale d'un noyau atomique (ce qui assure la stabilité de la matière dans les conditions usuelles). Dans la mécanique quantique d'aujourd'hui ceci se manifeste par le problème dit du "confinement des quarks"; à partir de là la notion de propriété locale apparait, on peut travailler réellement, en chimie par exemple. La symbolisation est l'apparition de structures formelles localisées aboutissant à la pensée.

DEL RE: Your lecture has been very rich and I may have missed something, but I have not caught a clear distinction between the word "structure" and the word "organization". I think this is something particularly important, especially in connection with biology. I often give, as an example, not a biological example but something much more familiar to us: a railroad network, or a railroad company, if you like. They have, at the same time, a structure and an organization, and I believe that the two terms do not denote the same thing, they are different. Could you please comment on this point?

THOM: Well, I think it is a question of linguistic use. For me a structure cannot be a continuous object. It is an abstract entity, discrete, fundamentally discrete. We may go up to dimension one, a graph for instance. A two-dimensional, or three-dimensional object can hardly be said to be a structure, except metaphorically, when for instance we design a building as a structure. I suspect it is a question of linguistic feeling.

DEL RE: Just a moment, organization in a living being is some kind of dynamic cooperation intended to fulfill a certain task like survival, for instan-

ce, in a changing environment. When I was giving the example of a railroad network, I meant that the organization of railroads, at least hopefully, should ensure that regardless of changes in number of passengers,
in number of trains and so on, the service should continue to be provided, whereas the structure is the junctions, the stations, the lines and so
on. So, in this sense I see an essential difference between the two things.
So I wonder if it is just a linguistic point.

THOM: Would you say that a structure has essentially a finalistic motivation?

DEL RE: An organization, not a structure, a structure is just a sort of correlation.

THOM: I would myself accept the statement that an organization is a structure with a finality, yes. But will you state the distinction between structure
and organization?

DEL RE: À mon avis, une structure, c'est quelque chose de statique, qui n'a
pas nécessairement une relation avec une tâche que l'organisme ou
l'objet quelconque dont on s'occupe est censé réaliser. C'est plutôt une
relation plus ou moins statique, peut-être essentielle, entre les différentes
parties. Voilà ce qui me gêne.

THOM: Autrement dit, pour vous, la structure est toujours un schéma appau-
vri de l'objet. C'est vrai qu'a mon sens une structure a toujours quelque
chose comme un aspect algébrique, on peut la décrire par un graphe ou
un schéma de ce type. Tandis qu'une organisation, à mon sens, c'est
quelque chose de beaucoup plus délicat à décrire.

DEL RE: Plus complexe?

THOM: Oui, on pourrait dire plus complexe si on n'avait pas peur du mot, ce
qui est mon cas, mais là, je pense qu'en biologie, une des grosses difficultés, c'est précisémente la définition de la notion de forme. La notion
de forme est une notion qui n'est pas susceptible d'une définition ni algébrique, ni mathématique. Là on bute sur une difficulté, par exemple, si
vous devez comparer un coquillage qui est en corne droite et un coquillage qui est en spirale, vous direz qu'ils n'ont pas la même forme, mais
néanmoins, du point de vue topologique, on peut dire qu'ils ont pratiquement le même plan d'organisation, n'est-ce pas? Chose plus compliquée,
il faut tenir compte des difficultés dues aux possibles ambiguités d'une
forme. Cela se voit à un niveau très élémentaire; un de mes exemples
favoris est le suivant: vous prenez un losange, avec une grande diagonale
et une petite diagonale, mettez la grande diagonale horizontale et la peti-

te verticale, et maintenant prenez le même losange tourné de π/2, est-ce qu'ils ont ou non la même forme? Je suis sûr que quelqu'un qui a une bonne culture de géomètre dirait immédiatement: ils ont la même forme, ils sont égaux. Mais l'homme naïf, l'*ageometretos* de Platon, sera sans doute d'un avis différent.

DEL RE: Le rapport avec le milieu ambiant.

THOM: Oui, c'est cela, il y a le rapport au milieu ambiant, il y a ce que le philosophe Ferdinand Gonseth appelait le référentiel. Par exemple dans le système de la signalisation routière, ou signalisation ferroviaire, les deux losanges n'auront pas la même signification, ce seront des symboles différents. Alors on a cette difficulté de la notion de forme mais même intrinsèquement il y a des problèmes à définir rigoureusement: quand est-ce qu'on peut dire que deux objets ont la même forme? Alors faites-vous intervenir la forme dans l'organisation, à votre sens?

DEL RE: Non, dans mon cas c'est essentiellement une coopération dynamique entre les parties.

THOM: Si vous abandonnez la forme, alors effectivement vous pouvez peut-être identifier l'organisation et la structure, c'est possible.

JAKI: Professor Thom, I would like to ask you a question about the very last phrase in your very interesting paper. There you say that the exclusivity of the absolute use of the verb "to be" should be left to metaphysicians and theologians. Before going any further I would like to propose three very brief phrases. One of them is, "There is matter"; the second, "There is a universe"; and the third, "There are other scientists". In what sense do you mean that such sentences should be left by scientists to metaphysicians and to theologians, in a sense which has a touch of contempt in it, or in a sense which has a slight admission of defeat about the scientific approach? Or is it a genuine concession on the part of scientists to philosophers or metaphysicians?

THOM: It is difficult to say in which sense I reached this conclusion which is perhaps irrelevant fot the subject of the Colloquium. I wanted just to stress the fact that for a sentence to have meaning, it has to have some kind of internal structure. And structure means at least two elements, if a sentence is reduced to one single element, it's trivial, and we may forget about trivial structures: hence a meaningful sentence needs at least two elements. Of these two elements, on is basically a *noun*, and the other is like an *adjective* or a *verb*. This verb may be intransitive (as "to sleep", "to think" an so on), but the verb "to be" is difficult to use in an absolute sense, it needs a predicate.

JAKI: But do you mean that the phrase, "There are other scientists", is a trivial statement?

THOM: I would say that such a sentence is semantically (if not grammatically) incomplete, it needs to be completed to qualify "other": "There are other scientists who do not think as I do."

JAKI: You have referred to Wigner's article, on the unreasonable effectiveness of mathematics and physics. There Wigner, I believe, makes the point that what he finds so unreasonably marvellous is that one of the very many mathematical or geometrical systems is far more applicable to the actual physical universe than all the other mathematical systems. Now, if one is a genuine Platonist, one is faced with the problem of why only one of the ideal geometrical theories translates itself into physical reality. There is an interesting article or essay by the young Immanuel Kant, dating from 1748 or 1749, his pre-critical stage, in which he voices his firm belief that to every set of dimensions, from the one-dimensional manifold to the dimensional manifolds, there has to correspond an actual physical universe. On the basis of your paper, how would you cope with this problem - because, obviously, we cannot prove that there is an infinite number of universes, each corresponding to any of an infinite number of sets of dimension.

THOM: Of course I cannot prove anything. The fact that mathematics is so fundamental for Physics reflects something relatively true, namely that the Universe in which we live, this sublunar world which surrounds us, is endowed with very good approximation by Euclidean Geometry. Now people (in ancient times) discovered that by playing with figures, it was possible to formalize their geometric properties in ordinary language. Euclidean Geometry provided a short connexion between the usual language, and our way of thinking with, say, the three dimensional Euclidean continuum. Of this miracle, I have no explanation, and I cannot really understand, except fictionally, a formally different universe in which living beings could exist.

LICHNÉROWICZ: Deux questions essentiellement. D'une part le rôle que tu fais jouer à l'analytique dans les rapports avec la physique. Eh bien, l'analytique, c'est évidemment la physique où il ne se passe rien. Et ce n'est pas cela qui m'amuse beaucoup... L'analytique, les équations aux dérivées partielles correspondent à une physique du repos, à une physique stationnaire. D'une manière générale, la Physique telle que nous l'avons est beaucoup plus hyperbolique qu'elliptique ou analytique. Cette physique se compose essentiellement de théories de propagation par ondes et rayons ... La physique pense en termes des mathématiques dont elle a besoin. De l'elliptique, et ce n'est pas le plus intéressant. Le deuxième problème que je voulais te poser, c'est: tu as parlé à propos des nombres de l'ontologie. J'aimerais savoir dans quel sens ceci est-il ontologique, car il y a beaucoup de sens pour le mot ontologie et l'emploi sans explication peut conduire à des quiproquos.

THOM: Oui, disons que cet emploi du mot ontologie est peut-être un peu métaphorique; j'ai voulu introduire en théorie générale des mathématiques, le schéma hylèmorpique d'Aristote.

LICHNÉROWICZ: Je pense un peu comme Epiménide le Crétois, il peut y avoir des ontologies du premier ordre, du second ordre, etc. Il s'agit ici, me semble-t-il, d'une ontologie d'un ordre très particulier.

THOM: Peut-être, peut-être... [En fait, comme on le voit sur l'exemple des entiers négatifs créés à partir des entiers naturels, la "matière négative" est créée à partir de l'opération (- 1) sur les entiers naturels. Or cette opération est la forme qui, en agissant sur cette matière, permet de la définir ... On est dans la situation (fort peu aristotélicienne) où c'est la forme qui crée la matière. En fait, la vraie matière, c'est le continu qui, par la Topologie, accède à une indépendance (ontologique?) vis à vis de la forme génératrice (et là, nous retrouvons Aristote...)].

LICHNÉROWICZ: Le calcul linéaire, ou quadratique, je veux bien que ce soit de l'analytique, c'est analytique, mais ce n'est pas cela qui épuise le caractère véritablement analytique.

THOM: Fais-tu allusion à l'emploi d'outils comme la distribution de Dirac? Est-ce qu'on sort alors de l'analytique?: peut-être même pas...

LICHNÉROWICZ: Ce n'est pas là pas véritablement le problème à mes yeux. Ici on est dans le différentiel, non pas dans l'analytique, trop rigide.

THOM: Oui, on est dans le C∞. C'est justement le cas où on l'utilise à plein avec les "test-functions".

LICHNÉROWICZ: Le vrai cadre est l'hyperbolique, on n'est pas dans le C∞ qui sert pour des tests, non pour les objets mêmes utilisés en Physique.

THOM: L'analytique est nécessaire parce que seules les fonctions analytiques peuvent être définies par un petit nombre de symboles, et que seules elles admettent une forme canonique (et calculable) d'extrapolation: le prolongement analytique. De plus, par leurs singularités, elles peuvent permettre la localisation d'accidents d'origine physique, par exemple les chocs.

LICHNÉROWICZ: L'un des cas qui m'intéressent concerne précisément les ondes de choc en Relativité Générale. Certaines discontinuités des dérivées premières des potentiels de gravitation ne peuvent correspondre à aucun choc physique. Il faut pouvoir faire des changements de cadre qui absorbent ces discontinuités (C^2 par morceaux). C'est pour cela que l'on détruit la Physique en imposant une structure $C\infty$.

THOM: Oui, enfin, il faudrait concevoir clairement la situation , je ne la conçois pas clairement. Mais je suis prêt à croire qu'en effet, dans une variété hyperbolique, si tu prends n'importe quelle fonction [comme potentiel], et si tu étudies la dynamique de son gradient, eh bien, tu auras en général de très mauvais accidents.

LICHNÉROWICZ: Mais ces accidents arrivent dans le réel. J'ai besoin qu'ils existent pour décrire des phénomènes réels.

THOM: Oui, oui, cela je veux bien le croire.

LICHNÉROWICZ: Et tu le sais mieux que moi.

THOM: Oui, pour ça que ça doit être comme ça.

MALU: Mon cher Thom, je me permets de vous demander un peu de m'expliquer ce que vous entendez exactement par la phrase suivantes que j'ai lue dans votre texte à la page 8: "in biology we have to accept the existence of the fact of finality".

THOM: Par finalisme, j'ai en tête l'affirmation un peu étroite de mon maître Aristote, qui a dit: "la Nature ne fait rien en vain", et c'est un peu cette idée que je crois on doit conserver toutes les fois qu'on s'efforce d'expliquer l'organisation ou la structure biologique, comme vous voudrez. Il faut essayer de justifier les choses par une contrainte préexistante, qu'on doit safisfaire, et essayer de démontrer le caractère optimal — ou suboptimal — de la solution pratiquée par la vie.

INFORMATION AND COMPLEXITY
(HOW TO MEASURE THEM?)

LÁSZLO LOVÁSZ

Dept. of Computer Science
Eötvós Loránd University, Budapest, and
Princeton University, Princeton, NJ

Computer science, also called *informatics*, is often defined as the theory of storing, processing, and communicating information. The key notion in the theory of computing is that of *complexity*. The basic tasks of computer science (and their variations) lead to various measures of complexity. We may speak of the complexity of a structure, meaning the amount of information (number of bits) in the most economical "blueprint" of the structure; this is the minimum space we need to store enough information about the structure to allow its reconstruction. We may also speak of the algorithmic complexity of a certain task: this is the minimum time (or other computational resource) needed to carry out this task on a computer. And we may also speak of the communication complexity of tasks involving more than one processor: this is the number of bits that have to be transmitted in solving this task (I will not discuss this last notion in these notes).

It is important to emphasize that the notions of the theory of computing (algorithms, encodings, machine models, complexity) can be defined and measured in a mathematically precise way. The resulting theory is as exact as Euclidean geometry. The elaboration of the mathematical theory would, of course, be beyond these notes; but I hope that I can sketch the motivation for introducing these complexity measures and indicate their possible interest in various areas. Complexity, I believe, should play a central role in the study of a large variety of phenomena, from computers to genetics to brain research to statistical mechanics. In fact, these mathematical ideas and tools may prove as important in the life sciences as the tools of classical mathematics (calculus and algebra) have proved in physics and chemistry.

Like most phenomena in the world, complexity appears first as an obstacle in the way of knowledge (or as a convenient excuse for ignorance). As a next phase, we begin to understand it, measure it, determine its laws and its connections to our previous knowledge. Finally, we make use of it in engineering: complexity has reached this level, it is widely used in cryptography, random number generation, data security and other areas. Some of these aspects are discussed by Adi Shamir in this volume.

Some examples

As a computer scientist, I will consider every structure (or object) as a sequence of 0's and 1's; this is no restriction of generality, at least as long as we study objects that have a finite description, since such objects can be encoded as sequences of 0's and 1's (every computer uses such encoding). For example, a positive integer is a finite sequence of 0's and 1's when written in base 2 instead of the usual base 10. Rational numbers can be encoded as pairs of integers, with some notational trick to show the sign and the element where the first integer ends etc.

There is a sequence which, in a sense, contains the whole of matematics. Imagine that we write down every conceivable mathematical statement (whether or not it is true or false). We start, say, with the equality $0 = 0$; second is the "equality" $0 = 1$; somewhere we write down the (true) identity $(xy)^2 = x^2y^2$, and then also the (false) identity $(x+y)^2 = x^2+y^2$; Pythagoras' Theorem appears and then also Thales' Theorem; Fermat's Last Theorem is listed (even though we don't know whether it is true or not), etc.

Details of how we do this are irrelevant; it is enough to know that in this way every mathematical statement, true or false, gets a number (its position in the list); given a statement, we can compute its position, and given a position, we can write up the corresponding statement.

Now we write down a sequence of 0's and 1's. The first element in 1, because the first mathematical statement in our list is true; the second in 0 because the second statement in the list is false etc. Anybody who knows this sequence knows the whole of mathematics!

So let us ask the question: how complex is a sequence of 0's and 1's? For example, consider the following sequence:

0000000000000000000000000000000000000000000000000000000000000000000000000
0000000000000000000000000000000000000000000000000000000000000000000000000
0000000000000000000000000000000000000000000000000000000000000000000000000
0000000000000000000000000000000000000000000000000000000000000000000000000
0000000000000000000000000000000000000000000000000000000000000000000000000
0000000000000000000000000000000000000000000000000000000000000000000000000
0000000000000000000000000000000000000000000000000000000000000000000000000
0000000000000000000000000000000000000000000000000000000000000000000000000

This is perhaps the simplest possible object, and not very interesting; the only thing that can be said about it is that it consists of 576 0's. The following sequence is only slightly more interesting:

0101010101010101010101010101010101010101010101010101010101010101010101010
0101010101010101010101010101010101010101010101010101010101010101010101010
0101010101010101010101010101010101010101010101010101010101010101010101010
0101010101010101010101010101010101010101010101010101010101010101010101010
0101010101010101010101010101010101010101010101010101010101010101010101010
0101010101010101010101010101010101010101010101010101010101010101010101010
0101010101010101010101010101010101010101010101010101010101010101010101010
0101010101010101010101010101010101010101010101010101010101010101010101010

This also consists of 576 entries, alternatingly 0 and 1 and is still very non-complex. Now consider the following sequence:

10110100100110110010110100110110100100110100101101100100110100100110110
00101100100110100100110110010110100100110110100110110010110110010011010
1011001001011011001001101001011011001011010011011010010110010011010010
1011010011011001001011010011011001011011001001011010011011010010011010
0110110010011001001101001001101100101101001001101001011001001101101001
0011010010110010010110100110110010010110110010110100100110110010110110
0010110100111101100100101100100110100100110110010110100110110110001011
101101100100101100110011010010110110010011011001001101101001101110010010110

This looks much more complicated! You really have to be a wizard to memorize it: there is no periodicity, no obvious regularity or pattern in the succession of 0's and 1's, and one feels that the sequence is very complex. But in fact it is generated by a very simple program:

```
main ()
{ int n, m=1, a[576];
a [1]=1
for (n=1; m<576; n++)
{ if (a [n]==0) {a[m+1]=1-a[m]; a[m+2]=1-a[m]; m=m+2;}
else {a[m+1]=1-a[m]; m++;}}}
```

It is not important what the details are: this is a simple rule to generate as many elements of the sequence as we wish. Note that it would not take a substantially larger program to generate 1,000 or 1,000,000 elements; one would only have to replace the number 576 in the program by 1,000 or 1,000,000.

Let us look at another strange sequence:

10110101000001001111001100110011111110011101111001100100100001000101101001
01111101100010011011001101110101010010101011110100111110001110101101111
01100000101110101000100100111011101010000100110011101101000101111010110 0
10000101100000110011001111001100100010101010010101111110010000001100000100
00111010101110001010001011000011101010001011000111111110011011111111011100
10000011110110110011100100001111011101001010100000101111001000011100111 00
01111011010010100111100000000100100001110011011000111101111111101000100111
01101000110100100010000000010111010000111010000101010111100011111101001110

If possible, this looks even wider that the previous sequence; yet, it is even easier to define: these are the first 576 digits of √2, written in base 2. Finally, look at the sequence

10101001000101110010110010010001100101010101010001011110000110001110 0000
10101110010111010100111011010001110101100101001101001101011100110011 0000
10011100101111100010000000101011000000111110010000011001010000000011 01010
00001001011001100011011110000110111001011001010011101111000011110101 0111
11101101010000000000111101010010011101011011010100111101101011110111 01100011
000101110010011111011100101010000010100101010110000011001001000100010 1010100
100010001101011011100100010010011000000000010010101011110110011011011 000
011001101100011110111111101111001010111100110111101010000110111010100 0000

Again, this looks chaotic, and it is: this is a random sequence, obtained by tossing a coin repeatedly and recording head and tails. There is no way to describe or memorize it other than bit-by-bit.

Now how complex are these sequences? As mentioned in the introduction, there are several ways of measuring this complexity, and we start with (perhaps) the simplest one: the *informational* or *Kolmogorov complexity*, introduced by Kolmogorov[1] and Chaitin[2]; see Chaitin[3] for a monograph on the subject. Informally, this defines complexity as the information content, or equivalently, as the length of a definition of the sequence. One has to make this precise; careless use of these words leads to paradoxes. So we proceed as follows: we fix some programming language (say C, which I was using above), and define the Kolmogorov complexity of a sequence x of 0's and 1's as the length of the shortest program which prints out x. We denote this number by $K(x)$.

It is fairly obvious that every finite sequence *can* be generated by some program; e.g. 011001... is printed out by

```
main ()
{
      printf ("011001...");
}
```

This also shows that if a sequence consists of n symbols then its Kolmogorov complexity is at most $n + 160$ (to be precise, we have to count everything in bits; a character is then 8 bits in the usual systems). The constant 160 is irrelevant, and depends on the programming language chosen; but it is not difficult to see that apart from such an additive constant, the specific language plays no role.

Looking at our examples, we see that the first one has, as expected, very small complexity: a sequence of N zeros can be printed out by

```
main ()
int n;
{
      for (n=0; n<N; n++) printf ("0");
}
```

---

[1] A. KOLMOGOROV, *Three approaches to the quantitative definition of information*, Prob. Info. Transmission 1, 1-7 (1965).
[2] G. CHAITIN, *On the length of programs for computing binary sequences*, J. Assoc. Comp. Mach. 13, 547-569 (1966).
[3] G. CHAITIN *Algorithmic Information Theory*, Cambridge Univ. Press 1987.

and a sequence N symbols, alternatingly 0's and 1's (where N is, say, even) can be printed out by

```
main ()
int n;
{
    for (n=0; 2*n<N; n++) printf ("01"):
}
```

These programs contain less than $\log N + 50$ symbols, so they are much shorter than the original length of the sequence. Such a program may also be viewed as a very compact *encoding* of the original sequence. Our third and fourth example have, in spite of the apparent complexity of the sequence, a similarly compact encoding: I have given the program above for the third example, and one could also write a program in a rather straightforward if more lengthy way to print out the digits of $\sqrt{2}$ (in base 2). In these programs, it is again the length of the sequence to be printed that is variable; the rest is constant. So from the point of view of the Kolmogorov complexity theory, there is no substantial difference between the complexities of the first four examples.

The fifth example, the random sequence is different: there is no better way to print it out than specifying its elements one-by-one. At least, it can be proved that if I choose a sequence randomly, say by flipping coins, this will be the case with very large probability. There is always a chance, if very remote, that we "hit the jackpot", and obtain a sequence with low complexity; even the sequence 000000... has a positive, if negligible, chance. It would be very useful to check if this is the case, and verify that we have not found one of these remote and unlikely sequences.

More generally, it would be very good to be able to compute the complexity $K(x)$ of a given sequence x. Unfortunately, this is impossible: *There is no algorithm to compute the Kolmogorov complexity of a finite sequence.*

I cannot resist including the proof of this here, since it is based on one of the classical logical paradoxes, the so-called *typewriter paradox,* and should not be too difficult to follow. (This is perhaps the simplest of the undecidability proofs in mathematical logic, going back to Gödel and Church, which have played such an important role in the development of the foundations of mathematics.)

First, the classical paradox. Suppose that we have a typewriter (or, rather a keyboard) containing all the usual symbols, and we want to write up a definition for every natural number, as long as we can fit the definition on one line (80 characters). We can start with the trivial decimal forms 1,2,3,... But after a while there may be more compact forms. For example, instead of 1,000,000,000 we can write "one billion" (slightly shorter) or 10^10 (even shorter). Since there is only a finite number of ways to use 80 characters, sooner or later we get to a number that cannot be defined by 80 characters. Look at the first such number; once again this is the least natural number not definable by 80 characters.

But this line is just the definition of this number, so it is definable by 80 characters!

To solve this logical contradiction, we have to analyze what "definable" means. A natural notion would be to say that it means that there is a program, containing at most 80 characters, that prints out this number; in other words, the Kolmogorov complexity of the number is at most 80. Is the line above a definition in this sense? To write a program which prints out this number, all we need is a subroutine to compute the Kolmogorov complexity. The paradox we formulated is then nothing but an *indirect proof* of the fact that the Kolmogorov complexity cannot be computed by any algorithm.

So this is bad news. Let us hasten to emphasize that there are algorithms known that compute a good approximation of the Kolmogorov complexity for most of the sequences. We cannot go into the details of these algorithms; I only remark that they are closely related to the notion of *entropy* known from physics and information theory.

The notion of Kolmogorov complexity has many applications; let me mention just a few. First, it is the key to a general logical framework for *data compression*. The program printing out a given sequence can be viewed as a compressed encoding of the sequence. Therefore, the Kolmogorov complexity of a sequence is the limit to which this sequence can be compressed without losing information. This is perhaps the most important property of Kolmogorov complexity from a technical point of view, but some others are more relevant for scientists working in other sciences. Kolmogorov complexity can be used to clarify some questions concerning the notion of *randomness;* I will return to this aspect in detail. It is also closely related to the notion of *entropy,* which is an important measure of "disorder" in thermodynamics and elsewhere. To state just a very simple case of this connection: if the elements of a 0-1 sequence of length n are independently generated from some distribution, then the Komogorov complexity of the sequence is approximately n times the entropy of this distribution.

Computational complexity

Let us now look at a more difficult notion of complexity. As far as Kolmogorov complexity goes, our first four examples were about equally complex. Yet, we feel that the first two examples are simpler, the last two are more complex. Is there a way to quantify this difference?

The trick is to ask the following question: "What is the 7,000,867,764th element of the sequence?". (We imagine that the sequence is extended to this in the natural way.) In the case of the first two, the answer is immediate: for the first, it is always 0, for the second, it is 1 (since 7,000,867,764 is even). To answer this question does not require any more effort than reading the question itself.

In the case of the third example, however, there is no immediate way to determine the 7,000,867,764th element. By definition, we can do this by determining all the previous elements, which takes about 7,000,867,764 steps (I have not carried this out). In the case of the fourth example, following the standard procedure for extracting square roots would take time about $7,000,867,764^2$. Of course, there is always a possibility that some trickier, less direct procedure gets the result faster; but no algorithm is known for either case that would do substantially better than computing the whole sequence.

So we get the following new complexity measure: how difficult is it to determine the nth element of the sequence? What amount of computational resource (time, space) must be used?

This notion is more subtle than Kolmogorov complexity is many ways. First, there does not seem to be any way to answer it by a single number. Rather, we have to consider the minimum amount of time (or other resource) used by an algorithm *as a function of n*. Further, strange as it may sound, there may not be a best algorithm for a given sequence; it could happen that for every algorithm to compute its members there is a much better one (taking, say, only the logarithm of the time the first algorithm takes, at least for large values of n), and then again, there is a third one much better than the second, etc.

Therefore, we ask the question in the following way: given an infinite sequence, and a function $f(x)$ defined on positive integers (this could be x^2, or $\log x$, or 2^x), can the nth element of the sequence be computed in time $f(n)$? (There are further subtleties: the answer depends on the machine model we use, and also on the way the sequence is specified, but I am ignoring these questions now.) I should also remark that it is customary to express the bound on the running time as a function of $\log n$ (which is the number of bits needed to write down n) rather than as a function of n. This is, of course, a matter of convention.

The most complex sequences are those of which no algorithm exists at all to compute their elements. Such a sequence is called *undecidable*. The sequence "Whole Mathematics" introduced at the beginning of the paper is such. One could easily transform the negative result about the computability of the Kolmogorov complexity into the undecidability of an appropriate sequence.

The most important question to ask turns out to be the following: Is the nth element of the sequence computable in a time proportional to $(\log n)^{const}$, i.e. in *polynomial time*? This question has initiated very active research in various branches of mathematics. It has turned out that very classical notions are quite unexplored from the computational complexity point of view. An outstanding example is the notion of prime numbers (positive integers that cannot be written as products of two smaller positive integers, like 2,3,5,7,11,...; 1 is not considered a prime by convention). It has been known since Euclid that there are infinitely many prime numbers, and that every positive integer can be written in a unique way as a product of primes. Yet, we do not know whether it can be decided in polynomial time whether a given positive integer is a prime; in other words, no algorithm is known that would decide in time k^{const} whether a given integer with k digits is a prime. Nor can we prove that such an algorithm does not exist. (If you want to translate this problem into computing the nth element of a sequence, consider the sequence 0110101000101000010..., in which the nth element is 1 if n is a prime and 0 if it is not).

Here we come to the central, and yet sometimes hopelessly difficult, request in complexity theory: prove that certain problems (sequences) are computationally hard, i.e., no algorithm whatsoever can solve them in the allotted time. Some results in this direction have been obtained, but most of the really fundamental questions are unsolved. Can the product of two n-digit numbers be computed in at most $1000 \cdot n$ steps? Can the prime factorization of an n-digit integer be computed in at most n^{100} steps? Can the shortest tour through n given cities be found in at most n^{100} steps?

There is one successful approach to the issue of showing that certain computational tasks are difficult. We can say only a few words (*cf.* Garey and Johnson[4] for a comprehensive treatment). Many important computational problems have the form "find something". The difficulty lies in the fact we have

---

[4] M.R. GAREY and D.S. JOHNSON, *Computers and Intractability: A Guide to the Theory of NP-Completeness.* San Francisco: Freeman 1979.

to search for this thing in a very large set; once found, it is easy to verify that we have indeed found the right thing. For example, to decide if a number N is a prime, we want to find a divisor. The trouble is that we have to search for a divisor among all integers up to $\sqrt{N}$; once a divisor is found (or guessed correctly) it is very easy to verify that we have indeed a divisor; it takes a single division of integers. The class of computational problems with this logical structure is called NP.

Now Cook[5] and Levin[6] have independently proved that there are problems in this class which are *complete* or *universal* in the sense that every other problem in the class NP can be reduced to them. So if we can solve such a complete problem efficiently (in polynomial time), then we have in fact solved *every* problem in the class. It is generally believed that such a single extremely powerful algorithm does not exist; hence, complete problems cannot be solved efficiently. The interest of this result was much enhanced by the work of Karp[7], who showed that many very common computational problems (for example, the famous Travelling Salesman Problem) are complete. This provides a way to show that certain problems are hard.

Unfortunately, there is an unproven hypothesis behind these arguments, and to replace it by an argument based only on usual axioms is perhaps the most challenging question in the theory of computing nowadays. This is the famous P=NP problem, which can be transformed into problems in logic, combinatorics, and elsewhere, leading in each case to very interesting and fundamental unsolved questions.

The difficulty of this problem, as well as its potential influence on mathematics, is best exemplified by an analogy. The first exact notion of an "algorithm" was formulated by the Greek geometers: the notion of a geometric construction by ruler and compass. Analogously to (computer) algorithms, there are both solvable and unsolvable construction problems. The design of construction algorithms has proved stimulating in geometry for a long time, and has contributed to the development of important tools (very useful also independently of construction problems) like inversion or the golden ratio. But the proof of *unsolvability* of basic construction problems (trisecting an angle, squaring a circle, doubling a cube, constructing a regular heptagon etc.)

---

[5] S.A. COOK, *The complexity of theorem proving procedures*, Proc. 3rd Annual ACM Symposium on Theory of Computing (Shaker Heights, 1971), ACM, New York, 151-158, 1971.

[6] L. LEVIN, *Universal'nyie perebornyie zadachi* (Universal search problems: in Russian), Problemy Peredachi Informatsii 9 265-266, 1972.

[7] R.M. KARP, *Reducibility Among Combinatorial Problems*, in: Complexity of Computer Computations, (ed. R.E. Miller and J.W. Thatcher), Plenum Press, 85-103, 1972.

illustrates this influence more dramatically. Such negative results were inacessible to Greek mathematics and for a long time after; the proof of them required the notion of real numbers and substantial part of modern algebra. (In fact, modern algebra was inspired by the desire to prove such negative results.)

Computational complexity has proved to be a very fruitful notion. Its basic framework was developed in the late 60's and early 70's, and the year that followed witnessed a fast spreading of its ideas to various branches of mathematics. The reason was that this theory led to an exact definition of the difficulty of a problem, and this definition matched the intuitive feeling of mathematicians about these problem. By now, it is customary in many branches of mathematics to start the study of a new notion or problem by a complexity analysis; the answer to the question whether a problem is polynomial time solvable determines pretty much the direction of further research. In the case of "easy" (polynomial time solvable) problems, one looks for efficient implementation, clever data structures. In the case of "hard" problems (complete problems in the class NP, for example) one aims at heuristic algorithms, approximate solutions, methods making use of hidden structure, etc.

A complexity analysis of an algorithm often shows where it can be improved; the solution of questions posed by complexity considerations often leads to theoretical and practical breakthroughs in algorithm design.

In the issues mentioned above, the large complexity of a problem appears as a disadvantage, a fact that makes the solution of the problem more costly, the understanding of the underlying structure more difficult. Some of the most exciting applications of complexity theory make use of complexity: these are *cryptography* (which is discussed by Adi Shamir) and *random number generation,* which I will discuss in the next section.

Complexity and randomness

Complexity theory offers new approaches to the notion of *randomness.* To understand what randomness means, how it arises and what its relations are to the notion of information, complexity, chaos, and entropy, is one of the most challenging tasks in mathematics, computer science, and (on a more general level) in the philosophy of science.

Randomness is a crucial issue in quantum physics and statistical mechanics; statistical descriptions (based on modelling the phenomena as random processes) are widely used in the social sciences. Many computational procedures (simulations, Monte-Carlo) need random numbers to run.

The two examples from physics underline the crucial question: is there true randomness, or is a random behavior of a system always a consequence of a very complex underlying structure? In statistical mechanics, the system (say, a container filled with gas, or a piece of ice heated to melting point) is, in principle, completely described by the Newtonian equations; the probabilistic description is an approximation of the combined behavior of an extremely large number of particles. On the other hand, physicists generally assume nowadays that events in quantum physics are truly random, there is no hidden underlying structure.

In computing, one usually resorts to pseudorandom numbers, i.e., one uses a carefully chosen, complex deterministic procedure whose output is "sufficiently random". ("Truly random" numbers could be generated using quantum mechanical devices, but these are impractical). Just what "sufficiently random" means is a difficult issues, and we will try to discuss it here.

First, a couple of examples. Sequences 1 and 2 above are obviously not random. Sequence 3 is definitely more "chaotic", but a little inspection reveals that it does not contain three consecutive 0's or 1's; a random sequence of this length should contain many more!

It is not so easy to make a distinction between examples 4 and 5, even though we know that example 4 was obtained by the completely deterministic (and in fact quite simple) procedure of extracting the square root of 2. In fact, the digits of $\sqrt{2}$ (in base 2) could be used in place of a random sequence in some applications. But one would face a serious danger: if we ever multiplied this number by itself, we would get exactly 2 (i.e., the sequence 10.0000000...), which is so obviously non-random that our computation would be all wrong.

Both Kolmogorov complexity and computational complexity offer ideas to help us understand the notion of randomness. In fact, the aim to define when a single sequence of 0's and 1's is truly random was the main motivation for the development of the idea of informational complexity. To understand this idea remember the point that if a sequence is truly random (obtained, say, by recording coin-flips or some quantum mechanical events), then there is no regularity in it, and one cannot expect to be able to describe it in any form more compact than the sequence itself. Martin-Löf[8] turns this around and suggests using this property as the definition of randomness. Consider a sequence x = 01101011000... of length n. This sequence is called *informationally random*, if

---

[8] P. MARTIN-LÖF, *On the definition of random sequences,* Inform. and Control 9, 602-619, 1966.

the Kolmogorov complexity of x is almost n. (This "almost" has to be quantified; this is technical but not difficult.)

To justify this definition, we state two facts.

Fact 1. *If we generate a 0-1 sequence at random, then with very high probability, it will satisfy the definition of informational randomness.*

Fact 2. *Every informationally random 0-1 sequence satisfies all the usual properties of random sequences (laws of large numbers, central limit theorems, statistical tests).*

Let me illustrate this last fact by the example of the simplest basic property of random 0-1 sequences: *the number of 0's and 1's is approximately the same.* Let me argue that an informationally random sequence also has this property; in other words, if a sequence x of length n has substantially more 0's than 1's (say, $2n/3$ 0's and $n/3$ 1's) then it can be printed out by a program substantially shorter than n (assuming that n is very large). The number of sequences of length n is 2^n; but the number of those containing only $n/3$ 1's is much smaller, only about 1.8^n. We can order these sequences lexicographically, and see where x is; say, x in the kth. Then we can print out the sequence x by the following program:

Take all sequences of length n in lexicographic order.

Discard those containing more than $n/3$ 1's.

Print out the kth of the remaining sequences.

When written as a formal program, the length of this statement is

$$\log_2 n + \log_2 k + \text{const.} < \log_2 n + 0.9n + \text{const} < 0.98n$$

(if n is lange enough). So indeed x can be printed out by a program shorter than n, and so it is not informationally random.

The above notion of informational randomness has two drawbacks. First, it is uncontrollable: since it is not possible to compute the Kolmogorov complexity by any algorithm, it is also impossible to determine whether a given sequence is informationally random. Second, it is *by definition* impossible to generate a random sequence by a computer; also, the behavior of a system in statical mechanics would again *by definition* be non-random. Now computer-generated (pseudo)-random sequences are widely used in practice, and statistical mechanics correctly describes a number of complicated physical phenomena. A satisfactory theory of randomness should be able to explain this success.

The theory of computational complexity offers some help here. Roughly speaking, we say that a sequence is *computationally random*, if given all but one of its elements, it is not possible to compute the missing element. One should, of course, make the phrase "not possible" more precise. A *random number generator* takes a "seed" (a truly random sequence) of some lenght n and uses it to generate a much longer pseudorandom sequence. We say that this random number generator is "secure", if every algorithm that works in polynomial time, and that predicts the ith entry knowing the previous entries, will err in approximately half the time. Such an algorithm may be viewed as a *randomness test*. Yao[9] proved that this randomness test is universal: if a random number generator passes it then it will pass every other statistical test that can be carried out in polynomial time.

In this model, it is not impossible any more to generate such a computationally random sequence by a computer; it only follows that then we have to allocate more resources (allow more time) for the generation of the sequence than for the randomness test. This is, however, only saying that there is no obvious reason why an algorithm working (say) in quadratic time could not generate a sequence that would be accepted as random by every test working in linear time. To construct such random number generators is difficult, and to a large degree unsolved; this is one of the most important unsolved problems in computational complexity theory. Answering it would have profound implications in the philosophy and practice of complexity theory.

This model, and its refinements, have worked very successfully in various applications of the notion of pseudorandomness, e.g. in cryptography. A glimpse of these applications is provided by the paper of Adi Shamir presented at the meeting.

Complexity and nature

I would like to close with some more speculative remarks about possible applications of these mathematical notions in the study of nature. Any time we encounter a complicated structure, we may try to measure how complex it really is. As we have seen, this is not merely the issue of size.

There are several alternatives to quantifying the notion of complexity, and none is "better" than the other; they merely consist in formalizing different aspects of what is meant by *complex*. It is quite natural, for example, to ask for the informational complexity of the genetic code, or of the brain. Such

---

[9] A.C. YAO, *Theory and applications of trapdoor functions*, Proceedings of the 23th Annual IEEE Symposium on Foundations of Computer Science, IEEE Computer 1982.

considerations may help understand the redundancy of the genetic code, and its role in inheritance and evolution. But it might also be interesting to study the computational complexity of the task of translating the genetic code into anatomical details of a living being, or the computational complexity of tasks performed by the brain.

The impossibility of certain tasks, observed, formalized and established, has repeatedly led to conceptual breakthroughs in science. The impossibility of constructing a perpetuum mobile led to the notion of energy and to the development of mechanics and thermodynamics. The impossibility of travelling (or sending information) faster than light leads to relativity theory; the impossibility of measuring the speed and position of a particle simultaneously is a basic fact of quantum mechanics.

The theory of computing offers some impossibility results of similar, or even greater, generality. First, the *Church thesis*, mentioned above, asserts that if a computational problem is algorithmically unsolvable (in the formal mathematical sense), then no device can be constructed to solve it, no matter what phenomena in nature are used. Some consequences of this thesis are explored by Penrose[10].

But one can go further and postulate that no phenomenon in nature can speed up algorithms by more than a constant factor. To be more precise, we have to note that one can speed up certain algorithms by using parall processing, but then the *volume* of the computing device must be large. One can postulate: if a problem cannot be solved in time less that T (in the formal mathematical sense), then no device can be constructed to solve it, using up a total of less than const $\cdot T$ of the space-time. We refer to Vergis, Steiglitz and Dickinson[11] for an elaboration of this idea; here we confine ourselves to a single example. It is known that to compute the minimum energy state of an alloy is computationally hard (non-polynomial, if some widely accepted hypotheses are used). So if we cool it down within non-astronomical time, it ought not to find its minimum energy state; thus complexity predicts the existence of different states with locally minimal energy.

Finally, it can be hoped that a better understanding of what randomness means (and I think complexity theory is crucial for this) could contribute to the answer to questions like the existence of true randomness in nature.

[10] R. PENROSE, *The Emperor's New Mind. Concerning Computers, Minds, and the Laws of Physics*, Oxford-New York: Oxford University Press 1989.
[11] A. VERGIS, K. STEIGLITZ and B. DICKINSON, *The complexity of analog computation*, Mathematics and Computers in Simulation 28, 91-113, 1986, Society Press, Washington, D.C., 80-91.

LOVASZ:

I would like to make a remark, just a little bit with the eye of a complexity theorist on the use of the verbs "can" and "cannot" in the definition of reduction. I had a slide in my talk where I put the word "cannot" in quotation marks, because there are many different senses in which you can use it, and many different senses in which you might want to reduce the behavior of a complex structure to its constituents. I think there are many examples of a complex structure where, although its behavior is completely determined in some sense by its constituents, you need entirely new notions, and a new approach and a new phenomenology to describe the behavior of the structure. There can be complexity-theoretic reason behind this (for one thing just to write up and solve the equations would be impossible, and I think this impossibility is very serious in any sense). There are nice examples particularly in statistical mechanics where the behaviour of certain systems depends a lot more on their structure than on the particular properties of their constituents. So I would like to see philosophers dealing with this "can" a little more carefully — or maybe giving us various interpretations.

INFORMATION, COMPLEXITY AND CRYPTOGRAPHY

ADI SHAMIR
Applied Mathematics Department
The Weizmann Institute of Science
Rehovot, Israel

In this paper we informally survey the interrelationship between Information Theory, Complexity Theory, and Cryptography. We describe several proposed foundations for the notion of cryptographic security, analyse their advantages and disadvantages, and describe some of the major research achievements in this area.

1. Cryptosystems based on information theory

The historian David Kahn, in his definitive study "The Codebreakers", states that the first monograph on cryptography was written by Leon Battista Alberti in 1466. It was motivated by a conversation he had in the Vatican Gardens with Leonardo Dato, the pontifical secretary who was in charge of the Vatican's secret communications. The monograph formalized the notion of frequency analysis, described the first polyalphabetic cryptosystem, and introduced a mechanical device called a cipher disk to implement it efficiently.

In spite of such early achievements, cryptography remained an art rather than a science for almost 500 years. The successes and failures of military cryptosystems had a major impact on world history, but until the middle of the 20-th century the field consisted of a loose collection of ideas and tricks, with few unifying ideas.

In 1949 Claude Shannon published his seminal paper "Communication Theory of Secrecy Systems", which was an extension of his 1948 paper "A Mathematical Theory of Communications", and was based on his experience as a military cryptographer during World War II. In this paper he introduced a mathematical model of an abstract cryptosystem, in which a set of cleartexts is randomly mapped to a set of ciphertexts under the control of a cryptographic key. The set of cleartexts contains both plausible and implausible messages, where the latter consist of grammatically incorrect, semantically incorrect, or out-of-context messages. The entropy of the actual message source is denoted by E, the entropy of the uniform source (which generates all the plausible and implausible messages with equal probability) is denoted by U, and the entropy of the source of cryptographic keys is denoted by K. In his paper, Shannon proves that such a cryptosystem is secure as long as the number of ciphertext symbols obtained by the cryptanalyst is smaller than the unicity distance of the cryptosystem, which is defined as $N=K/(U-E)$. The quantity U-E is called the redundancy of the source language, and indicates the compressibility of the language (e.g., English is about 75% redundant, and thus English text can be compressed to about 25% of its original size without losing intelligibility).

The formula made it possible to formally prove the security of several paper-and-pencil cryptosystems. The best known among them was the one time pad, invented in 1917 by Gilbert Vernam of ATT. In this cryptosystem, each letter in the English alphabet is treated as a number in the range 0 to 25 The cleartext is an arbitrary English text, and the key is a sequence of random English letters of the same length as the cleartext. Encryption is carried out by adding the numerical values of corresponding cleartext and key letters modulo 26, and reinterpreting the numerical result as an English letter. Decryption is carried out by subtracting key letters from their corresponding ciphertext letters. Each key letter is used only once, and thus to encrypt a new cleartext the user has to use a new segment of key letters. This cryptosystem is absolutely secure: A cryptanalyst (who obtains the ciphertext but has no knowledge of the key) cannot obtain any information whatsoever about the cleartext, regardless of how much ingenuity, expertise, or computing power he may have. This provable security made the one time pad the favorite cryptosystem of spies and diplomats. For example, when the "red line" was established between the White House and the Kremlin in order to discuss international crises, American and Russian experts agreed to protect the messages on it with the one time pad.

The difficulty in using the one time pad (and the reason it is not universally used) is the need to generate, distribute and store in a secure way a large number of random letters, which exceeds the expected length of all the messages which may be sent on the communication line. One possible alternative is to use another English text as the key. If the sender and recipient agree in advance to use a particular book which is widely available but whose choice is unknown to the opponent, they can start their messages with an indication of the point in the book where they start extracting key letters, and proceed to encrypt and decrypt in the same way as in the one time pad. However, since the entropy of the key in this case is only 25% of the entropy of a key which consists of truly random letters, this scheme is breakable in theory as well as in practice. On the other hand, if the key consists of four independent English texts, whose numerical values are all added (modulo 26) to the cleartext, the scheme becomes unbreakable since the entropy of the key is restored to its one time pad level.

Information theory provides a solid mathematical foundation for cryptography, but its conclusions tend to be overly pessimistic. A cryptosystem which is information-theoretically secure cannot be broken even by infinitely powerful opponents using infinitely powerful computers for an infinite amount of time. However, a cryptosystem which is information-theoretically insecure may provide excellent protection against real opponents. For example, information theory labels as insecure any cryptosystem which encrypts large cleartexts with a fixed size key, even if the best available attack on it takes longer than the age of the universe.

2. Cryptosystems based on physical assumptions

Several cryptosystems whose security is based on physical assumptions have been proposed in the last 15 years. In this section we demonstrate this approach with three examples.

2.1 Cryptosystems utilizing a deck of cards

The problem of achieving cryptographic security can be reduced to the problem of generating common secret bits. If two communicating parties A and B can generate a single random bit which is known to A and B but unknown to the cryptanalyst C (who eavesdrops on all the communications between A and B), then A and B can repeat the protocol arbitrarily many times, and thus generate the equivalent of the one time pad discussed in the previous section. We now show that this seemingly impossible feat is achievable if A, B and C have access to a physical deck of three cards, whose faces are marked by 1, 2 and 3, and whose backs are indistinguishable. We assume that the three parties are located in a single room, that A and B do not have a chance to talk privately with each other (e.g., in order to establish the meaning of concealed gestures), and that the cards can be randomly shuffled with their faces down.

Each party takes one card from the deck, and notes the number marked on its face. Without loss of generality, let us assume that A got 1, B got 2, and C got 3.

In particular, the cryptanalyst C can deduce that A and B have cards 1 and 2, but he has no information whatsoever about who has which card. A now randomly chooses a number of a card which is different from the one he has, and publicly states: "I do not have card number x" (where x can be either 2 or 3 in our hypothetical example). If B also states that he does not have this card x, the protocol is considered a success. Otherwise, the protocol is considered a failure.

We can now analyse the protocol in the following way. If the protocol is successful, x must be the card held by C. Since the cryptanalyst C already knows which card he has, he has learned absolutely no new information from the public statements about the distribution of cards to the other participants. A, on the other hand, now knows everything about this distribution since he knows that his card is 1, C's card is x, and B's card is the remaining third card. B can also deduce the complete distribution in an analogous way. A and B can now agree on a common bit by using the publicly declared convention that this bit is 0 if A's card is numerically smaller than B's card, and 1 otherwise. The result is an unbiased random bit which is known to A and B but unknown to C. If the protocol was a failure (because B stated that he had the card x declared by A), the cards are collected and reshuffled, and the protocol is repeated with the new distribution of cards and a new random choice of x by A. Since the probability of success is 1/2 per iteration, A and B can expect to generate about $n/2$ common secret bits by executing the protocol n times.

2.2 Cryptosystems based on physical padlocks.

The process of encryption can be modelled by the physical process of placing a written message in a strong metal box, and locking the box with an external padlock. We assume that the box and the padlock are so strong that they cannot be defeated, and thus the only way to read the message is to unlock the padlock with the appropriate key.

The only difficulty in implementing such a scheme is the key distribution problem. Since the locking and unlocking are carried out by different parties in far away locations, we have to decide who makes and who gets the cryptographic keys, and how they are transported.

The classical solution (which was used in all the cryptosystems developed before 1976) was that the sender chose a random encryption/decryption key, and sent it to the recipient via a secure key channel. The key could not be sent over the same communication channel as the message, since this channel was supposed to be controlled by the cryptanalyst. The key could not be sent encrypted in another box with another padlock, since this would merely change the problem from the secure distribution of message encrypting keys into the secure distribution of key encrypting keys. It was thus necessary to assume the existence of another channel, which could not be eavesdropped on by the cryptanalyst. In practice, keys were carried by trusted couriers in diplomatic pouches, or memorized by spies in face-to-face meetings with their controllers. However, this key distribution problem created a major obstacle to the widespread adoption of cryptographic techniques in personal and business communications.

In 1976, Diffie and Hellman from Stanford University proposed a radically different solution to the key distribution problem, known as "public key cryptography". It is motivated by the observation that open padlocks can be locked without using a key by pushing the shackle in until it engages the internal spring mechanism. In other words, the processes of locking and unlocking the padlock are quite different, and only the latter requires the posession of a key. Assume now that B, the intended recipient of the message, distributes to all his acquaintances identical copies of unlocked padlocks, and keeps their common key safely in his office. The padlocks can be distributed via insecure channels, since we assume that their keys cannot be deduced by inspection. When A wants to send a message securely to B, he uses one of these distributed padlocks to lock the message in a strong metal box, and sends it via the insecure channel. When B receives the box, he unlocks it and reads the message. Since the unlocking key never left B's office, there is no need to send it via a separate secure channel.

The mathematical implementation of this idea is based on pairs of seemingly unrelated cryptographic keys, Ek (for encryption) and Dk (for decryption). Each user picks a random pair, publishes Ek, and keeps Dk secret. All the Ek's can be published in a publicly available key directory, which lists the name, address, telephone number and encryption key of all the users in the system. If the computation of Dk from Ek is unfeasible, the publication of the encryption keys does not endanger the secrecy of the corresponding decryption keys. When A wants to send a message securely to B, he can find out B's encryption key by consulting this directory. This solution eliminates the key distribution problem, and reduces the number of encryption keys in a large network of n users from $n(n-1)/2$ to n. The best known and most widely used cryptosystem of this type is the RSA scheme, developed in 1977 by Rivest, Shamir and Adleman.

In 1978, I pointed out that the padlock example can be further refined. In fact, it is possible to send messages securely without any prior exchange of keys even if we assume that keys are necessary both to lock and to unlock all the padlocks in the system. To achieve this, we slightly modify each metal box by installing a second hole through which we can hang a second padlock. The message is secure as long as the box is locked by at least one padlock.

Assume now that A wants to send a message securely to B, with whom he had never talked before. A locks the box with padlock PA, using the key KA which only he has access to. The locked box is sent to B, but the message cannot be read. Instead, B adds to the box a second padlock PB, using the key KB which only he has access to. This "doubly encrypted" box is now returned to A via the insecure channel. Upon receipt, A unlocks and removes PA, and sends the box back to B. Since it is now protected only by padlock PB which B can unlock, the message can be read!

2.3 Quantum cryptography.

A very different approach to cryptographic security was developed in the last decade by Weisner, Bennet, Brassard, and Crepeau. They decided to base their cryptosystems on quantum theoretic effects, and in particular on the behaviour of polarized photons. Consider a system in which the sender can send to the recipient a single photon polarized in a desired direction.

The recipient can test whether this photon is polarized in a particular direction, but unless the actual and measured directions are either parallel or perpendicular to each other, the result will be probabilistic and the actual polarization of the photon will change to the measured direction.

To send a single bit, the sender chooses at random one of two coordinate systems: In the first system a zero bit is represented by a single photon polarized in a vertical direction and a one bit is represented by a single photon polarized in a horizontal direction, and in the second system the two directions are rotated clockwise by 45 degrees. The recipient does not know which coordinate system was used by the sender, and chooses at random one of the two coordinate systems to measure the polarization. If he guesses correctly, the answer is correct. If he measures the polarization in the wrong coordinate system, he gets a random answer.

The process is repeated many times (with new photons and random choices), and the recipient ends up with a long sequence of bits, half of which are correct.

The sender and recipient then initiate a public discussion of the coordinate systems they chose or each bit, and thus they know which bits were received correctly and which bits were destroyed by the measuring process. With an appropriate protocol, they can extract from this process a common random sequence of bits which can be used as a one time pad when actual messages are sent over the telephone network.

The main property of this process is that any attempt by the cryptanalyst to measure the polarization of the transmitted photons will corrupt the polarization of at least some of them, and thus will become known to the communicating parties. As a result, quantum cryptosystems make passive eavesdropping impossible, and active eavesdropping detectable.

In the last couple of years, researchers have constructed experimental devices which can send information securely over short distances with polarized photons. They are now trying to send information securely over longer distances with a fiber optic cable, but it is too early to decide whether such devices can be practical.

3. Cryptosystems based on complexity theory

The most promising approach to cryptographic security is to use techniques derived from complexity theory. This theory, developed by theoretical computer scientists since the mid 1960's, attempts to classify computational problems by the amount of time and memory required to solve them on a digital computer. Due to the differences between various computational models, the theory is asymptotic in nature: Instead of assigning a concrete complexity value to each instance of the computational task, the theory deals with the rate of growth of this complexity as the size of the instance grows to infinity. Feasible computations are defined as those whose complexity grows at most polynomially with the size of the instance.

A one way function is defined as a function whose evaluation is a feasible computation, but its inversion is an unfeasible computation. While we cannot prove at this stage that such functions exist, we have a number of candidate functions which are widely believed to be one way. The best known example is the function that takes two large prime numbers and multiplies them together. It is very easy to carry out this multiplication, but there is no known feasible algorithm which can reverse this process and factor the result. The problem of factoring large numbers had been investigated in the last three centuries by some of the best mathematicians (including Euler and Gauss), but all the algorithms discovered so far for this problem require superpolynomial time.

Complexity-based cryptosystems are designed to withstand feasible attacks, rather than attacks by infinitely powerful cryptanalysts using infinitely powerful computers. If the complexity of the best attack on a cryptographic scheme with a k bit key grows as 2^k, then even the ultimate parallel computer which uses each elementary particle in the universe as a processing element would not be able to break the case of $k=200$ in the next trillion years.

A particularly promising way of using complexity in cryptography is to utilize a complex computation in order to generate a pseudo-random sequence (of bits, digits, or characters) for the one time pad cryptosystem. A pseudo random generator is a deterministic algorithm which stretches a short truly random seed of length k into a longer sequence of length n, where n is some polynomial in k. The set of possible outputs of such a generator is only a negligible subset of the set of all sequences of length n, but there should be no feasible way to tell them apart. Note again that infinitely powerful opponents will be able to determine that the pseudo random sequences are compressible (and thus nonrandom) by trying all the possible seeds of length k. However, for sufficiently large k this process will take an unfeasible amount of time.

To formalize this notion, we use the concept of "distinguisher". A distinguisher D is any polynomial time algorithm whose input is a sequence of size n, and whose output is 0 or 1. There are two probabilities associated with such a distinguisher: The probability PR of outputting 1 when its inputs are truly random sequences of length n, and the probability PG of outputting 1 when its inputs are the outputs of the generator G applied to a truly random seed of length k. The generator G is called perfect if for all feasible distinguishers D, (PR-PG) decreases faster than the inverse of any polynomial when the sizes of k and n grow to infinity. In other words, there is no feasibly computable measure which can significantly differentiate between truly random and pseudo random sequences. Any generator which passes this very strict test can be safely used as a source of sequences for the one time pad cryptosystem, since any successful attack on the cryptosystem could be used as a distinguisher between the presumably breakable pseudo random version of the cryptosystem and the provably secure truly random version of the cryptosystem.

The main disadvantage of the original one time pad cryptosystem was the large key which had to be sent in advance via the secure key channel and stored in a secure way by both the sender and the recipient. By using a secure pseudo-random generator, the communicating parties can replace the long random sequence by a short random seed. To encrypt a new binary cleartext of length m, the sender generates the next m pseudo random bits, adds corresponding bits modulo 2 and transmits the result. To decrypt the ciphertext, the recipient regenerates locally the same m pseudo random bits, subtracts corresponding bits modulo 2, and reads the result. This greatly simplifies the key management problem, reduces the storage requirements, and eliminates the need to estimate in advance the likely size of the cleartexts.

The remaining problem is to develop secure pseudo random generators, which can defeat any feasible distinguisher. This problem received a lot of attention in the research community during the 1980's, which culminated in 1989 with a proof by Impagliazzo, Levin, Luby and Hastad that any one-way function can be turned into a secure pseudo-random generator. This proof linked two seemingly unrelated fundamental concepts, and had a profound impact on the foundations of cryptography.

DISCUSSION
(A. Lichnerowicz, chairman)

ARECCHI: The subject of this week is the emergence of complexity in mathematics, physics, etc. Do you think that quantum cryptography schemes are complex in any way? I am referring to papers by Bennett and Tipster.

SHAMIR: I think that quantum cryptographic schemes are not purely based on complexity theory; it is a different branch. I brought it up just because I was showing all kinds of nifty ideas about how to construct cryptosystems...

DEL RE: I should like to repeat a question I asked before. In your context, what exactly do you mean by theory of complexity and by complexity itself?

SHAMIR: I am only trying to give meaningful answers to a very restricted class of complexity questions, those which are well defined in the context of computations on finite strings. We define a computation as a transformation of input strings to output strings. Then the concept of complexity is defined as the number of steps required to do it, or, in alternative, the number of memory cells required to measure time, space, etc.

THIRRING: What is meant by a function being invertible or not, is clear. But what exactly do you mean when you speak of a function that is hard to invert?

SHAMIR: As a first approximation, the following definition can perhaps be given: no polynomial-time algorithm exists which will compute x given $f(x)$.

THOM: Just a Bourbakist question precising the last theorem. If x is one-to-one, what is the result?

SHAMIR: One-to-one functions can still be hard to invert. In the information theoretic sense, when you ignore computation aspects, the problem of going back in easy. But in complexity theory, the unique existence of x means nothing, it may still be very difficult to get to it.

THOM: If you take the string and associate another string in a one-to-one way I would say that you just get a permutation between strings, that is a random generator.

SHAMIR: A computer scientist will look at things differently, and will say that, even though you might claim that the description of two prime factors, P followed by Q, each of 500 bits, giving a thousand-bit string is isomorphic with the thousand-bit string which is the product, new mathematical content is added by the multiplication. I claim that from a complexity theoretic point of view going in this direction is easy, the reverse is still hard, in spite of the isomorphism.

LAMBO: You mentioned security, safety of messages, military codes, etc. I come from a university in Nigeria, where almost one out of every five students is doing computer science. This is probably true of many parts of Africa and the third world. Those students are all going to banks, government services, etc. In the last few years there has been almost an epidemic of frauds all over the place most of them made possible by the use of computers, so much so that you are almost afraid even to put your money in the bank. Is there any way of turning this flood of students going into computers towards more positive activities?

SHAMIR: Cryptology has two aspects. The theoretical one and a practical one. I have told you about some of the applications, and I can tell you that many of the concerns that you have raised, like bank security, are actually addressed by a concrete system that we are developing today. But I want to warn you about one aspect which is impossible to solve: if the bad guys are in the management of your bank, and they are the guys to whom you give all the cryptographic keys, then I do not know of any cryptographic solution to that problem.

DOBEREINER: I still have some difficulty with your definition of "hard" and "simple". It seems to me that being "hard" or "simple" depends on how your computer is constructed.

SHAMIR: You have indicated a very important reason why we work with asymptotics rather than specific numbers and sizes. For the sake of concreteness we refer to a Turing machine. A Turing machine does not have any primitive operation of multiplication. The only operations it knows are how to read one bit, how to write one bit, how to move the head right of left. Nevertheless, within a polynomial-time factor it can mimic the operation of any computer.

RUBBIA: Let me first congratulate you on the ability with which you have been defending your case. You are not only an excellent cryptographer but you might become a remarkable lawyer as well! Coming back to serious questions, as you correctly say, a computer can simulate the functions of another computer. But I am confused when you distinguish between "easy" and "hard". I cannot give a definite, mathematical meaning to such a concept. What you call a "simple" function for a given computer may become a "hard" function for a different computer and vice versa. "Simple" and "hard" depend on the architecture of the computer. I can build conceptually with the help of appropriate hardware a new computer for which a given function which for my computer is "hard" becomes "simple". Something which was difficult for one computer is then going to become easy for another computer. So, concepts of "easy" and "hard" are in my view computer-dependent. I find it difficult to make them "objective".

SHAMIR: The difference between us it that you consider one hundred steps on a computer to be an easy computation, one billion steps to be a hard computation. I define things in a different way. According to my definition an "easy" computation in one which for inputs of size N asymptotically requires N, N^2, ... N^{100}, ... operations; a "hard" computations is one which, as N increases, requires 2^N steps. We made the conscious decision not to take the fuzzy intuitive notions of "easy" and "hard", but to give them a precise meaning: "easy" is polynomial-time computable, "hard" is *non* polynomial-time computable.

LICHNEROWITZ: Je voudrais compléter la réponse précedente. Je pense que c'est une distinction asymptotique en fonction de N; c'est le comportement. Quel que soit l'ordinateur que vous utilisez, du moment qu'il fonctionne en binaire, vous aurez un invariant qui est ce comportement asymptotique. Le type de facilité ou difficulté, c'est exactement cela.

RUBBIA: Hard and not so hard, easy in the ordinary language and hard and easy in the mathematical language are not at all the same. Mathematical concepts need a rigorous definition.

DOBEREINER: Your presentation illustrates a process which is exactly opposite to the process we are supposed to investigate, that is, the emergence of complexity. You do exactly the opposite in a very interesting way, namely you reduce complexity to noncomplex propositions and methods. But there is a price to pay, there. You have repeatedly referred to the Turing machine, and to the view that we must not ask what understanding is, of how to define it, as if understanding could be defined in any terms other than understanding. That is to say, the very interesting methods you are proposing cannot themselves claim to be intelligent.

SHAMIR: At variance with your remarks, I think that I am among the few lucky ones who are exploiting complexity. Most people are unhappy with the emergence of complexity, they would prefer it if the world were very simple, but then it would be doom for a cryptographer like myself.

LEJEUNE: Monsieur le Président, je voudrais poser la même question qu'hier: d'où vient le mot "random"? What is the etymology of that word? Who introduced it? And what does it mean really? Is it just a flap of swatches or what? Is randomness chosen at random? Je voudrais faire remarquer que nous avons deux grands spécialistes de "randomness", et que ni l'un ni l'autre ne sait ce que veut dire le concept qu'ils emploient. Je m'explique. Dans l'utilisation, ils savent parfaitement la fonction de ce mot, ils nous l'ont expliqué de façon parfaitement claire. Mais ce qui m'intéresse, c'est que les mots du vocabulaire sont reliés les uns aux autres, et en fait, l'étymologie nous permet de comprendre la consistance de la pensée humaine. Et quand on rajoute un concept qui ne vient de nulle part, au sens étymologique au moins, on aimerait connaître son histoire.

THOM: The etymology of "random" is very well known. It is the old French word "randon", it was a way of hunting, "chasse à la randon", and the word "randon" has later given the word "randonnée", which means "a long hike".

DAY TWO
28 OCTOBER 1992

THE EMERGENCE OF ORDER AND COMPLEXITY
IN PHYSICS

W. THIRRING

Institut für Theoretische Physik

Universität Wien

The search for order in nature has a long history. The first significant landmark in this unending quest are Newton's laws. He realized that the trajectories in phase space of mass points are determined by a first order differential equation. They define a flow, that is a one parameter group on one to one mappings of phase space onto itself. The parameter is the time and the flow describes the time evolution of arbitrary initial points. Furthermore for Newton's law for the gravitational force the flow is particularly simple since the trajectories are either closed or go to infinity. This simplicity was later found to be due to a hidden symmetry of the problem. Since the force is a central force, the equations are invariant under rotations in configuration space but in addition also under a combined transformation of coordinates and velocities. Altogether this gives an invariance group isomorphic to 0 (4), the group of rotation in 4-dimensional space. The trajectories when projected into donfiguration space are ellipses as empirically found by Kepler for the motion of the planets. These ellipses caused some uneasiness because in Aristotelian philosophy the most perfect curve is a circle and only this was felt to be worthy of the earth. Today we do not attach a deeper significance to this notion of order, but it is amusing to note that the trajectories are actually perfect circles when projected on velocity space. In spite of its simplicity Newtonian mechanics was superseded by more fundamental notions. The forces at distance turned out to be transmitted by fields which followed their own laws.

In the quest for the correct field equations, another point of view turned out to be instrumental, namely mathematical beauty. As in art beauty cannot be well defined scientifically but I shall illustrate it by the history of Maxwell's equations for the electromagnetic field.

Table 1
Maxwell's Equations in the Course of History

The constants c, μ_0, and ε_0 are set to 1, and modern notation is used for the components.

The Homogeneous Equation   **The Inhomogeneous Equation**

Earliest Form

$$\frac{\partial B_x}{\partial x} + \frac{\partial B_y}{\partial y} + \frac{\partial B_z}{\partial z} = 0 \qquad\qquad \frac{\partial E_x}{\partial x} + \frac{\partial E_y}{\partial y} + \frac{\partial E_z}{\partial z} = \rho$$

$$\frac{\partial E_z}{\partial y} + \frac{\partial E_y}{\partial z} = \dot{B}_x \qquad\qquad \frac{\partial B_z}{\partial y} + \frac{\partial B_y}{\partial z} = j_x + \dot{E}_x$$

$$\frac{\partial E_x}{\partial z} + \frac{\partial E}{\partial x} = \dot{B}_y \qquad\qquad \frac{\partial B_x}{\partial z} + \frac{\partial B_z}{\partial x} = j_y + \dot{E}_y$$

$$\frac{\partial E_y}{\partial x} + \frac{\partial E_x}{\partial y} = \dot{B}_z \qquad\qquad \frac{\partial B_y}{\partial x} + \frac{\partial B_x}{\partial y} = j_x + \dot{E}_x$$

At the End of the Last Century

$$V \cdot B = 0 \qquad\qquad V \cdot E = \rho$$
$$V \times E = - \dot{B} \qquad\qquad V \times B = j + \dot{E}$$

At the Beginning of This Century

$$F^{\beta z}{}_{,z} = 0 \qquad\qquad {}^{*}F^{\beta z}{}_{,z} = j^{\beta}$$

Mid-Twentieth Century

$$dF = O \qquad\qquad \delta F = J$$

The early form looks somewhat complicated with space (x, y, z) and time t, the electric and magnetic fields E and B interwoven in a strange way. Nevertheless Boltzmann saw in it such awe-inspiring beauty that, quoting

Faust, he exclaimed: "Was it a god who traced these lines?" Later the notation was made easier by the introduction of symbols for the differential operations appearing in these equations. It was then recognized by Lorentz that the equations are invariant under transformation mixing (x, y, z) with t and E with B. Einstein understood the far-reaching consequences of this invariance. The deepest understanding of the structure of these equations was reached when one freed oneself from the limitation to Cartesian coordinates and admitted the theory of differential manifolds into any suitable coordinate system. There one finds that there is only one differential process not referring to a particular coordinate, the exterior derivative d. If the manifold has a metric as additional structure, one can define a scalar product and with the adjoint operator to d the coderivative δ. Thus from this point of view the most natural differential equations are the specification of the exterior derivative and the coderivative of the quantity F which contains E and B. The fact Maxwell's equations assume the simple form $dF = O$, $\delta F = J$, is not just a suitable shorthand. None less than Dirac himself believed that this aesthetics had a deeper significance.

"The Lorentz transformations are beautiful transformations from the mathematical point of view, and Einstein introduced the idea that something which is beautiful is very likely to be valuable in describing fundamental physics. This is really a more fundamental idea than any previous idea. I think we owe it to Einstein more than to anyone else, that one needs to have beauty in mathematical equations which describe fundamental physical theories" (Dirac).

The way to the equations for the gravitational field was even harder and when Einstein found them finally he had to admit that under the present condition the change for Newtonian mechanics is minuscule. Nevertheless, now it is firmly established and what it did for the Kepler problem was to reduce the symmetry group from 0 (4) to 0 (3).

As a consequence the orbits are no longer closed but the axis of the ellipse rotates slowly. Nevertheless the invariance group 0 (3) of rotations in coordinate space insures that the system is still what is called integrable. This implies that the motion is still orderly in the sense that uncertainties in the initial conditions grow only linearly in time. The worst case compatible with group structure of a flow is that they grow exponentially and they do that if there is no further symmetry in the problem. This is exactly what happens when the gravitational attraction between the planets is taken into account. Naively one might think that the characteristic time $1/\gamma$ for the exponential increase ~ $e^{\gamma t}$ of uncertainties is of the order 1000 years since the mass and therefore the gravitational effect of Jupiter is 1/1000 times the mass of the sun. Actually the situation is far more complex and depends on the irrationality of the ratio of the frequencies of the orbits. If they are sufficiently irrational, resonance effects are averaged out and the perturbation of other planets by Jupiter causes only a small deformation of the orbits. Nevertheless the stability of the planetary system over times ~10^9 years is still something a miracle.

Since in practice (and in quantum mechanics in principle) the initial point in phase space cannot be determined with arbitrary precision, classical mechanics loses its predictive power for $t > 1/\gamma$ and one can talk only about probabilities. Therefore systems with $\gamma \neq 0$ are felt to be disordered and chaotic and the question arises of how to give a mathematic definition to these vague notions. Kolmogorov has defined two related quantities, the complexity and the dynamical entropy. The former is the length of the shortest computer program which can cope with the system. Since this will be discussed in the mathematical sections I shall concentrate on its physical counterpart, the dynamical entropy. This latter gives the maximal asymptotic information gain by periodically measuring a subsystem. In chaotic systems in the course of time information gets lost so that new information can always be gained. Mathematically these quantities are defined as follows. Suppose you have a probability distribution $\rho(z)$ for, say, finding a particle at a point z. By measurement you will not be able to determine its position exactly but you can say in which of the boxes A_i the particle was. The A_i are supposed to cover the space $M = \cup_i A_i$ and $\mathcal{A} = \{A_i\}$ is called a partition. The probability for finding a particle in A_i is $\mu_i = \int_{Ai} dz\, \rho(z)$ and the information gained by knowing in which of the boxes the particle is given by the entropy

$$S\,\rho(\mathcal{A}) = -\sum_i \mu_i \ln \mu_i.$$

More information is gained if the grid is refined for instance by considering the intersection of two partitions

$$\mathcal{A} \vee \mathcal{B} = A_i \cap Bk, \quad \mu_{ik} = \int_{A_i \cap B_k} dz\, \rho(z)$$

$$S\,\rho(\mathcal{A} \vee \mathcal{B}) = -\sum \mu_{ik} \ln \mu_{ik}.$$

Now assume that a flow $\emptyset : M \to M$ is defined on M which means that at a time t a particle is at $z \in M$ at the time $t + 1$ it is at $\emptyset(z) \in M$. Measuring at the time $t, t+1, t+2... t+n$ in which box the particle is equivalent to measuring at t where it is in the partition $A \wedge \emptyset(A) \wedge ... \wedge \emptyset^n(A)$. Then the information gained by the last measurement is

$$S(A \wedge \emptyset(A)...\wedge \emptyset^n(A) - S(A \wedge \emptyset(A) \wedge...\wedge \emptyset^{n-1}(A)).$$

It turns out that for $n \to \infty$ this converges to a definite limit

$$h_\rho(A, \emptyset) = \lim_{n \to \infty} [S(A \wedge \emptyset)(A) \wedge...\wedge\emptyset^n(A)) - S(A\wedge\emptyset(A) \wedge...\wedge \emptyset^{n-1}(A))].$$

The dynamicalentropy is now defined as the upper limit of the values which this quantity can assume by varying the partition A:

$$h_\rho(\emptyset) = \sup_A h_\rho(A, \emptyset).$$

For a measure of the indeterminism of a deterministic system this quantity contains a surprising amount of information. For a certain class of systems it contains in fact all information in as much as they are isomorphic if they have the same dynamical entropy. It turns out that generally

$$h_\rho(A, \emptyset) \leq S_\rho(A)$$

and from this point of view the most chaotic systems are so-called K-systems for which

$$\lim_{n \to \infty} h_\rho(A, \emptyset^n) = S_\rho(A) \; \forall A.$$

This means that after many time steps all partitions A have a complete memory loss such that by remeasuring one gains the full information. In spite of being completely chaotic they have a well-defined structure in as much as there are three more seemingly different characterizations. One might object that the simple classical model of a flow $\emptyset$ on M is too poor a model for a reality which is quantum mechanical. It turns out that all this (including the Liapunov exponent γ) can be generalized to quantum systems and that the real systems we find in nature are presumably K-systems in this generalized sense.

HARMONY AND COMPLEXITY
ORDER AND CHAOS IN MECHANICAL SYSTEMS

PETER H. RICHTER

University of Bremen, Germany

Introduction

The quest for harmony was Johannes Kepler's guiding principle when he struggled to discover the divine plan in the order of our universe. As you all know, he saw the *Mysterium Cosmographicum* in 1596 (fig. 1) as an orderly arrangement of the five Platonic solids, each touching two planetary spheres and thereby defining their successive distances. According to this model the six planets Mercury, Venus, Earth, Mars, Jupiter, and Saturn are held in their respective spheres by octahedron, icosahedron, dodecahedron, tetrahedron, and cube — an order for which he had a number of special reasons, following the Pythagorean tradition. It is remarkable that this picture "explains" the actual relation in our solar system to within some 10% (see table 1). Kepler was eager to confirm his model using the data that Tycho Brahe had collected throughout his life as a meticulous observer of the sky. He suffered a terrible disappointment when he discovered that the planets do not follow circles but ellipses; this seemed to be a fatal blow to the vision of his youth. But gradually he found order of a higher kind, and praised the Creator for this...

I am not referring to the laws to which his name became attached later, the laws of the celestial two-body problem: elliptical motion, angular momemtum conservation, and 3:2 scaling of spatial versus temporal units. This was an admirable achievement to be sure, but there was more, which is little known. In the first place Kepler was not interested in just two bodies. He (a)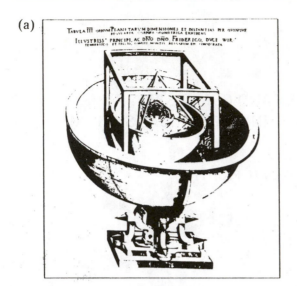

(b)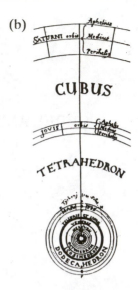

Fig. 1 - Two versions of Kepler's world model. a: The six planetary spheres are assumed to obtain their respective radii from the five Platonic bodies. The outermost sphere is Saturn's. Inscribed is a cube which Jupiter's sphere touches from the inside. As a consequence, Saturn's sphere is $\sqrt{3} = 1.732$ times the size of Jupiter's. The actual value fluctuates in the range 1.84 ± 0.20 because the orbits are not circular but rather elliptic. A tetrahedron is seen to lie between the spheres of Jupiter and Mars, then a dodecahedron between Mars and Earth. b: After Kepler found the elliptic nature of the planetary orbits, he modified his model by attributing the spheres a thickness that takes into account the difference of maximum and minimum distance from the sun.

looked at the solar system as a whole, and tried to detect regularities in the particular values of eccentricities. In *Harmonices Mundi*[1], his most mature opus of 1619, he reports his findings, insisting that geometrical harmony be related to that of musical intervals. Until quite recently there was no basis to appreciate this kind of "chimériques spéculations" (Laplace, 1821) within the framework of theoretical mechanics. For those, however, working today in the field of nonintegrable classical mechanical systems, Kepler's observations are surprisingly near the mark. The musical harmonies that he identifies in the frequency ratios of planetary motion (at perihelion, aphelion, or average position) are nothing but *resonances* in modern terms. Based on the geometrization of mechanics as worked out by Poincaré and his followers, we

---

[1] J. KEPLER, *Harmonices Mundi*, Linz, 1619.

	Octah.	Icosah.	Dodecah.	Tetrah.	Cube
Platonic Solids					
ratio of radii	1.732	1.258	1.258	3.000	1.732

Planets	☿	♀	⊕	♂	♃	♄
	Mercury	Venus	Earth	Mars	Jupiter	Saturn
mean radius (a.u.)	0.387	0.723	1.000	1.524	5.203	9.546
ratio of radii	1.868	1.383	1.524	3.414	1.834	

Table 1

can now relate to "more or less rational" and to "more or less irrational" frequency ratios, and we see the two extremes as particularly conspicuous types of dynamical behavior on which natural evolution may operate.

The key to an understanding of how harmony in Kepler's sense arises, is the observation that number theory is an important ingredient of stability analysis in complex dynamics. Nobody suspected that, for at least two hundred years, after Newton had developed his theory of classical mechanics, and of celestial mechanics in particular. In a continuous world where *natura non facit saltus,* how could it matter whether or not a number was rational, i.e., could be expressed as a ratio of two integers. We know that almost all numbers are irrational, i. e. cannot be written as a ratio of integers (remember the classical Greek argument for $\sqrt{2}$ to be irrational). Nevertheless, there are sufficiently many rational numbers to approximate any real number to arbitrary precision; take, for example, a decimal number of sufficient though finite length. Eventually the difference of $\sqrt{2}$ and 1.41421356... ought to be irrelevant for all practical purposes, and so, it was felt, the distinction of rational versus irrational should not be of interest in physics.

But then there was the notorious difficulty of proving the stability of our solar system. Inspite of Laplace's bold assertion to the contrary, rigorous proof is lacking to this very day. In fact, when Poincaré addressed this problem a hundred years ago, developing his famous Méthodes Nouvelles de la Mécanique Céleste[2], it appeared the system might not be stable because for all *rational* frequency ratios (in a sense to be explained later) it turned out there was an instability. It took mathematicians some 70 years to recover a slightly more optimistic point of view: the so called Kolmogorov-Arnold-Moser theorem of 1963 established the notion that *sufficiently irrational* situations may be "robust" against the onset of *Chaos*, and thereby represent an element of stability. Still this is far from directly applicable to our solar system with its 10 celestial bodies. Definite knowledge exists only for special cases of the three body problem. My lecture will therefore focus on such simple systems, and demonstrate a few aspects of their dynamic complexity. Among these, perhaps the most fascinating is that with a certain degree of universality, special numbers turn out to be particularly important.

Chaos

If Kepler laid the foundation of modern astronomy, Galilei was the father of physics. Newton stood on these two giants' shoulders when some 70 years later he formulated his *Principia Mathematica*[3]. Common to all three, and to physicists ever since, has been the desire to discover Nature's order and express it in mathematical terms, with a tendency to accept simplicity as an indicator of truth. But there are two sides to Nature: one is the set of laws, the other the set of phenomena that they govern. And while the laws seem to be simple indeed, it becomes more and more obvious that the time course of events is typically very complex This is by no means a new discovery. Newton knew it well, and many scientists including Maxwell, Boltzmann, Einstein, Born have expressed it very clearly. It must be confessed, however, that for purely opportunistic reasons, when physicists turn to explaining phenomena, they tend to select the simple ones, those that can be handled with available tools. What else could they do? Nobody can be blamed for being silent about what he does not understand.

In mechanics, the simple problems are called *integrable*, and they have occupied students ever since Newton, Euler, and the Bernoullis provided the first solutions. As these grew more and more impressive, they nourished the

---

[2] H. POINCARÉ, *Les Méthodes Nouvelles de la Mécanique Céleste*, Gauthiers-Villars, Paris, 1899.
[3] I. NEWTON, *Philosophiae Naturalis Principia Mathematica*, London, 1687.

prejudice that everything might be integrable for a sufficiently powerful mind. Laplace's demon is but the symbol of its extreme form. Unreal as it always was, it did not succumb to better insight: its exorcism had to wait for powerful computers to be universally available.

... and for Jim Yorke, mathematician of the University of Maryland, to introduce the catchword *Chaos* into the field of Complex Dynamical Systems, in the title of a short article in 1975 (*Period 3 Implies Chaos*[4]). He gave it a precise mathematical meaning, but of course consciously played with the many resonances that this notion elicits. Activity of gigantic proportions grew out of this jocular expression and spread throughout the world. Any serious mechanics course these days tells the students that integrability is the exception, chaos the rule.

I hope you do not object too much to the liberty taken in stealing a well established term from the context of genesis. Of course we are talking about something very different. But considering that applications to the evolution of the solar system, and even the cosmos as a whole, have been worked out, Yorke's terminology may not be altogether unjustified.

So what do we mean by *chaos*? We mean dynamical behavior of the state of a system that can be formulated as a set of simple deterministic rules, and yet is unpredictable in the long run. We assert that this behavior is ubiquitous. But what do we mean by *unpredictable in the long run*? It means that, given the slightest uncertainty in the initial state of a system, there comes a time when the rules are no longer of any help in predicting the state of the system.

The essence of this behavior can be grasped by a very simple mathematical example. Let the state of a system be described by the set of numbers x in the interval $(0,1)$, and let the rule be: multiply by 2 and ignore the integer part,

$$x \rightarrow 2x \bmod 1. \tag{1}$$

The prediction we might be thinking of is to say whether x is smaller or larger than 0.5. Now let the number x be given in their binary representation: $x = 0.x_1x_2x_3...$, the bits $x_1, x_2, x_3...$ being either 0 or 1. Then the rule (1) can be stated more explicitly:

$$0.x_1x_2x_3... \rightarrow 0.x_2x_3x_4...; \tag{2}$$

[4] T.-Y. LI, J. A. YORKE, *Period Three Implies Chaos*, Amer. Math. Monthly 82, 985 (1975).

the bits are shifted to the left by one unit, and the leading bit is lost. Now think of an initial uncertainty. It means that only a certain number N of bits $x_1,...x_N$ can be specified. Our rule then allows prediction up to the N-th step. After that, we cannot tell whether the number will be smaller or larger than 0.5. Observing the system after the N-th step will reveal more detailed information about the initial state than we had at the beginning. We say that this process generates information.

This behavior is often referred to as the *butterfly effect*, and illustrated by the butterfly in Florida that prevents or provokes a thunderstorm in New York. I do not like this particular illustration because it does not correspond to our daily experience, even though our weather system is undoubtedly chaotic. The truth about real physical systems is that only parts of them are of type (1), and that it may be difficult to identify those parts.

My favorite physical example to demonstrate this kind of chaos consists of ingredients that Galileo Galilei introduced into physics. I call it *Galilean chaos*, and am happy to present it to this Academy to which he was so closely related[5]. Consider an inclined plane in front of a vertical wall, and a mass point that is released over it. Assume that the mass point is reflected elastically both at the plane and at the wall (fig. 2). Between any two bounces it will follow a Galilean parabola, so everything is very simply described and computed (though difficult to demonstrate in a real experiment because friction and non-elasticity interfere strongly). And yet this elementary physical system is prototypically chaotic!

How do we analyze this? First we might just look at particular orbits and collect some experience about their behavior. Three typical examples are given in figs. 2. Part (a) shows a *periodic orbit*. It so happens that it is reflected back into itself at two points: at the end of its vertical rise, and when it meets the inclined plane at a right angle. This is obviously a very special situation, and even though there are infinitely many different periodic orbits, most orbits are not periodic. The next level of complication is a *quasiperiodic orbit* as shown in fig. 2b. It bears some of the regularity of periodic orbits in that it comes close to its initial situation at regular intervals, but it never quite returns into itself. The *chaotic orbit* of fig. 2c is clearly different. It looks erratic at a first glance, and there is obviously no way to predict its time course in the long run.

It may be fun to observe a number of such orbits, playing with the initial

---

[5] P. H. RICHTER, H.-J. SCHOLZ, A. WITTEK, *A Breathing Chaos*, Nonlinearity 3, 45-67 (1990).

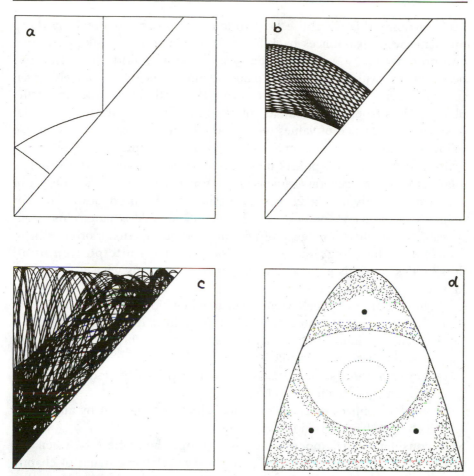

Fig. 2 - Galilean chaos. An inclined plane meets a vertical wall at an angle $\theta = 40°$. A
mass point bounces back and forth inside this wedge, with elastic reflections at
the walls. Its orbit may be periodic (a), quasiperiodic (b), or chaotic (c),
depending on the initial conditions. The Poincaré section (d) combines the
essence of the three pictures. The periodic orbit (a) appears as a set of three
points (at the centers of three islands); the quasiperiodic orbit (b) generates the
line around the main island, and the chaotic orbit (c) fills the grey area. The
coordinates plotted are the $v_{\parallel}$ and $v_{\perp}^{2}$.

conditions. In the end, however, we get tired and wonder whether there is a
way to survey all possible orbits in a single picture.This is precisely what
Poincaré achieved when he suggested throwing away most of the
information about those orbits and only retaining the essence. Of the many

ways of doing this we chose here to look at an orbit at precisely those moments when it bounces back from the inclined plane. Taking the two components $(v_{\parallel}, v_{\perp})$ of its velocity at these moments, parallel and perpendicular with respect to the plane, identifies the orbit uniquely, given the total energy which is constant during the motion: the location on the plane derives from the available potential energy. Poincaré's idea was to observe the sequence of values $(v_{\parallel}, v_{\perp})$ which an orbit generates when we follow it from bounce to bounce at the plane. A periodic orbit generates a finite number of such points because it returns to where it started. A quasiperiodic orbit generates a dense line of points, or perhaps a finite set of such lines. Finally, as fig. 2d shows, a chaotic orbit produces a cloud of points. This cloud does not, in general, cover all possible situations; its extent is restricted by the coexistence of periodic and quasiperiodic orbits. But in contrast to these regular cases, it does cover a finite portion of all possibilities, and this is in somewhat quantitative terms what we mean by long term unpredictability.

Chaotic and orderly behavior is typically highly interwoven. The picture of fig. 2d is simpler than most such Poincaré sections. It contains
  • a central island of order, with quasiperiodic orbits of type (b) sur-
    rounding a simple periodic orbit;
  • a cycle of three smaller island of order, with quasiperiodic orbits sur-
    rounding a periodic orbit of type (a);
  • one big connected region of chaos which is generated by any initial
    condition from within it.
Now consider what happens when we change the angle θ between the vertical wall and the inclined plane (fig. 3). The relative amount of chaotic versus regular behavior is strongly dependent on θ — an observation that came as a surprise but is not untypical. Though not in all details, it may be understood on the basis of a systematic stability analysis of periodic orbits (op. cit.).

The sequence of fig. 3 starts with two completely chaotic situations. Computer experiments indicate that this behavior is found for all angles $\theta > 45°$. It should be possible to prove this as a mathematical fact, but so far no proof has been found, for reasons I shall not go into. The special case $\theta = 45°$ is not shown in fig. 3 because it contains no chaos whatsoever: in this and only this case the system is *integrable*, i.e., contains only periodic and quasiperiodic orbits. Then, slightly below 45°, there is very little chaos and lot of different quasiperiodic regions with periodic centers.

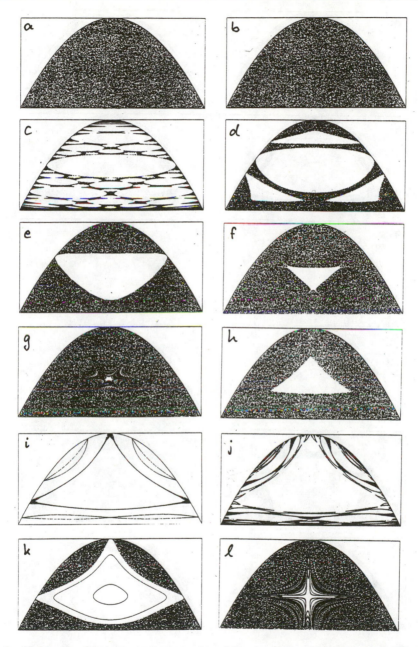

Fig. 3 - Sequence of Poincaré sections of 24 different angles θ. (a) $\theta = 60°$; (b) $\theta = 47°$; (c) $\theta = 44°$; (d) $\theta = 40°$; (e) $\theta = 37.5°$; (f) $\theta = 35.5°$; (g) $\theta = 34.26°$; (h) $\theta = 32°$; (i) $\theta = 30°$; (j) $\theta = 29.5°$; (l) $\theta = 25.91°$.

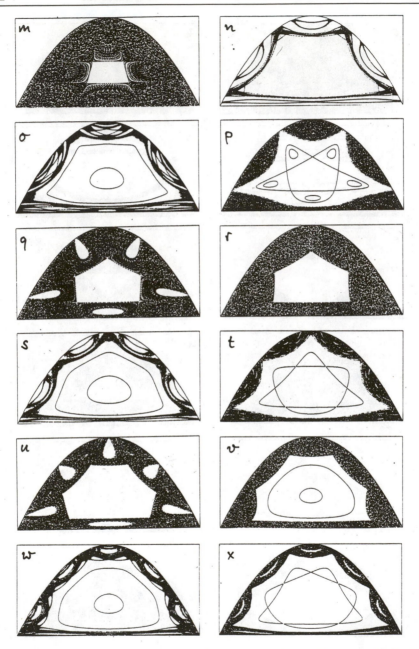

Fig. 3 (continued) - (m) $\theta = 25.5°$; (n) $\theta = 22.5°$; (o) $\theta = 21.5°$; (p) $\theta = 20.5°$; (q) $\theta = 20.25°$; (r) $\theta = 20°$; (s) $\theta = 18°$; (t) $\theta = 17°$; (u) $\theta = 16.6°$; (v) $\theta = 16°$; (w) $\theta = 15°$; (x) $\theta = 14.5°$.

As θ decreases further, chaos spreads, and order gets compressed into a small triangular region before it disappears completely at $\theta = 34.26°$. The sequence continues with order recovering around the center, and taking over almost completely at $\theta = 30°$. Then comes a repetition of this scenario with decreasing amplitude in the oscillation of chaoticity. We have termed this kind of parameter dependence the *breathing chaos,* and found that it is characteristic of other systems as well. Fig. 4 gives a quantitative account of this breating. Part (a) shows the amount of chaos in terms of relative area of the chaotic part of the Poincaré section. Part (b) exhibits the so-called *Lyapunov exponent* of the main chaotic region. This is a number which characterizes the chaos dynamically; it tells us how fast the chaotic orbit spread over the accessible region of phase space.

Take two initial conditions, close together, from the chaotic region, and follow their time course. Observation tells us that in the long run, their separation grows exponentially. The Lyapunov exponent is the rate of this growth, its inverse gives the time scale on which prediction is possible. Large Lyapunov exponents indicate fast mixing, i.e., rather quick loss of predictability. Small Lyapunov exponents, on the other hand, indicate slow mixing; predictability will be lost eventually, but only after a very long time. The range of angles $45° < \theta < 90°$ may be completely chaotic from the point of view of the sheer amount of chaos: nevertheless the Lyapunov exponent indicates that there are differences from the dynamical point of view. The time development of chaos is fast around $\theta = 60°$, and slow near the integrable limits $\theta = 45°$ and $\theta = 90°$.

To summarize, chaos is an elementary property of mechanical systems. It can be characterized by the extent to which it covers phase space, and — in more detail — by the time scale on which prediction is possible in each chaotic region. A survey on how phase space is partitioned into orderly (periodic or quasi-periodic) and chaotic regions, can be obtained from Poincaré sections. (In practice, this method is limited to cases with only two essential degrees of freedom.) In our example of Galilean chaos, it appears as if there is only one chaotic region into which a more or less complicated web of orderly behavior is embedded, depending on the parameter θ. Closer inspection would show that the situation is even more involved, with many small chaotic bands foliating the seemingly regular zones. In other examples, the various chaotic regions may be comparable in size, but typically this partitioning of phase space depends strongly on certain parameters.

While the size of a chaotic region tells us to what *extent* mixing occurs in

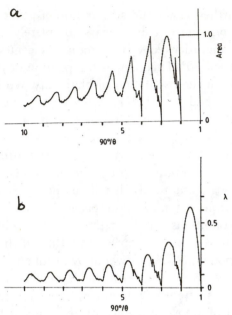

Fig. 4 - a: relative area of chaos in the Poincaré section, as a function of angle θ. The area was determined by pixel counting on an 8192 x 4096 grid, with 50 million iterations of an initial point inside the chaotic region. b: Lyapunov exponent of the main chaotic region, computed from 10 million interations of a chaotic initial point. — The abscissa $90°/\theta$ was chosen so as to show up the similarity of intervals $90°/(n + 1) < \theta < 90°/n$, $n = 1, 2,...$

the long run, the Lyapunov number is a measure of the *speed* of this mixing. Whatever the precision to which an initial state is given, the unavoidable uncertainty will be blown up exponentially at this rate. After a length of time of the order of the inverse Lyapunov exponent, no prediction is possible within the given chaotic region. Observing an orbit for longer times means unfolding the initial randomness. Though deterministic in its rules, the process turns into a random number generator. (By the way, in the very long run, the distribution of points in a chaos band will be homogeneous due to the general properties of Hamiltonian processes. However, this statement may not mean much in practice because a typical chaos band is punctured by holes of arbitrarily small sizes, and therefore is difficult to identify in detail.)

Note that I have restricted mechanics to systems without friction. This is, to be sure, in the tradition that Galilei established in opposition to Aristotelian physics, and that paved the way for Newton. But of course, in real nature there *is* friction, and we know it plays an important role in driving towards a final state. In the case of a mass point bouncing up and down, the introduction of friction leads to a trivial final state: the orbit will come to rest at the lowest accessible point. But this is different in celestial mechanics, the case of interest to Kepler...

Celestial Mechanics

Newton derived Kepler's ellipses from his law of gravitational attraction between any two celestial bodies, using the framework of analytical mechanics that he had established. However, this first great triumph of theoretical physics missed Kepler's point of trying to understand the solar system as a whole. Why are the planetary orbits the sizes they are; why are the periods of Jupiter and Saturn so nearly the ratio 2:5? Such questions have no place in Newton's physics. They are matters of initial conditions which may be arbitrarily set. From this point of view, the harmony that Kepler saw is pure chance, and thus irrelevant.

In our modern perspective, we attempt to reconcile both these attitudes. Galilei and Newton are right on relatively short time scales while Kepler's questions concern the long term evolution. This is where chaos come into play, and where Aristotelian dynamics begins to play a role in Darwinian phrasing.

The fact that celestial mechanics is plagued by chaos has been known for a long time. Poincaré (op. cit.) made it very clear that no analytical integration of Newton's equations can be expected in the case of three or more celestial bodies. So what *can* be done? Well, we can combine Poincaré's geometric thinking with numerical integration, and perform computer experiments to become familiar with the problem. We can identify regions of stability and regions of chaos, and thereby characterize the phenomenology on which evolution must have operated. This kind of work was initiated by E. Strömgren around 1920[6], and among the numerous investigations with electronic computers I mention those of Hénon[7] and Wisdom[8].

---

[6] E. STRÖMGREN, *Connaissance actuelle des orbites dans le problème des trois corps*, Bull. Astron. 9, 87-130 (1935).

[7] M. HÉNON, *Exploration numérique du problème restreint. I.* Ann. d'Astrophys. 28, 499-511 (1965), *II.* Ann. d'Astrophys. 28, 992-1007 (1965), *III.* Bull. astron. 1, 1, 57-79 (1966), *IV.* Bull. astron. 1, 2, 49-66 (1966), *V.* Astron. Astrophys. 1, 223-238 (1969).

[8] J. WISDOM, *Urey Prize Lecture: Chaotic Dynamics in the Solar System*, Icarus 72, 241-275 (1987).

The problem in its simplest non-trivial form is the so called restricted three body problem. It consists of two main celestial bodies (which I will call *Sun* and *Jupiter* for easy identification), and a third body so small that it does not influence the motion of the other two. Thus Sun and Jupiter behave according to the rules of two body dynamics. Their motion takes place in a fixed plane, and is assumed to be circular (circles are of course special cases of ellipses, the eccentricity being zero). The third body is assumed to move in the same plane, attracted by the gravitational forces of the main moving bodies. Jacobi noticed long ago that the time dependence of these forces can be eliminated by taking the point of view of a coordinate system that corotates with Sun and Jupiter. The scales of lengths, times, and masses can be chosen such that the distance between Sun and Jupiter, their angular velocity, and their combined mass are unity. The third body's motion, viewed in the rotating system, is then taking place in the constant potential shown in fig. 5, and under the influence of Coriolis forces.

The Jacoby potential consists of the gravitational attraction of Jupiter (mass μ, position $(1 - \mu, 0)$) and Sun (mass $1 - \mu$, position $(-\mu, 0)$), and of the centrifugal potential associated with the rotating frame. Its value approaches minus infinity at the position of the main bodies and towards the far outside. It has two maxima at the so called Lagrange positions L_4 and L_5 which form an equilateral triangle with the main bodies (irrespective of Jupiter's mass μ!) And there are three saddle points L_1, L_2, L_3, called Euler points, on the line connecting the main bodies.

To this very day, the test body's motion in Jacobi's potential has not been grasped in its entire complexity. In the following I shall attempt to give a pictorial survey in terms of Poincaré sections. As usual, such sections may be obtained in many ways, and our choice is purely a matter of convenience.

We consider small values of μ ($\mu < 0.05$), and Jacobi energies corresponding to the maximum of the potential (so that the test body is not energetically excluded from any part of the plane). A typical orbit will then resemble a Kepler ellipse surrounding the Sun, but its perihelia and aphelia will not be fixed in space. Rather will they be moving in a more or less orderly way, and it is this motion that we shall follow in our Poincaré sections. All the rest of the motion will be ignored because it does not contain new information with regard to the question whether the motion is periodic, quasiperiodic, or chaotic. In other words: specifying the Jacoby energy and the position of peri- or aphelion — in heliocentric polar coordinates r and φ, Jupiter lying at $(r, \varphi) = (1,0)$ — the test body's motion is completely defined.

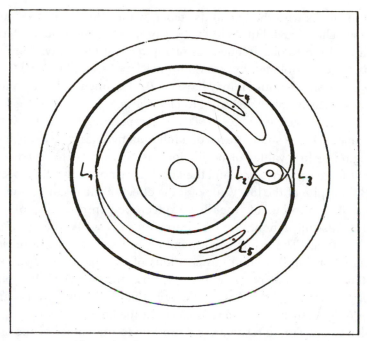

Fig. 5 - Equipotential lines for the motion of a test body in the gravitational field of two main bodies, of masses μ and $1 - \mu$, at positions $(1 - \mu, 0)$ and $(-\mu, 0)$, in a coordinate system that rotates with the main bodies. The potential attracts towards these bodies, and also towards the outside, due to the centrifugal force associated with the rotating frame. it has maxima at the Lagrange points L_4 and L_5 which form an equilateral triangle with the positions of the main bodies; the saddle points L_1, L_2, L_3 are called Euler points. In addition to the forces that derive from this potential, there is a Coriolis force deflecting the test body to the right in proportion to its own velocity. As a consequence of this peculiar force, the positions at the Lagrange points turn out to be stable provided the second main body is sufficiently light ($\mu < 0.0385$).

(Strictly speaking, there are two possibilities, depending on whether the angular velocity φ at peri- or aphelion is positive or negative. To have a unique orbit, we select the more interesting case $\dot{\varphi} > 0$.)

So the rules of our game are as follows. Given an orbit of our test body, we monitor its distance r from the Sun and wait for r to be a minimum or a maximum, $\dot{r} = 0$. If $\dot{\varphi} > 0$, we plot the coordinates (r, φ) of this event, and wait for the next. What do we get in the limiting case $\mu = 0$ where Jupiter's perturbation is absent? Clearly, the test body moves along a Kepler ellipse, with peri- and aphelion fixed in space. Looking from our coordinate system which now rotates with the Sun at its center, we see the distances r_p and r_a of peri- and aphelion fixed, but their angles φ proceed in equal steps, depending on the ratio of the Kepler frequencies of Jupiter and test body. The Poincaré section is therefore foliated by invariant lines $\{r = r_p\}$ and $\{r = r_a\}$ (of which one may be missing because it may have $\varphi < 0$). The angular increments along these lines are easily calculated from Kepler's laws.

Now consider the case of $\mu = 0.0001$, corresponding to one tenth the real mass of Jupiter. Fig. 6 shows traces of a foliation by lines $\{r = \text{const}\}$, especilly in the region of low values of perihelion distances r_p. (The corresponding orbits have large eccentricities, and the aphelia are not shown here because of $\dot{\varphi} < 0$.) But in spite of the low value of μ the effect of the perturbation is quite pronounced. A number of lines have developed into chains of islands, indicating resonant behavior at their centers, with quasiperiodicity surrounding them. The orbits corresponding to these islands have perihelia oscillating ("librating") around the positions of a stable resonance. In doing so they tend as much as possible to avoid close encounters with Jupiter, the source of perturbation here. A special case is that of the two islands around $r = 1$, $\varphi = \pm 60°$. These are the two resonances at the Lagrange points L_4, L_5 (see fig. 5), also called Troian resonances according to the names of the asteroids found there.

Outside the islands there is an extended region of chaos, apparently radiating out from Jupiter which occupies its center. What is the long term fate of an orbit in this region of phase space? In contrast to the elliptic orbits, these orbits tend to have close encounters with Jupiter where they pick up erratic changes of their perihelion positions. To illustrate this observation we have marked, in the upper part of fig. 6, those points in the Poincaré section whose orbits run into collision with Jupiter before their next or second return to the section. (The time reversed situations are also shown.) It is obvious that chaos, in this particular context, is intimately related to collisions. In addition, the computer experiments show that many points in the chaotic region eventually leave the system.

This indicates how the insertion of chaotic elements into a system's dynamics may contribute to order. A perturbation with relative mass $\mu = 0.0001$ added to Sun surrounded by cosmic dust, creates the structure of fig. 6. The islands mark regions where test particles can survive in the long run, while the space in between will be cleared by collision or ejection. As a result, Jupiter will accumulate mass and become an even stronger perturbation of the system. At $\mu = 0.0005$ the system's structure is that of fig. 7. The biggest

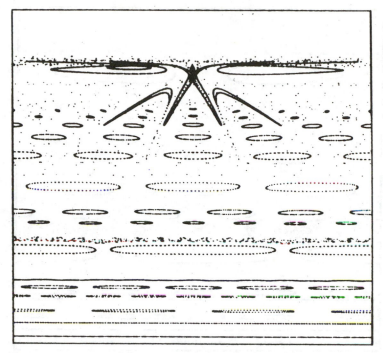

Fig. 6 - Poincaré section of the restricted three body problem for $\mu = 0.0001$ and
Jacobi energy equal to the peak value of the potential. The picture records
perihelion and aphelion positions (r, φ) of orbits if $\dot{\varphi} > 0$. The angle φ is plotted
along the abscissa ($-180° \leq \varphi \leq 180°$); the radial distance from the Sun is plotted
along the ordinate ($0 < r \leq 1.2$). The islands surround regions of order where test
particles may survive forever. The chaotic regions in between will eventually be
cleared of particles because of collisions with Jupiter or ejection from the
system. The batman like figure in the upper part marks the set of orbits that run
into collision with Jupiter before the next or second return to the Poincaré
section, both forward and backward in time.

islands have grown even bigger while the small ones have disappeared into
the chaotic region. This makes the distinction between order and chaos more
clearcut than when the perturbation is small.

Let us finally consider the sequence of pictures in fig. 8 where we increase
the value of μ from 0.005 to 0.03852. Linear stability analysis of the
Lagrangian libration points L_4, L_5 indicates they ought to be stable in this
range of perturbing masses. However, in a Hamiltonian system stability
cannot be established by linear analysis only. It needed the formidable
development of Kolmogorov-Arnold-Moser theory before it could be shown in

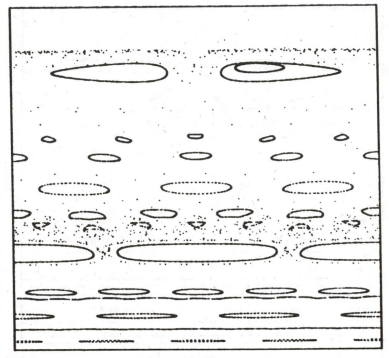

Fig. 7 - Poincaré section of the restricted three body problem for μ = 0.0005 and
    Jacobi energy equal to the peak value of the potential. In comparison to the case
    of μ = 0.0001, the prominent islands have grown in size, and their number has
    become smaller. The chaotic region has also grown. The distinction between
    order and chaos has become more clearcut.

1967[9] that stability in fact holds — except for perhaps three special values
where the conditions of Moser's theorem do not hold. Fig. 8 illustrates that
for two of these cases, viz. μ = 0.01352 and μ = 0.02429, the stability region
around the Troians is indeed compressed, and expands again in the
intermediate mass ranges. This is reminiscent of the breathing chaos that we
observed in the preceding section. Loss of stability of a resonance is to be
expected when the libration around the periodic orbit is itself resonant to
order two, three, or four. In that respect the case θ = 25.91° of our Galilean
chaos (fig. 3l) corresponds to μ = 0.01352 (fig. 8d). The angle θ = 34.26° (Fig.
3g) corresponds to μ = 0.02429 (fig. 8f). In both systems stability is lost
finally beyond the resonance of order two, at θ = 45° or μ = 0.03852.

    [9] A. DEPRIT, A. DEPRIT-BARTHOLEMÉ, *Stability of the Triangular Lagrange Points*, Astron. J. 72,
173-179 (1967).

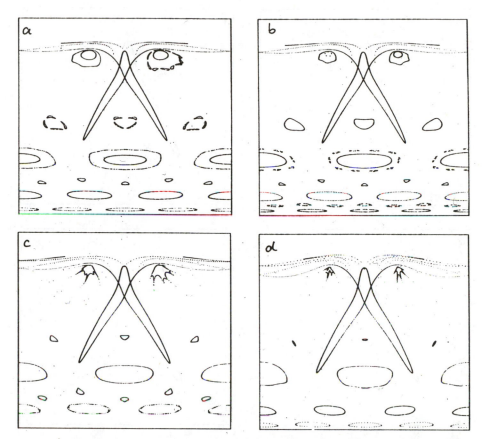

Fig. 8 - Poincaré sections of the restricted three body problem for eight different values of μ. The Jacobi energy is always equal to the peak value of the potential. Each picture shows the main resonances (in particular those around the Lagrange points L_4, L_5) and the set of points whose orbits collide with Jupiter before the next return to the Poincaré plane. (a) $\mu = 0.00500$; (b) $\mu = 0.00551$; (c) $\mu = 0.00827$; (d) $\mu = 0.01352$.

The Troian resonance is certainly an interesting feature of these pictures, but not the most important. The main observation concerns the general behavior of chaos versus order. The sequence of fig. 8 illustrates clearly how under the influence of increasing perturbation both chaos and a few strong resonances expand while all the minor features gradually disappear. If it is true that we may interpret the chaotic orbits as being condemned to disappear in the lung run, then what remains are just a few stable islands whose motion, if compared to Jupiter's, exhibits frequency ratios of the

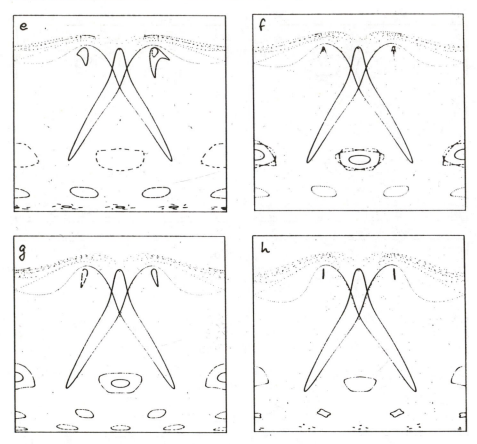

Fig. 8 (continued) - (e) μ = 0.02000; (f) μ = 0.02429; (g) μ = 0.03000; (h) μ = 0.03852.

highly rational kind 2:1, 3:2, or 4:3. In musical terms, this is octaves, fifths, ot fourths. And we are back to Kepler's harmonies.

Conclusions

What can we conclude from these observations? We have seen that the simplest possible three body system in celestial mechanics contains all the complexity of chaotic behavior, notably the long term unpredictability of individual orbits. But on the other hand, surveying the development of chaos as we leave the integrable limit μ = 0, we recognize a definite pattern in the emergence of structure. As the strength of the perturbation grows, the relationship between chaotic and regular regions in phase space loses complexity. Fewer and fewer islands of order survive, and these islands become the prominent feature of phase space (before eventually they may also be flooded by chaos).

All this is pure Newtonian mechanics, presented in terms of Poincaré's geometric ideas. The difference from Newton's way of thinking is that the focus is on whole classes of orbits and their stability rather than on single orbits. This allows us to think of the problem's global features: given the two main bodies and a dust of test particles to start with, how will the system evolve? Clearly it will matter for a dust particle whether or not its orbit is in resonance with the main motion, and in what kind of resonance it is. If it happens to move in one of the major islands of stability, chances are that it will stay there for a long time. We may speculate that under the influence of some slight dissipation (due to collisions or tidal friction) the orbit is drawn towards the centers of its islands. In this way different dust particles may approach each other and combine to form larger bodies. But then the three body problem gets more complicated...

Almost four hundred years after Kepler conceived his *Mysterium Cosmographicum,* we have not solved his problem of understanding the global dynamics of our solar system. But at least it appears there *is* something to understand: that the arrangement of the planets is not pure chance but the result of an evolutionary process in which the respective stability properties of regular versus chaotic types of obits have played an essential role.

DISCUSSION
(G. Puppi, chairman)

THOM: You mentioned the ratio of masses that distinguishes chaotic and non-chaotic behavior of the Troian asteroids. Can you calculate or explain why just this particular value distinguishes the two cases?

RICHTER: This calculation was made by J. J. Lagrange some two hundred years ago. It is a standard linear stability analysis about the critial point L_4 of the potential, called Lagrange point. The non-intuitive part of this problem is that L_4 is a *maximum* of the potential function; the stability is therefore an effect of the Coriolis forces. This effect depends on the mass ratio of the two main bodies, and linear stability analysis shows that L_4 is unstable when this ratio exceeds the value $(1 - \sqrt{69/9})/2 = 0.038521$). I must confess I have no intuitive explanation for this result.

MOSHINSKY: The Kepler picture that you showed us, and, more generally, the fact that there are integrable systems, related to the presence of symmetries. My question is this: does the algebra of symmetry groups play a role in chaotic phenomena?

RICHTER: Thank you for mentioning symmetries. We make extensive use of symmetry properties when we look for periodic orbits. It was Birkhoff's idea that when the system possesses a discrete symmetry including time reversal invariance, the Poincaré map can be written as the product of two involutions. You may than consider the invariant lines of these involutions, and their iterates under the map. Any intersection of two such lines gives a periodic orbit. As periodics orbits are the backbone of the dynamical picture, it is very important to have these symmetries.

MOSHINSKY: But not specially Lie algebras?

RICHTER: No, Lie algebras don't play an important role in chaos, except in the negative sense that some continuous symmetry is broken by the perturbation that leads into chaos. Lie algebra may have something to say in connection with perturbation theory, in the near vicinity of integrable limits. But it is our experience that pure theory does not help us very much in this context. Let me take this opportunity to express my point of view, as a physicist, with respect to mathematical theorems. I am very grateful for the security that rigorous results like the Kolmogorov-Arnold-Moser theory provide. I take them as a source of ideas for our computer experiments, even where their assumptions do not hold any more. The observations made on the computer screen can then be interpreted in the light of mathematical clarity, but they take us beyond established theorems, and so may stimulate further mathematical thinking.

RUBBIA: First of all let me thank the speaker for this very clear and extremely illuminating presentation. I'd like to add a comment to this, regarding the operational experience one needs to understand these phenomena where chaos plays a role together with some regularity of the orbits. You mentioned astronomy as a field of application, but of course there are very few examples in real astronomy; you replace that by computer simulations. I would like to point out the existence of another system which is very useful for these studies. Indeed, we are observing, in our own field, all the same phenomenology: in the behavior of cyclic particle accelerators. Let me tell the audience that in a particle accelerator, or a storage ring, we have one turn every few microseconds, and we can keep particle for days. So we reach 10^{10} to 10^{11} cycles without any appreciable friction, and that corresponds to the number of cycles the earth has made around the sun since the beginning of the history of our planetary system. So it's long, long time, and there is strictly no friction. You can then realize situations similar to the one you mentioned, of several particles interacting with each other, or interacting with a perturbative force that you can introduce from outside. And indeed, we do see the same type of phenomena: we see periodicity with the magic numbers, associated to certain islands of stability, we see chaotic phenomena occuring around these islands. Many of the phenomena you indicated are now emerging in practice from accelerators. I think this is a very interesting tool for understanding the fuzzy borderline between order and chaos.

RICHTER: Thank you. It's probably fair to say that the impact that boosted this field in modern times comes from elementary particle physics, and also from plasma physics. But the mathematical foundations were laid in celestial mechanics a long time ago.

THIRRING: Your last result was very remarkable, that for large values of μ it's exactly the rational frequencies which give you some stability. When you apply perturbation theory to small μ you find these are just the values where you cannot show there is any stability. And usually what happens for large perturbations is that the islands around frequency ratios which are completely irrational, like the golden mean, are the ones which survive longest. Also in reality, just at frequency ratio 2 there is a gap in the distribution of asteroids — of course for a smaller value of μ than you considered.

RICHTER: I did not have time to speak about the golden mean and its special role in this kind of system. It is one of the most fascinating aspects, and there are systems where it can be shown very neatly that orbits are particularly stable when their frequency ratio is the golden mean (because this is the most irrational of all numbers). I have studied one such system extensively, namely the planar double pendulum. However, in the case of the restricted three body problem, that scenario does not seem to be particularly important. If you want to observe it as a global feature of a system, you need two strong chaos bands that grow under increased perturbation and squeeze a last KAM torus between them before it breaks up into a "cantorus" and gives way to large scale chaos. In our situation the chaos emanates from only one center, and that's Jupiter the perturber; it radiates into the system and spreads out gradually as μ increases. In such a case the golden mean does not play a conspicuous role. Granted, if you look at the system under a microscope, then you will find areas where the golden mean scenario takes place; it is a "universal scenario" after all.

RUBBIA: I'd like to ask about the possibility to compute the behavior of the whole solar system, the ten body example you mentioned. The reason why this comes to my mind is associated with the asteroids which are part of our planetary system. This is a very practical problem: can we predict an asteroid hitting our planet? Is the motion of the asteroids, or these stoned flying around in our system, a chaotic movement, or is it regular? We know that in past history, a large number of collisions have occured; we also know the dinosaurs presumably disappeared because of that. It is important that modern science should be able to predict such an event with precision. It's a different issue to prevent it, but we should try at least to foresee it. There is no doubt in my mind that sooner or later in the future such a thing will occur on earth, because it did occur many times in the past, so it's a real problem. Can your calculation help in that respect?

RICHTER: The person who has done this kind of calculation with highest precision so far is Jack Wisdom at the Massachusetts Institute of Technology. He has a special purpose computer with one processor for each of the ten bodies of our solar system, and he has simulated its motion for the next five hundred million years or so. He finds with some confidence that taking only those ten bodies into consideration, the system is essentially stable, except for small scale chaos in Pluto's motion. With respect to the asteroids, Jack Wisdom has also done a lot of careful computations, taking into account the influence of Jupiter and Mars. He finds that when an asteroid comes close to Mars it may get deflected into an orbit that approaches the Earth. He claims that many meteorites that hit us originate from the asteroid belt. So, yes, there is a chance that small celestial bodies will collide with our planet every now and then. But prediction will be very difficult because of the sensitive dependence of the motion on initial conditions that we do not know well enough.

COMPLEXITY IN SCIENCE: MODELS AND METAPHORS

F. T. ARECCHI

Università di Firenze and
Istituto Nazionale di Ottica

1. Introduction and synopsis

Many recent inquiries have been devoted to the old problem of the power of human knowledge. Today there are two opposing attitudes, namely, the scientific one, in which in line with Galileo and Descartes, we start from self-evident notions and deduce all possible necessary consequences, and the rhetorical one, in which we rely on argumentations. The two approaches, that we may call "top-down" and "bottom-up", are illustrated respectively by Carnap[1] and Perelman[2]. Both aspects were treated in Aristotle's logic, but the triumph of the scientific attitude has made the second one neglected up to recently (Perelman, Boyd and Kuhn[3]). However, argumentation is based on analogies, and the anological approach appears as a connection between different cultural backgrounds stored in man's mind, rather than having a direct grasp on reality.

Two crucial questions arise, namely, i) are the two approaches really in conflict? and ii) rather than examining historically what has been produced in the two areas, can we start from one side and recover some aspects of the other?

As a working scientist, I shall take the latter attitude. Nowadays the average scientific community in unfamiliar with the notions and viewpoints

---

[1] R. CARNAP, *Philosophical foundations of Physics: an introduction to the philosophy of science*, Basic Books, New York 1966.

[2] C. PERELMAN and L. OLBRECHTS-TYTECA, *Traité de l'argumentation. La nouvelle rhétorique*, Presses Univ. de France, Paris 1958.

[3] R. BOYD and T.S. KUHN, *Metaphor and thought*, Cambridge Univ. Press, Cambridge 1979.

of traditional philosophical debate. On the other hand, the success of the scientific programme has brought about a large amount of investigation into its procedures. Most scientists are aware of what they are doing and take good care of it, even though often in a language not equivalent to that used by philosophers.

I shall try to point out the power as well as the intrinsic limitedness of the scientific approach with special reference to investigations in physics in which I am currently involved.

A crisis analogous to that which exploded a few decades ago in mathematics with questions of undecidability and intractability has been brought about in physical sciences by the investigation of non-linear dynamical phenomena (chaos and complexity). Coping with this crisis implies re-discussing the role and power of human knowledge, with anthrolopogical and ontological consequences.

The main result is that our cognition of reality can not be fully expressed within a single formalised procedure (that we call "model"), but it requires a network of different models, corresponding to different languages (Arecchi 1992)[4]. "Metaphors", or bridges not strictly formalised among different models are an essential constituent of this linguistic network.

Furthermore, nonlinear dynamics introduces a new paradigm in cognitive sciences. At variance with a Turing machine , a biological cognitive system does not get trapped in undecidable or intractable problems, but it decides, and does it within a (reasonably) short time. This happens because it does not obey fixed rules, that is, it does not "speak" a single language, but it continuously readjusts, adapting to the intrinsic evolution of the event under observation. In other words, knowledge is intentional since it consists in a continuous "adaequatio" of the mind to the observed reality.

In particular human knowledge is not only an efficient survival strategy, but — and this is unique among living beings — it includes a reflection on the structures, the natural forms which oblige us to re-shape our cognitive strategies. Thus, a hierarchy of orders emerges as an intrinsic property of reality, not as superimposed by the trascendental activity of the human mind which imposes its own "values".

In Sec. 2, I analyse the historical construction of physical language as a way out of the ambiguities of ordinary language. The effectiveness of physics

---

[4] F.T. ARECCHI, *Why science is an open system implying a meta-science*, in The science and theology of information (ed. by C. Wassermann), Editions Labor et Fides, Genéve, 1992, pp. 108-118.

fostered some pretensions which had an impact on our philosophical ideas. Nowadays we are able to criticise those pretensions from within the scientific programme, and thus recover the right to coexistence of many independent languages. Sec. 3 distinguishes between "models" (or intra-linguistic procedures) and "metaphors" (or inter-linguistic procedures). In Sec. 4, I highlight nonlinear dynamics as it applies to the passage from individuals to collective phenomena, and criticise the term "self-organization" showing why it is more appropriate to replace it with "hetero-organisation". Sec. 5 is a critical review of the current definitions of complexity. Sec. 6 extracts the ontological implications of complexity. Sec. 7 summarises the main theses which emerge from this study.

2. The language of Science and its limitations

Physical language was introduced by Galileo as a way out of the ambiguities typical of ordinary language (fig. 1a). Rather than trying to pick up the "essence", that is, to formulate words which fully describe the "nature"

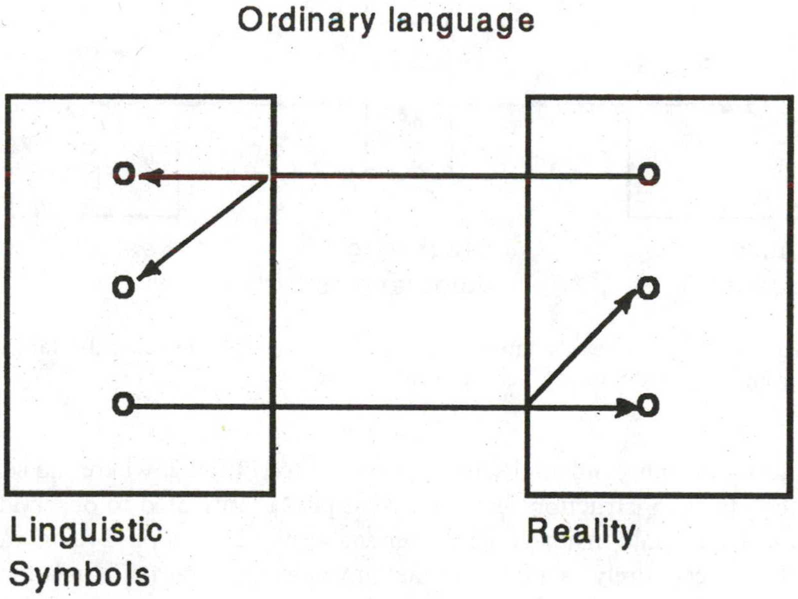

Ordinary language

Linguistic
Symbols   Reality

Fig. 1a - Mapping of reality into the symbol-words of ordinary language.

of an event X, physics limits itself to assigning some quantities related to ("quantitative affection of") X, as e.g. the length of X, the weight of X, etc.

As a consequence, the approach of the observer to reality R is mediated by some measuring apparatuses M (fig. 1b). The new language in the space of symbol S is made of symbols- numbers which are the outputs of M and whose "semantics" depends upon the operations performed by M, while the "syntax" connecting those numbers is the mathematics. To complete physics as a formal language we fix a set of primitive relations among the numbers that we take as the "physical laws" and that become the "axioms" of the language. Whether those relations are extracted from reality by trial and error (guessed and then tested) or whether they are a trascendental human activity ordering otherwise raw data is a matter of philosophical debate.

Physical language

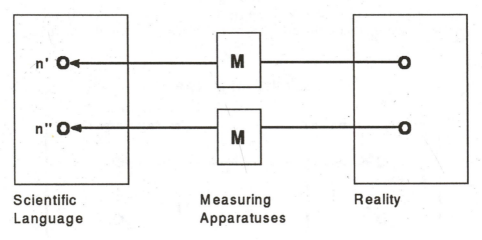

Scientific Language   Measuring Apparatuses   Reality

Fig. 1b - Mapping of reality into the symbols-numbers of the scientific language, through the mediation of measuring apparatures M.

A Kantian transcendental scheme is illustrated in fig. 2, where the role of the theoretical construction seems purely a priori, unrelated to observation. Naive realism would instead take the scheme of fig. 1b as a passive operation whereby M acts purely as a filter. In fact, my working experience as an active physicist suggests that observations play a twofold role, namely, 1) in motivating the selection of the set of definitions and axioms, and 2) in testing the correspondence as shown in fig. 3, which is a synthesis between the Carnap scheme (fig. 2) and the Galileo scheme (fig. 1b).

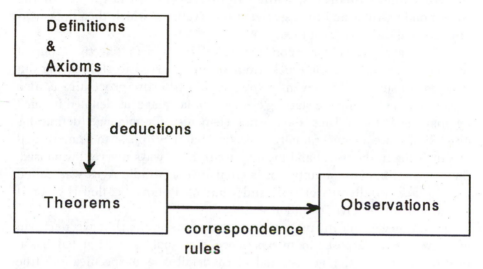

Fig. 2 - The Carnap scheme of a scientific theory.

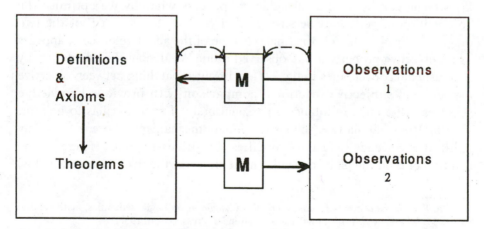

Fig. 3 - The scheme of critical realism. Observation 1 motivates the choice of definitions and axioms. Observation 2 is either a prediction or a verification of the model. M is suggested by the pre-formalized aspects of the observed reality (dashed lines from R to M) but it is also influenced by the observer's activity (dashed line from S to M).

The Carnap scheme of fig. 2 is after all a representation of what Tarski[5] calls a model, that is, a possible semantic interpretation of an abstract theory. In our case, the theory is the collection of definitions and axioms and the consequent theorems, while the interpretation is the set of the corresponding rules and the associated observations, which then appear as physical realisations of that theory.

What Carnap calls "correspondence rules" are in fact a selective group of recipes which pick up some behaviour of an observed world, leaving out what is considered "irrelevant". Now, such a filtering procedure is also performed in selecting the starting elements of language, as denoted in fig. 3 by applying M twice. Thus, the starting elements of a language (definitions plus axioms) are no longer totally subjective, but they appear to be motivated by our living in the world and making use of it, in ways not yet formalised. Such a pre-linguistic adaptation is what in the Middle Ages was called "adaequatio intellectus et rei", and what M. Polanyi[6] called the "tacit dimension" and A. Livi[7] "common sense".

In fig. 4 we qualitatively sketch the adjustment that we perform whenever we are exposed to an experience. We usually have a global classification parameter (the "genus") and a local detailed set of specifications (the "species"). As we increase the resolution, that is, the number of details, we specify our experience better, but we reduce the extension of the genus. Any knowledge can be seen as an adaptive process whereby we optimise the competition of genericity versus specificity. The intersection E of the two diagrams (fig. 4), which is the final outcome of the adaptive process, appears as an objective property of the observed event. We denote the adaptation by the dashed feedback lines in fig. 3. They imply a matching between objective aspects and subjective reactions (comparison with previous knowledge schemes). Of course in infancy a personal body of expertise has to be built starting from "tabula rasa" plus some innate mechanisms (Lorenz)[8] and any cultural experience (learning) modifies the successive ones (Piaget)[9]. The position of E on the plane depends on the slopes of the two diagrams, that is,

---

[5] A. TARSKI, *The concept of truth in formalized languages*, in Logic, semantics, mathematics: papers 1923-1938 by A. Tarski, Engl. transl., Clarendon Press, Oxford 1956.

[6] M. POLANYI, *Personal knowledge: towards a post-critical philosophy*, Routledge and Kegan Paul, London 1958.

[7] A. LIVI, *Filosofia del senso comune*, Ed. Ares, Milano 1990.

[8] K. LORENZ, *Kant's doctrine of the a priori in the light of contemporary biology*, in: General systems, Eds. L.V. Bertelanffy and A. Rapaport, Soc. Gen. Syst. Research, Ann Arbor 1962.

[9] J. PIAGET, *Introduction à l'épistémologie génétique*, Presses Univ. de France, Paris 1950.

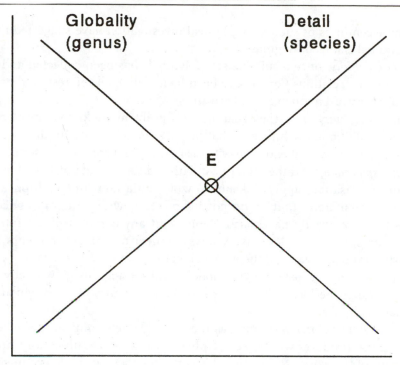

Fig. 4 - Trade off between genericity (globality) and specificity (detail) for different resolution, leading to an optimal appreciation (the crossing point) where the two competing requirements find the best compromise E. The slope (indicative) of the diagrams and hence the position of E depends on the individual, through innate and learned skills.

on the individual, through inherited and learned skills. For a hypothetical external "supervisor" the position of E will be different for different observers, but any individual will always recognise that particular reality which has turned him (her) to a particular E.

To sum up, the scheme of fig. 3 includes the "naive" realism of fig. 1b as well as the subjective aspects of fig. 2. I call this generalised attitude "critical realism".

Scientists are generally realists in the sense that they take as strong evidence what "common sense" suggests to them before any linguistic formulation, but they are critical realists, since they are aware of the strong filtering action of M, in virtue of which they always selected specific aspects of reality.

Three centuries of success in physical investigation have suggested three unwarranted claims or dogmas.

i) *Decidability:* once a suitable set of symbols has been collected and the corresponding relations (laws) have been formulated, all the rest is a matter of straightforward mathematical derivation.

ii) *Predictability:* once the equations of a problem are known, given a set of initial conditions the future of that problem is univocally fixed, since the solution of a differential equation with assigned initial conditions is unique.

iii) *Explainability:* rather than postulating different sets of laws for any level of organisation, apply Ockham's razor (Entia non sunt multiplicanda praeter necessitatem, that is, do not introduce new explanatory concepts unless they are strictly necessary). Hence split any complex object into its components, find the elementary laws connecting the components and reconstruct the behaviour of the whole object.

The first dogma fosters a "methodological reductionism" whereby any scientific language has to be reduced to the set of formal rules which are peculiar to physics.

The second dogma was taken as the basis for determinism in the sense that a finite (and knowable) set of physical laws determines any future behaviour of the world. Notice that this is not the theological determinism of monotheist religions which attributes to God the knowledge of any event whatsoever, but this is a kind of anthropological determinism, since it fixes its roots into a set of first principles that we can grasp.

The third dogma fosters an "ontological reductionism", that is, a tentative to consider the elementary constituents into which we analyse a composite body (molecules, or atoms) as a sufficient and complete cause of the macroscopic forms in which the body manifests itself.

The above three dogmas have found "in-principle" limitations from within the same scientific language. The three limitations are respectively:

i) The Gödel undecidability theorem (1931)

ii) Deterministic chaos, i.e. sensitive dependence on initial conditions (Poincaré 1890, Lorenz 1963 etc.).

iii) Complexity, i.e. emergence of new information in a large system, not included in the separate constituents (Kolmogorov, Chaitin, Bennett, Atlan, etc. from 1960).

We consider these three limitations as new paradigms of science which introduce crises into the very foundations of mathematics (i)) and experimental physics (ii) and iii)). We wish to show that these crises are healthy. On one side they provide new thrust to scientific investigation, on the other side they destroy the claim of science to be the only language qualified to describe reality (scientism), so that science becomes open to interdisciplinary relations with other languages.

i) *Undecidability*

Once we have organised a group of observations in a set of statements, we presume to be able to do scientific research in a deductive way, taking those statement as axioms and deducing all possible consequences (figs. 2 and 3). The Gödel theorem hampers such a claim. It says that, given a set of axioms, there is eventually a statement which is true because it is expressed by the rules of formal language, but for which we can prove neither that it is true nor that it is false without falling into a paradox. We must consider such a lack of univocity as a limit to the reliability of a long list of grammatical constructions. Thus, we must get off the space S of linguistic symbols and fish in the real word R for further relevant input. Truth can no longer be considered as the correctness of a formal procedure but it must be given a semantic value with reference to R.

ii) *Chaos*

Determinism was based on the uniqueness of the solutions to dynamical equations, once initial conditions have been assigned. Although a trajectory may be unique in starting from certain initial conditions, it is enough for it to have a minimal uncertainty to lose the predictability of its future path. Now, these minimal uncertainties are intrinsic to the method of measurement itself. In coding events by numbers, we can assign with accuracy only the rational numbers, but by far most numbers are irrational, such as the square root of 2, that is, they consist of an unlimited sequence of digits. As infinity can neither be encompassed using our systems of measurement nor recorded in our memory, the truncated version of an infinite number introduces a tiny initial uncertainty, the effect of which become enormous when we try to extend our prediction beyond a certain time. Deviation from the "unique" path can be expressed diagrammatically (fig. 5). Let us compare two equal paths but with different surrounding "landscapes": the first on a valley floor, the second on the ridge of a hill. The initial "exact" position A gives the required path; a slightly wrong position B gives a path which in the first case converges on the correct one, but in the second diverges (away) from it.

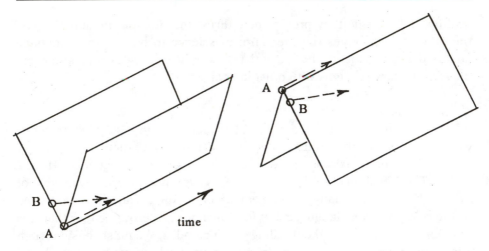

Fig. 5 - Energy landscape around a dynamical trajectory.
a) valley floor means regular motion, that is, convergence of nearby trajectories;
b) ridge of a hill means irregular motion, i.e., rapid divergence of nearby trajectories: this is chaos.

Even simple physical systems like Poincaré's three body problem have critical paths which run along a ridge and can give rise to deterministic chaos. Similar effects, called *deterministic chaos,* are observable today in various different situations: chemical reactions, the motion of fluids, lasers, cardiac rhythms, the movements of asteroids, economic and social trends etc.

Why did it take two centuries to discover that paths are not always stable (at the bottom of a valley), but are often unstable (on the ridge of a hill)? The fact is that physicists have limited their considerations mainly to stable equilibria and have then only examined small perturbations around these. Now these tiny movements obey *linear* dynamics which always produces trajectories with valley floors, i.e., ensure future predictability.

Thus, the simplified models used to study nature excluded certain pathologies. Nonlinear dynamics is, however, the way in which nature normally behaves.

iii) *Complexity: a heuristic approach*
Entering nonlinearity means discovering complexity. The fact that one's starting-point is never a geometric point from which a single line emerges into the future but is in general a small blob from which lines fan out in all directions, can be seen as a case of dynamic complexity, i.e., as an

infringement of the simplicity requirement which forms the basis of the Galileian method.

In addition to the dynamic complexity of deterministic chaos we can see the emergence of structural complexity which consists of the impossibility of satisfactorily describing a large object by reducing it to an interplay of its component parts with their elementary laws. In this sense, complexity is associated with descriptions rather than being an intrinsic property of objects. A classification of the describing processes should then provide different complexity measures.

Let us consider three sequences of letter of our alphabet: the first one is random, the second is regular, that is, a chain of letters repeated in the same order, and the third one is part of a poem, or any other literary text. We define as complexity the cost of the effort which enables us to realise one of the three sequences. In the first case the programme is simple because it can be associated with the outcomes of a random number generator. For the second case the instruction is very short. In the third case no programme exists which would be shorter than the poem itself. This last sequence will be called complex, the other two simple.

We can say, in anthropomorphic terms, that complex systems are those in which a choice has to be performed, and hence a unique description can not exist. Once we have recognised that the reductionist's dream leads to complexity, the only way out is to give up insistence on working with a single set of elementary laws. In respecting the various, irreducible levels of the description of reality, physics is on the one hand respecting the scientific organization of other areas of science (from biology to sociology) without trying any longer to reduce these to "applied physics", on the other hand it is rehabilitating certain aspects of the Aristotelian "organicism" for which "a house is not the sum of its bricks and beams inasmuch as the architect's plan is an integral part of it".

3. Models and methaphors

From the consideration of Sec. 2 it emerges that there is no unique science, but rather different scientific approaches dealing with different aspects of reality. An example is shown in the following scheme (taken from

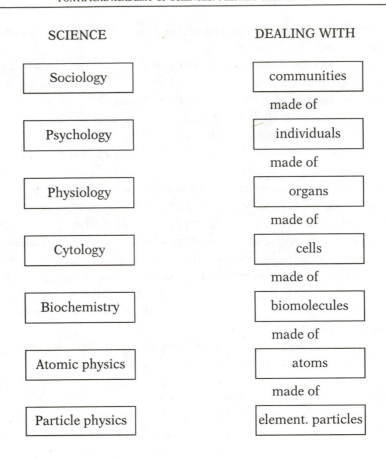

Fig. 6 - Different levels of scientific description.

the book by F.T. Arecchi and I. Arecchi)[10] where some areas of investigation follow the criterion that the object at each level can be analysed into elementary components (atoms) belonging to the lower level (fig. 6).

By no means can we reduce any higher level to the lowest one. We are far from Democritus' reductionistic dream still current in the early decades of modern physics or molecular biology. Indeed, the study of complexity has shown that the amount of information at a given level is larger than the sum of the information of its components. Think of a literary text, where a large part of the information is in the mutual correlation of words, and not in their

---

[10] F.T. ARECCHI and I. ARECCHI, *I simboli e la realtà*, Jaca Book, Milano 1990.

individual information as provided by a dictionary. Similarly, even though living individuals are made of cells and cells by atoms, neither cytology nor atomic physics can fully describe the behaviour of a living being.

We call "model" a finite set of symbols (concepts and laws) which provide "fair" predictions for a given class of events. By this definition, a model is co-extensive to a language, since a model is characterised by the set M of measuring rules which specify concepts and laws. "Fairness" is not the result of an engraving of reality upon the mind like that produced by a camera, but a correspondence between model and observation within the rules selected by M. Such a point of view is developed by E. Agazzi[11] and by M. Artigas[12]. Fairness means a truth, even though a partial truth i.e. projected in the subspace of events allowed by M. Comparison of fig. 2 and fig. 3 shows that Carnap's scheme missed the realistic foundation of the starting elements of the language.

Notice that the adaptation scheme of fig. 4, being pre-linguistic, applies to ordinary language (fig. 1a) as well. Thus the knowledge process leads to a satisfactory compromise for any experience.

However, the language of physics with the artificial insertion of M, on one hand enhances our precision providing reliable numbers and avoiding ambiguities, on the other hand extracts in general a finite set of elements with which to start the deductive procedure. Notice that the artificial separation into a set of initial elements and a deductive procedure, typical of formalised languages, opens those problems that Cantor tried to solve with his set theory and which lead to all recent limitations of mathematics.

Since the set of symbols is finite, it is impossible to extract an unlimited number of consequences in order to predict all possible behavior. In fact we come across undecidability questions, in line with the Gödel criticism of the axiomatic formulation or arithmetic. Thus a closed science, based on a finite set of rules, fails even in trying to describe a single layer of reality.

There are two possible ways out of this limitation. The first one is to break up the unlimited number of deductions which eventually ends by a Gödel limitation and block the deductive procedure to a finite number of theorems, all decidable. This finistic approach (Webb)[13] is not tenable within

[11] E. AGAZZI (ed.) *L'objectivité dans les différent sciences*. Ed. Universitaires, Fribourg CH 1988, pp. 41-54.
[12] M. ARTIGAS, *Scientific creativity and human singularity*, in *The science and theology of information* (ed. by C. Wassermann et al) Editions Labor et Fides, Genéve, 1992, pp. 319-326.
[13] J.C. WEBB, *Mechanism, Mentalism and Metamathematics* (An essay on finitism), D. Reidel, Amsterdam 1980.

a formalised science, since the set of language elements pre-assigned by M may not fit some of the observations. Already within the physical theory (Wolfram)[14] we resort to non finitistic procedures (the thermodynamical limit in statistical mechanics, the path integral formulation in quantum mechanics) which do not fit Webb's finitistic self limitation.

Another finitistic approach (Basti and Perrone)[15] consists in not fixing the set M of measuring procedures, which delimits the language, but in exploiting the adaptive process of fig. 4 in order to adjust to any observed event. The intersection of fig. 4 is the best compromise between a finite resolution and a finite set of properties specifying the species.This way corresponds to a rejection of the Galileian scheme of fig. 1b and to a return to fig. 1a. It is also related to the theory of argumentation (Perelman) as we shall discuss in Sec. 7. Since Basti and Perelman seem — though in different ways — to give up the Galileian scheme of fig. 1b, I intend to introduce an alternative approach which recognises the power and specificity of each different science.

We admit the intrinsic limitation of any scientific language, and carry a double line of reasoning, a deductive one within each language and an analogic one, consisting of heuristic correlations among different languages.

Let us then introduce into a science A the notion of "metaphor", where a metaphor is a symbol borrowed from another science B (e.g. a psychological concept used at a physiological level). A metaphor is strictly not usable in building a model, because it is not formulated via M. However it is rich in heuristic, since it brings correlations with other levels of reality. In other words, a linguistic term, formulated at a given level, is the outcome of its own M which extracted a specific characteristic. Beyond that characteristic the term is still "rich in ontology", since it carries the flavour of other characteristics which were filtered out by M but were anyway part of the pre-formalised experience of the scientist. This extra meaning becomes useful when approaching other aspects of reality not captured by M.

As shown in the scheme of fig. 7, taken from the same book by Arecchi and Arecchi, a given science (e.g. cytology) builds a legal model description out of legal symbols (concepts and laws specific to that level of description).

---

[14] S. WOLFRAM, *Undecidability and intractability in theoretical physics*, Phys. Rev. Lett. 54, 735 (1985).

[15] G.F. BASTI and A. PERRONE, *A theorem of computational effectiveness for a mutual redefinition of numbers and processes*, Proc. Int. Symposium of information Physics (ISKIT '92), Kyushu Inst. of Technology, 1992.

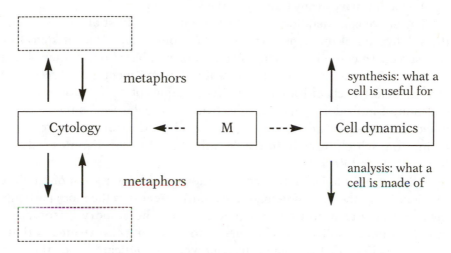

M: measuring procedures which characterize the model

Fig. 7 - Role of metaphors in bridging different linguistic levels.

However it can also borrow symbols belonging to other sciences: these are metaphors. If metaphors enter a model, then the model is a "hybrid", that is, less formalised, but in general more powerful, insofar as it overcomes the Gödel limitations.

To conclude this section, all hybrid models (as e.g. Landau models in phase transitions: see Sec. 4), which include a given amount of metaphoric content, make science more effective, i.e. to say open to other languages, but as a result it is no longer strictly formalised.

Extending these considerations from one science to the overall scientific programme, we realise that natural sciences have to interact with other linguistic formulations of reality in order to be effective, insofar as the Gödel limitations forbid any finite set of sciences from being a complete self-consistent description of reality, as claimed by some scientistic ideologies which have tried to limit relevant knowledge only to scientific statements.

On the other hand, even though limited to partial connotations, each science implies an ontology, since the adaptive process of Fig. 4 is controlled by reality itself. Hence the question whether statements about reality not formalised within the procedure of fig. 3 are relevant or not, that is, whether metaphysics, even though not being a Galileian science, discloses aspects of reality as an affirmative answer, once we acknowledge that the adaptation of fig. 4 is preliminary to any filtering of the kind of fig. 3.

The above considerations raise further questions. First of all, where does the subdivision of different sciences (fig. 6) come from? Does it stem from intersubjective consensus among scientists or is it based on "matter of fact" differences? In the first case we have a Kantian argument, in the second we have an objective classification of different sections of reality. Notice that the question of individuals versus a collectivity (molecules versus the cell) goes back to Socrates and it is the most crucial question in our concept formation, which already in the Middle Ages caused disputes between Nominalists and Realists.

I have used "model" in the sense of a formalised description of an object of science (e.g. the solar system, or the atom), but also in the more particular meaning of interpretation of a theory in another theory, insofar as a homomorphic application of a theory to another context means that a careful reading of that context has shown the applicability of the same formal tools developed on the previous occasion. In such a case we will say that one description is reducible to the other, that is, the two are included in the same level.

Our general definition of model includes also pragmatic models such as numerical simulations and mock-ups. Indeed a more or less extended set of measurements can filter out some subsets of behavior which are fit either by a numerical simulation (as Montecarlo simulations in a liquid state) or by a toy model built for demonstration purposes.

To conclude, the question whether models exist only in the human mind or are suggested by things, that is, the old controversy of subjectivism versus objectivism (fig. 2 versus fig. 3), has been answered in this Section in the objective sense. While on one hand, one could say that even the measuring procedures which extract data from things are subjective (why did I choose to measure e.g. weight rather than length?), on the other hand, we recognise that the choice of M is never arbitrary, but rather motivated by objective regularities.

4. Bifurcations and organisation

Most dynamical models until late 1800 were linear, that is, with a straight line diagram of force versus position, and consequent parabolic behaviour for the potential energy (fig. 8). If there is a strong viscous

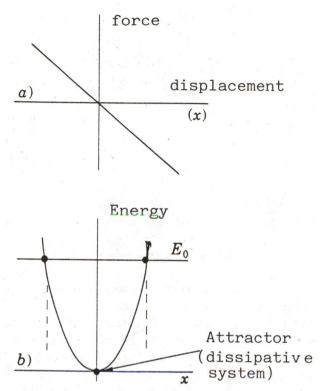

Fig. 8 - Linear dynamics.
The straight line is the force versus position. The associated potential energy
has a parabolic shape. In the dissipative case, the motion stops in the
bottom, which is then called "attractor".

damping (dissipative systems), the kinetic energy is released to the
surrounding medium, and the falling body stops at the minimum of energy
$x = 0$. We call this final point "attractor" and "basin of attraction" all initial
positions which merge at the bottom of this energy landscape.

By change of a "control parameter" we can distort the energy landscape,
as e.g. in fig. 9, and enter the realm of nonlinear dynamics.

Linear dynamics is universal in the sense that any initial condition will
converge on the unique attractor. There are many applications of nonlinear
dynamics characterised by different attractor numbers and positions, as in
the example of fig. 9. Fig. 10 shows how the energy landscape changes for
different λ. At a critical point λ_c the single valley is replaced by two separate
valleys this is a qualitative change. If we represent the position $\bar{x}$ of the

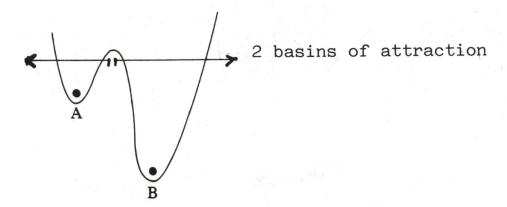

Fig. 9 - Nonlinear dynamics: example of energy landscape with two attractors.

attractors versus λ, the change at λ_c appears as a "bifurcation", as shown by the solid line of fig. 10b. The dashed line represents the hilltop.

    We have shown the simplest of all bifurcations. Accounting for geometric constrains, one can classify all possible shapes of bifurcations (this was called "catastrophe theory" by R. Thom, but I prefer the less conspicuous term of bifurcations).

    The previous figures refer to a simple system, that is, a marble sliding along the energy landscape. The ambitious programme of statistical mechanics is to extend this description from a single object, as an atom or a molecule, to a large collection of objects behaving as a single body. As far as the component particles do not interact, or interact by linear forces, like the atoms of an elastic chain, each of them is embedded in an energy landscape like that of fig. 8 with just one attractor and no bifurcations. It is immaterial whether we have 1 or 10^{22} atoms (such is the number in a cubic centimetre of solid matter), anyway, the single object is representative of all the atoms. That is why an ideal gas has no qualitative changes, no "phase transitions". If instead we consider nonlinear forces, a parameter change can induce bifurcations. We are no longer able to provide a "global" description which holds everywhere, but we attempt a "local" description, the so-called Landau model, which applies the above ideas to a collection of many interacting particles as if they were a single body. Such is indeed the approximate behaviour close to a phase transition.

    Consider a macroscopic magnet. Its magnetization is the sum of all the atomic contributions (the so called atomic spins). At a low temperature, they

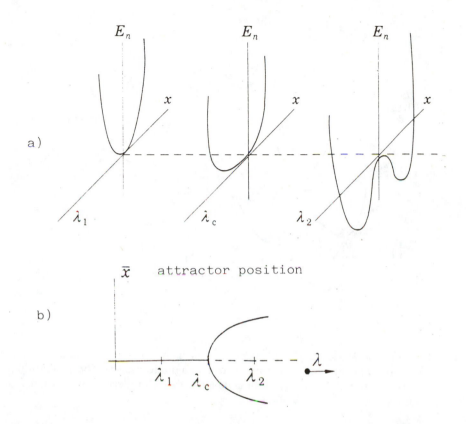

$\bar{x}$ attractor position

Fig. 10 - a) Modification of the energy landscape for a varying control parameter λ.
b) The position of the valley bottoms (solid lines) and of the hilltop (dashed line) are plotted versus λ. λ_c = bifurcation value. For each λ, the bottom is a stable fixed point, the hilltop is an unstable fixed point.

attract one another and hence they become parallel yielding a macroscopic sum. At a high temperature, the mutual attraction is contrasted by thermal agitation which scrambles the single spin orientations. Thus they compensate one another, giving a vanishing total sum.

The overall magnetization, that we may call the order parameter, acts as the position of the single particle of fig. 10. At a high temperature it is attracted by the zero value, at a low temperature it is attracted by two macroscopic values of opposite sign. Indeed the symmetric bifurcation does not privilege either choice. To point out that this is a collective and not just a

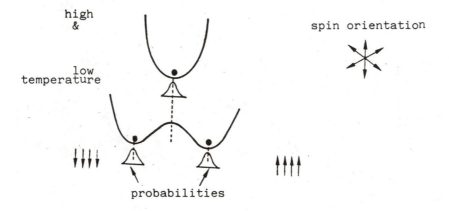

Fig. 11 - Collective Model of a magnetic system at high temperature (paramagnetic phase) and at low temperature (ferromagnetic phase). The probability curves around the energy minima denote the role of thermal fluctuations.

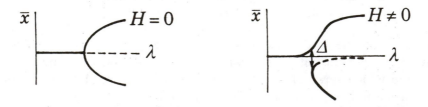

Fig. 12 - In the absence of external fields ($H = 0$) a temperature decrease yields equal probabilities of aligning the magnets upward or downward. An external field ($H \neq 0$) breaks the symmetry making one branch preferred to the other.

single particle description, we sketch also some bell-like curves which give the probability spread of magnetization around each attractor due to thermal fluctuations (fig. 11). Let us consider the probability at high temperature. This being symmetric around the zero value, as the system cools down we have equal probability of landing with magnetization up or down.

The presence of a tiny external field $H \neq 0$ induces a "symmetry breaking" which makes the two branches no longer equivalent (fig. 12). Thus an external perturbation is responsible for selection. Repeat this selection many times, through a sequence of bifurcations (fig. 13). The system reaches a state of "organisation", that is, it assumes just one among a large number of possible configurations. Some people call this "self-organisation", to stress

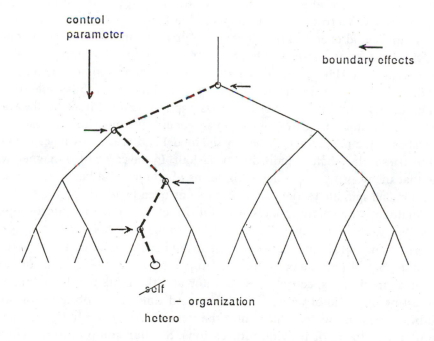

Fig. 13 - In a long chain of bifurcations (the control parameter here points downward) an external perturbation provides a selection at each bifurcation, thus the final state is unique, even though the dynamics allows for a very large number of final states: this is the meaning of organisation.

that the bifurcations depend on the inner dynamic relations among the system components. I do not find that word appropriate, because a specific path (dashed line of fig. 13) is the result of many selections imposed by external perturbations. Thus, it was not enough to have the right number of components (atoms, biomolecules, etc.) but it was crucial to put them in that specific environment or ecological niche which provided the appropriate selective push (symmetry breaking) at each bifurcation. For this reason I propose the new word a "hetero-organisation" as more appropriate than self-organisation.

Only if we could describe each system together with its ecological niche, and the niche of the niche, and so on, would we fulfil this endeavour. Thus, this programme requires the feasibility of a global description and we have just learned that the Landau models are just local approximations, close to a bifurcation.

The Landau model arises from heuristic considerations. However in some cases we are able to trace the whole passage from microscopic dynamics to a macroscopic description of the Landau type. This occurs for lasers, fluids, and chemical reactions, for which we can build a macroscopic model almost from first principles. Such an area of investigation was called "Synergetics" by Haken[16] to denote an identity of behaviour based only on dynamical arguments, independent of the physical nature of the problem.

In fact only under most restricting assumptions can we trace all the way from microscopic to macroscopic, and in general we must rather resort to heuristics of the Landau type, whereby the model is suggested by macroscopic symmetries rather than by individual behaviour. In the case of the magnet, we saw that the transition is explained in terms of competition between thermal agitation and spin interaction and that it is controlled by temperature.

Analogous heuristic models without direct reference to microscopic behaviour were applied by A. Turing[17] (1952) to the morphogenesis, that is to pattern formation in chemical reactions or embryonic development. In this case the control parameter is not the temperature, but the amount of two chemical species, respectively the activator and the inhibitor. To obtain an interesting dynamic of patterns the space and time scales of activator and inhibitor must be widely different, otherwise the relative change in their populations does not provide bifurcations. Similar competitions, which

---

[16] H. HAKEN, *Synergetics: an introduction*, 3d Ed. Springer-Verlag, Berlin, 1983.

[17] A.M. TURING, *The chemical basis of morphogenesis*, Phil. Trans. Rov. Soc. London, 237, 37, (1952).

recall the struggle for life of two competing biological species (Volterra-Lotka population models: see Haken) explain the convective instability of a fluid or the onset of the laser action.

The difference of time scales between the two competing processes has been called by Haken "slaving" of the fast variables, which adapt themselves to the conditions imposed by the slow variables. These latter which survive for a long time, are the order parameters, like the magnetization in the above example.

5. Complexity: attempts at formal definitions

We try to establish a quantitative indicator C of complexity. This has been done by many people, but all definitions proposed so far are subject to criticism.

5.1. Probabilistic definition

What is common to all the organisation phenomena is the emergence of space-time structures which seem "new" with respect to what is known of the constituents as shown in fig. 12b. Let me quote P.W. Anderson[18].

"The behaviour of large and complex aggregates of elementary particles, it turns out, is not to be understood in terms of a simple extrapolation of the properties of a few particles. Instead, at each level of complexity entirely new properties appear, and the understanding of the new behaviour requires research which I think is as fundamental in its nature as any other. That is, it seems to me that one may array the sciences roughly linearly in a hierarchy, according to the idea: The elementary entities of science X obey the laws of science Y.

But this hierarchy does not imply that science X is "just applied Y". At each stage entirely new laws, concepts, and generalisations are necessary.

Has a higher level to be necessarily described by different laws, or is organisation a recipe to guess the higher level once we know the rules at a lower level? Such a programme seems to be contained in the following statement by Atlan[19].

«... We cannot assume that the finality of natural organisation has been set up by some consciousness either from the outside or from the inside. In

---

[18] P.W. ANDERSON, *More is different*, Science 177, 393 (1972).
[19] H. ATLAN, *Self creation of meaning*, Physica Scripta 36, 563 (1987).

either case, this would amount to resorting to God as an explanatory principle, ... and no one knows what God's purpose was ... in setting up an organism or an ecosystem, as opposed to a man-made machine which has been designed with a given known purpose in mind. This is why various attempts have been made to understand mechanisms of self-organisation by which non-purposeful systems, not goal-oriented from the outside, can organise themselves in such a way that the meaning of information is an emerging property of the dynamics of the system."

My criticism is that Atlan's statement is based on a subjective attribution of a purpose, a "value", by part of the observer. Such is also the case of the tentative by S. Kauffman[20] to attribute an "emergent" complexity to an otherwise simple Boolen model.

Let me detail my criticism. The emergence of meaning, in Atlan's sense, is based on a comparison between the global information I_t of a compound system and the sum $\sum_i I_i$ of the information I_i of the individual components. If $I_t > \sum I_i$, the difference, which is the mutual information (M.I.) among the components, is the complexity C, that is,

$$C = M.I. = I_t - \sum I_i$$

In the case of a written text, M.I. is related to the word positions in the phase, while I_i are the single word information provided by the dictionary. Of course, a random collection of words or a collection put in some conventional order (e.g. alphabetic) has $C = 0$. Now information, following C. Shannon, is related to the "surprise" of a message, and hence to the inverse of its probability p, via a logarithmic dependence

$$I = < \log 1/p > = - \sum p_i \log p_i$$

Here the pointed brackets denote an average over the set of possible situations in which the event may show up, and this average is the sum of $\log 1/p_i = - log p_i$ with each term weighted by its probability p_i of occurrence.

The drawback of this approach is that assignment of a probability measure is subjective, related to the "value" we attribute to an object. In communication theory, where we worry only about information loss in a communication channel, it is sufficient to choose a conventional probability measure. When we refer to a piece of the real world, information has a

---

[20] S. KAUFFMAN, *Emergent properties in random complex automata*, Physica 10D, 145 (1984).

semantic connotation, and the probabilities must be assigned after a pre-scientific recognition which has attributed different weights to different events, depending on their use.

An apparently objective definition is the one based on "frequence" by J. Venn and R. von Mises. We partition the space of events by small equal boxes, and attribute to the i-th box a probability given by the number of times N_i the event falls into it, divided by the total number N ($N \geq N_i$) of events

$$p_i = N_i / N$$

While this definition is helpful to classify trajectories in chaotic dynamics, it is useless for classifying forms of objects. For instance, if we take a chip of a few cubic centimetres of glass and shape it as a Venetian artistic cup, the above criterion applied to the space occupied by the amorphous chip and by the formed cup yields the same information, provided that the glassy material has the same consistency. So, information provides criteria just for detecting sponge-like (fractal) structures in the interior of the glass. Only by attributing a higher weight ("value") to those boxes of the partition along the desired contour do we succeed in attibuting higher information to the cup.

"Semantising" the probability measure means that the object is not explainable "per se" but only in relation to a context, to an environment. Hence the open structure that I called "hetero-organisation" in Sec. 4.

5.2. Computational definition

Besides the semantic problems related to a probability measure, the previous definition requires knowledge of an ensemble of systems, in order to assign probabilities as relative frequencies of occurrence.

It seems reasonable that we should be able to describe a system as ordered or disordered irrespective of how much we know about it. What is needed is a measure of complexity that refers to individual states, rather than to ensembles. Mathematicians have proposed such a measure using a new branch of mathematics known as algorithmic complexity theory.

The algorithmic complexity of a state is the length (measured in bits of information) of the shortest computer programme that can describe the state.

Suppose that a particular string of information looks like 101010101010... We would assign this state a low algorithmic complexity because we can recover it by a very short algorithm, namely, «Print 10 n times". This regular binary array can be regarded as the arithmetic equivalent of a crystal. We cannot, however, generally display an arbitrary sequence of ones and zeros using a short algorithm. It can be proved that almost all sequences cannot be reproduced by algorithms significantly shorter than themselves. That is, the algorithm contains almost the same information as the sequence itself. In attempting to generate such a sequence, therefore, we can do little better than simply display a copy of the sequence. Only rarely is a sequence "algorithmically compressible", in order words, can be generated by an algorithm containing less information than the sequence itself.

This notion provides a formal definition of randomness: a random sequence is one that cannot be algorithmically compressed. This satisfies our intuitive expectation that random sequences are devoid of all patterns; the existence of any pattern would imply a more compact description because we could write a short computer programme to specify that random. Indeed, Chaitin[21] has shown in an extension to Gödel's incompleteness theorem of mathematics that one can never prove a given string to be random. On the other hand, one can show that a string is non-random, simply by discovering a short algorithm to generate it.

Algorithmic complexity is not a suitable measure of organisational complexity of a system, because it would assign a high complexity to a random state of a gas, for example. A gas is disordered, but it is not organised. What is needed is a definition that assigns low organisational complexity to systems that are either highly ordered or highly disordered.

C. Bennett[22] has proposed a different definition of organisation, called the *logical depth* of a system. This is the "difficulty of generating a description of the system from the shortest algorithm". A system with high organisational complexity, such as a living organism, would require a long and elaborate computation to describe it. This reflects the long and elaborate sequence of steps in the evolution of the organism. On the other hand, we can describe a crystal by a short computation from a single algorithm. But what about a random gas? As we have seen, a short programme cannot generate an algorithmically random state at all. We can do little better than take a description of the actual state, translate it into computer language and

---

[21] G.J. CHAITIN, *Algorithmic information theory*, Cambridge University Press, Cambridge, England 1987.

[22] C.H. BENNETT, *Dissipation, Information, Computational complexity and the definition of organisation* in Emerging Syntheses in Science (ed. by D. Pines) pp. 215-234 Addison Wesley-New York, 1988.

use as the algorithm "Print that". This is pretty brief, and it implies that a gas, like a crystal, has low logical depth.

Since "logical depth" is (roughly speaking) identified with time required to compute the message from its minimal algorithmic description, here again we find a difficulty: the choice of the minimal description is undecidable in the Gödel sense!

Identifying "logical depth" with the complexity of a physical entity, Bennett appeals to the computational view of physical processes, in which physical processes are viewed as computing equations specified by the laws of nature. The solar system can, in this view, be seen as an analogue computer solving Newton's equations. We may begin with a very elementary set of rules or algorithms to make such a computation (the "minimal algorithmic description"), like Newton's laws for the solar system or the rules of molecular combination in the case of living systems. The "logical depth" of an object — its complexity — is measured by how long it takes a computer to simulate the full development of that object beginning with the elementary algorithm and taking no short cuts. Complexity, in this sense, is a measure of how hard it is to put something together starting from elementary pieces.

And here we come across the difficulty of extremely long computation times, in which case the problem becomes practically unsolvable.

The behaviour of a physical system may always be calculated by explicitly simulating each step in its evolution. Much of physics has, however, been concerned with devising shorter methods of calculation that reproduce the outcome without tracing each step. Such shortcuts can be made if the computations used in the calculation are more sophisticated than those that the physical system can itself perform. Any computations must, however, be carried out on a computer. But the computer is itself an example of a physical system. And it can determine the outcome of its own evolution only by explicitly following it through: no shortcut is possible. Such computational irreducibility occurs whenever a physical system can act as a computer.

Computational reducibility may well be the exception rather than the rule: most physical questions may be answerable only through irreducible amounts of computation. Those that concern idealised limits of infinite time, volume, or numerical precision can require arbitrarily long computations, and so be practically undecidable (Wolfram).

Let me give an example. Suppose we want to classify the head-tail configurations of 100 identical coins put in sequence. The total number is $2^{100} \sim 10^{33}$. If a fast computer evaluates 10^6 configurations per second, it still needs 10^{27} sec $\cong 10^{20}$ years, that is 10^{10} times the age of the Universe!

Thus the formal procedures have not only to face the Gödel undecidability, but also intractability whenever the computing time increases exponentially with the size of the problem.

6. Metascience-toward an ontology of complexity

From the difficulties we have encountered trying out probabilistic and computational definitions of C, we have realised that a phenomenon is complex when it can not be described in a unique way. We can thus define C as the number of irreducible descriptions or models of that event.

In Sec. 3, I called "model" a procedure which, starting from an assigned set of measurements, defines a language (concepts, laws and syntactical consequences). This definition is different from Tarski's definition used in meta-mathematics, whereby a model is a possible realisation of an abstract set of rules. Starting from the consideration that in physical sciences the rules are formed by measuring procedures, it is these that define the model.

Is there any way of comparing different models, or are they incommensurable as sustained by Kuhn[23]? The question requires a meta-symbolic space MS (fig. 14) to compare the different models classified in S, depending on different sets of measuring apparatus M_i applied to the piece of the reality R under observation.

If the problem of confrontation of different models is not solved at the level MS, it may require a meta- meta-level and so on, "ad infinitum"! The comparison can be made by introducing extra-scientific criteria, but where do they come from? If they are imposed as "values", then our scientific endeavour is contaminated by a highly subjective procedure.

On the other hand we should expect that MS considerations be directly motivated by reality, through that pre-linguistic knowledge (dashed line from R to MS), that provides a realistic ground to metaphors.

The crisis of the scientific language outlined above, has questioned the claim that science be the only language able to speak about reality. But would the result be pure scepticism? In other words, do the limitations of Sec. 2 destroy the trust in the power of science and hence make us despair of our ability to grasp reality (nihilistic exit)?

Not at all! On the contrary, any cognitive strategy adapts successfully to

---

[23] T.S. KUHN, *The structure of scientific revolutions,* Univ. of Chicago Press, Chicago 1962.

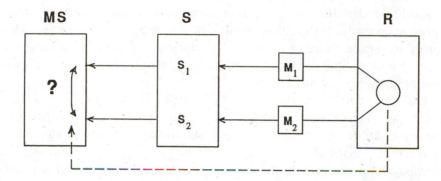

Fig. 14 - Mapping of a piece of reality R into the space of scientific symbols S through different models associated with different measuring procedures M. Successive mapping of S into MS, in order to compare different theories referring to the same reality.

the chain of changes of a complex event. We refer again to fig. 13, to be considered not only as a morphogenetic strategy but also as a cognitive strategy which justifies what was anticipated in the scheme of fig. 2. My criticism means in fact that science is an *open* language which has to readjust itself whenever this is required by reality, in order to avoid undecidable or intractable questions.

It has been objected (Webb and Hofstadter[24]) that finistic procedures, that is, handling of a finite number of possibilities, avoids the Gödel undecidability. Thus we might foresee a science and human knowledge as well, based on finitism, and hence not limited by Gödel. In fact, quantum and statistical physics are not limited to finitistic procedures, and furthermore, intractability, that is, a prohibitively long processing time, is peculiar to most problems.

On the contrary, we experience that man decides, and decides within a short time. To use the computer metaphor, man does not act as a "braid" of self-referential routines and is not trapped by undecidable questions, but he rather changes his rules, adapting them to the bifurcations of an external organising reality (fig. 13).

Neural models of the past decade considered the brain as an extended

---

[24] WEBB, *op. cit.;* D.R. HOFSTADTER, *Gödel, Escher and Bach: an eternal golden braid,* Basic Books, 1979.

dynamical system whose energy landscape had many valleys corresponding to stable fixed points. By a suitable codification, each of those points represents a useful pattern. Recognition of a new pattern means providing energy to escape from an attractor (codification of the old pattern) and entering another one (new pattern to be memorised). This requires a given amount of energy and time.

On the other hand, physiologists (see Skarda and Freeman[25]) have recently given evidence that the electric activity of neurons is chaotic in the absence of an input, and that it regularises in the presence of a stimulus. Notice that, informationally, deterministic chaos is not as hopeless as thermodynamic disorder. In this latter case a Maxwell demon can not make any further prediction out of the system, since he needs the same amount of information he wants to handle (Gabor)[26]. In the former case, even though the trajectory slides downhill (fig. 5), loss of information takes some time. Thus, there is room either to stabilise chaos (Ott et al.)[27] or to synchronise another system to a chaotic one (Pecora and Carroll)[28]: in the presence of chaos the Maxwell demon is efficient!

Combining physiological evidence with synchronisation strategy suggests a new cognitive paradigm: the most efficient neural network has a chaotic activity, and it synchronises itself to an external pattern. Synchronisation means that the rules are not fixed as in a Turing machine (which gets stuck in undecidable or intractable questions), nor are they arbitrarily modified by the observer (pure subjectivism of the trascendental orderings of Kantian derivation), but they are motivated by the observed reality. Thus, any symmetry breaking into the organisation tree of an external reality induces a corresponding rule modification in the recognising system.

One could object that such a superiority with respect to a Turing machine is common to any biological cognitive system (even the immune systems which build antibodies against a virus!). This is true, but, as Blaise Pascal said, we are the only living system which does not limit knowledge to a survival strategy, but that makes a problem of the fact that many levels of order exist (the different layers of a tree such as fig. 13) and hence builds an

[25] C. SKARDA and W. FREEMAN, *How brains make chaos in order to make sense of the world*, Behavioral and Brain Sciences, 2, n. 10, pp. 161-195 (1987).

[26] D. GABOR, *Light and information* in *Progress in Optics*, (ed. by E. Wolf), Vol. 1, North Holland, 1964.

[27] E. OTT, C. GREBOGI and J. YORKE, *Controlling chaos*, Phys. Rev. Lett. 64, 1196 (1990).

[28] L.M. PECORA and T.L. CARROLL, *Synchronization in chaotic systems*, Phys. Rev. Lett. 64, 821 (1990).

ontology of complexity. In other words, man is not the measure of all things, as Protagoras said but he is rather measured by things.

Let me cite Thomas Aquinas (In I Sent., XIX, V, ii): "Therefore our science does not measure things, but is measured by things, as said by Aristotle in book X of The Metaphysics".

7. Conclusive theses

In this essay, I have used my working experience as a physicist to show how the two logical approaches to a problem, the deductive one starting from necessary evidence, and the argumentation one starting from circumstantial evidence, are not fully unrelated, as it may appear from defenders of scientific and rhetorical methods. In fact, use of metaphors in science is necessary in order to overcome the limitations of a model.

We have discussed the realistic grounds of a measuring procedure M upon which a model is based, and have also shown that metaphors, or analogies, are not just linguistic tools, but they are based on real features, as they shape our knowledge through adaptive processes. This heuristic role is always used at a preliminary stage of scientific formalisation. It was inappropriately called "metaphysics" by Popper[29], on the wrong basis that only languages formalised as in fig. 3 are relevant. An example is the role that anthropic principles have in cosmology (see e.g. Dalla Porta[30]). The different role of deductions within a model and metaphoric bridges outside a model explains why I do not consider the "included third" principle as a logical tool (Nicolescu[31]) but rather as a metaphoric suggestion.

An insight into recent physical ideas (bifurcations, organisation and complexity) was necessary in order to discuss the above questions from the point of view of a physicist, without getting involved in the lengthy or questionable approaches developed by philosophers throughout the centuries.

From the previous considerations I extract the following theses.

1) *Against methodological reductionism*

Since any formal language has Gödel limitations, to overcome them it would require feeding with more input, thus changing the very structure of

---

[29] K.R. POPPER, *Conjectures and refutations*, Harper and Row, New York 1963.

[30] N. DALLA PORTA, *The heuristic role of the anthropic principle*, Charles University and Nova Spes Seminar on *Models: science and knowledge*, Prague, 20-21 November 1992.

[31] B. NICOLESCU, *Levels of complexity and levels of reality: nature and transnature*, Plenary Session of the Pontifical Academy of Sciences, Roma, 26-31 October 1992.

the language. Therefore we must recognise that different languages have the right to describe different aspects of the same reality independently, and thus provide partial truths.

2) *Against methodological anarchism*

The different descriptions cannot be bridged by formal rules, since each description has its own rules. However they can be bridged by "metaphors". In Thomistic language, these are analogies, predicated of things which have been considered from different points of view by different sciences.

3) *Against logical atheism*

If logical reductionism were true, then only scientific language would be relevant and one could speak scientifically only of directly measurable things. This would exclude a science of God, or natural theology, which should then be considered in terms of a private life of devotion, without any logical impact. The established independency of different scientific levels, plus the fact that any statement on reality results from a balance between genericity and specificity (fig. 4) which is preliminary even to fixing measuring procedures (fig. 3), gives legitimacy to metaphysics as a science with its own rules, and hence to a natural theology (Gonzales)[32].

4) *Against concordism*

We cannot apply univocally concepts or findings of the physical sciences (cosmology, biology etc.) to natural theology and viceversa. Any tentative of this kind would have the nefarious consequences of Galileo's trial. Instead, it is legitimate to use the analogy procedure, that is, to apply metaphoric bridges not as a poetic touch, but as an expression of a complex reality, which cannot be confined within the formal boundaries of a single language.

5) *Truths versus Truth*

We have seen that any description reveals partial truths. Since Truth is "adaequatio intellectus et rei" and since any reality is not exhausted by a single language, to comply with Truth is a non linguistic affair which involves all our being. Therefore the search for Truth can not be synthesised in linguistic formulae but is a life commitment.

---

[32] A.L. GONZALEZ, *Teologia natural*, Ediciones Univ. de Navarra, Pamplona 1982.

ASPECTS OF ORDER AND DISORDER
IN COMPLEX SOLIDS

C.N.R. RAO

Indian Institute of Science and

Jawaharlal Nehru Centre for Advanced Scientific Research

Bangalore - 560 012 India

Introduction

The crystalline state is the most ordered state of matter, exhibiting infi nite long-range order. The liquid state, on the other hand, shows only short-range order. The non-crystalline glassy state produced by quenching matter in the liquid state shows interesting structural features due to the presence of short and intermediate range order. Apart from the noncrystalline state of matter possessing positional disorder, we have the plastic state exhibiting orientational disorder. The newly discovered fullerenes, C_{60} and C_{70}, just like other spherical molecules, exhibit the plastic to crystalline state transition. By freezing the plastic state, we get the orientational glassy state. The liquid crystalline state is characterized by orientational order although there is no translational order. An important new state of solids is the asicrystalline state with icosahedral structures involving the "forbidden" five-fold symme-try. A quasicrystalline phase can transform to a more disordered glassy phase or to a more ordered crystalline phase. We shall discuss these various states of solid matter. In addition, we shall briefly examine the nature of disorder in crystalline solids and explore how one can quantify disorder in noncrystalline solids. We shall show how Nature avoids a high proportion of random point defects in crystalline solids, although there could be some disorder at the ultramicro level, often involving differences in composition and structure from those of the bulk crystal.

Defects and Disorder in Crystals

Crystals rarely attain a state of perfect order even when they are cooled close to absolute zero. At ordinary temperatures, crystalline solids generally depart from perfect order and contain several types of imperfections, which are, indeed, responsible for many solid state phenomena such as diffusion, electrical conduction, plasticity and so on. The types of disorder that occur in solids fall into one of the following categories: point defects, linear defects, planar defects and volumetric defects. Point defects arise from the absence of atoms (or ions) on lattice sites, the presence of atoms in interstitial positions, the presence of atoms in the wrong positions (not possible in ionic solids) and the presence of alien atoms. The presence of point defects results in the polarization of the surrounding crystal structure, giving rise to small displacements of neighbouring atoms or ions. For example, a cation vacancy in an ionic solid will have an effective negative charge and cause displacements of neighbouring anions. The energy of formation of a point defect depends mainly on the atomic arrangement in its immediate neighbourhood since the relaxation effects decrease rapidly with distance. Linear defects in crystals are dislocations, corresponding to rows of atoms which do not possess the right coordination. Planar defects are grain boundaries (boundaries between small crystallites), stacking faults, crystallographic shear planes, twin boundaries and antiphase boundaries. Segregation of point defects can give rise to volumetric (three-dimensional) defects.

Point defects order themselves, giving superstructures; point defects such as vacancies are eliminated by the formation of planar defects (shear planes) or structural singularities.

There is a wealth of information in the literature on the real structure of defect solids, but rarely has it been necessary to invoke point defects. In fact, much of the recent literature on the chemistry of defect solids does not pertain to the regime of point defects; instead, chemists are mainly concerned with superstructures, shear and block structures, and intergrowths. Defect behaviour in solids actually varies from the entropycontrolled point-defect regime at one end to the enthalpycontrolled systems such as crystallographic shear (cs) planes at the other, superlattice ordering and defect complexes coming in between.

There are many examples of solids exhibiting superlattice ordering of vacant sites or interstitials as a result of interaction between them. Defect complexes have been observed in many inorganic solids. Defects are assimilated into the structural elements of the lattice so that there are no longer any point defects. Crystallographic shear is a well known structural principle where instead of sharing corners metal-anion polyhedra share edges to accommodate anion deficiency (fig.1). An interesting phenomenon in solids is the occurrence of disordered intergrowths of lamellae of a different composition (and structure) in the host matrix of a solid. Such intergrowths are commonly seen in complex oxides showing thereby that uniformity in composition and structure at the ultramicro level is an idealized description of a solid. What is even more surprising is that there are solids exhibiting recurrent intergrowth of related structural/compositional units.

The point-defect regime in crystals is found only when the defect concentration is extremely small. In the case of disordered alloys (where the atoms occupy wrong sites, e.g. A on B site and B on A site), one can quantitatively treat the disorder to order transition, the simplest treatment being that due to Bragg and Williams.

Crystals undergo transitions from one form (structure) to another and the transitions can be thermodynamically of first or higher order. The high-

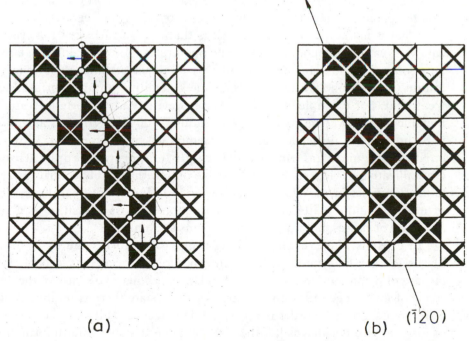

(a)   (b) $(\bar{1}20)$

Fig. 1 - a) Cubic ReO₃ structure showing oxygen vacancies which are eliminated by
b) crystallographic shear.

temperature phase in a transition is generally more disordered and has higher symmetry. Structural phase transitions were elegantly treated by Landau who proposed the concept of an order parameter. We can understand the nature of a phase transition by studying the behaviour of the order parameter in the transition region.

Noncrystalline Solids

Although much of the study of solids pertains to the realm of crystalline solids, there has been a tremendous increase in interest in non-crystalline (or amorphous) solids in the last two decades, probably because of the universality with which almost all materials can be transformed into the amorphous state as well as the diverse applications of these solids. While the term amorphous or noncrystalline solid is a general one, the term glass is used only with reference to amorphous solids prepared by slow-or fast-cooling of melts. Today we have metallic glasses, organic glasses and polymeric glasses. Glasses are characterized by the so-called glass transition. At the glass transition temperature, the glass melts or an undercooled liquid freezes.

What is not sufficiently recognized is that a large variety of materials - catalysts, catalyst supports, xerographic photoreceptors, optical fibres, large-area solar cells and many biominerals - are all noncrystalline. Most noncrystalline (or amorphous) materials become crystalline on appropriate heat treatment. For example, ordinary glasses devitrify at some temperature; metallic glasses become crystalline on heating (~700 K or so). Crystallization is by no means a universal feature as exemplified by fused quartz. Vitreous carbon is a material employed in high-temperature applications. Crystallization of vitreous B_2O_3 is possible only with severely contrived crystallization procedures. While crystallization is an important phase transition of non-crystalline materials, the phenomenon of glass transition is characteristic of glasses, but not of all amorphous substances.

In crystals, the existence of long-range order forms the basis of all discussion; crystal symmetry, Brillouin zones and associated aspects of crystals are of paramount importance in solid state chemistry because of the presence of long-range order. In the noncrystalline state there is no long-range order, but short-range order is very much in evidence. Although the absence of long-range order makes it difficult to obtain exact structural information from diffraction and other experiments, we are able to say a lot about the

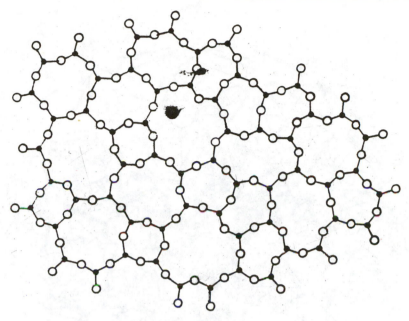

Fig. 2 - Random network model for a A_2B_3 glass

structure of noncrystalline solids. Some of the recent techniques such as
EXAFS and MASNMR are able to provide precise information on the structural features of these solids, but much has been gained in our understanding of these solids by the structural models of the amorphous state.

One of the early models to describe the amorphous state was by
Zachariasen, who proposed the continuous random network model for covalent inorganic glasses. We are now able to distinguish three types of continuous random models:

Continuous random network (applicable to covalent glasses as in fig. 2)

Random close packing (applicable to metallic glasses, see fig. 3)

Random coil model (polymeric glasses)

All these models involve a description of the amorphous state in terms of
statistical distributions.

Most materials, if cooled fast enough (and far enough) from the liquid
state, can be made into the form of glasses. The various techniques available
for the preparation of glasses are cooling of supercooled liquid phases,
vapour deposition, shock disordering, radiation disordering, desolvation and
gelation (fig. 4). All these techniques with the exception of the first produce

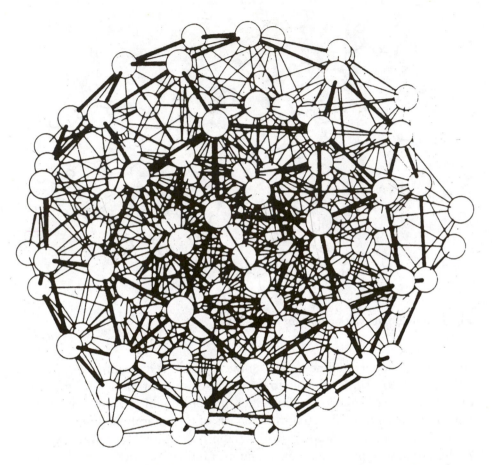

Fig. 3 - Computer simulation of a random close packing of atoms. Heavy lines indicate elements of pentagonal symmetry

materials which cannot easily be characterized, especially with regard to their entropies relative to those of the corresponding crystals.

A variety of glasses, possessing different types of bonding, have been prepared. The rate of cooling has to be very fast to prepare glasses from metallic materials; the temperature drop required is around 1000 degrees in a millisecond ($\sim 10^6$ Ks^{-1}). Melt-spinning, where the metallic glass is spun off a copper rotor as a ribbon at a rate exceeding a kilometre per minute is employed for the purpose. Most other glasses prepared by quenching melts require slower rates, SiO_2 requires a cooling rate of hardly 0.1 Ks^{-1} and this is attained by

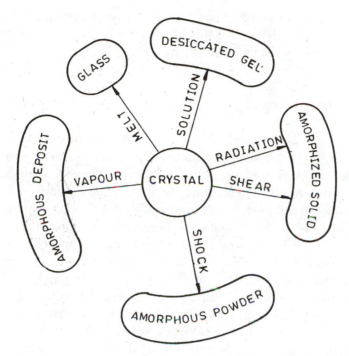

Fig. 4 - Ways of making non-crystalline (positionally disordered) solids

allowing the melt to cool freely. *Splat-quenching* techniques with a range of cooling rates (10^5 - 10^8 Ks^{-1}) have also been employed to prepare metallic glasses. Of special interest is the technique of *laser-glazing* where the surface of a crystalline material is made glassy by exposing it to a moving focussed laser beam so that the subjacent solid acts as a heat sink; the quenching rate in this technique is estimated at 10^{11} Ks^{-1}. Crystals can also be rendered disordered or glassy by subjecting them to shock or large doses of radiation.

The vapour deposition method is widely used to obtain amorphous solids. In this technique, atoms, molecules or ions of the substance in dilute vapour phase are deposited on to a substrate maintained at a low temperature.

Computer simulation studies help in the understanding of the nature of the glassy state vis-a-vis the liquid state. A question that arises is whether there is a way of quantifying disorder in such positionally disordered glasses. One approach based on graphtheoretical considerations makes use of the minimum spanning tree (MST). MST is characterized by the average edge length, m, and the standard deviation, σ. One can obtain m-σ plots for different atomic arrangements and find out how disordered a system is. We have found that in real molecular glasses the disorder can be higher or lower than in a liquid depending on the molecular shape.

The plastic crystalline state

Ordinarily, crystals possess both translational and orientational order. Amorphous or non-crystalline states discussed earlier have no translational order. Crystals of certain substances, especially those formed by spherical or globular molecules, exhibit orientational disorder while retaining the long-range translational order. Such an orientationally disordered crystalline state is referred to as the *plastic crystalline state*. The transition from the crystalline to the plastic crystalline state is exhibited by molecular crystals as the molecules acquire orientational degrees of freedom with higher temperature. The characteristic thermodynamic feature of plastic crystals, which led to the recognition of their existence by Timmermanns, is that the entropy change in the crystal-plastic crystal transition is far greater than the entropy change accompanying the melting of the plastic crystal. In addition to the low entropy of melting, plastic crystals usually melt at a relatively high temperature; for example, plastic crystalline neopentane melts at 257 K while n-pentane, which does not form the plastic state, melts at 132 K. In some cases, plastic crystals directly sublime without melting, camphor, being a typical example. The newly discovered supermolecules of carbon, fullerenes, C_{60} and C_{70}, show plastic-crystal transitions since they are spherical molecules (fig. 5).

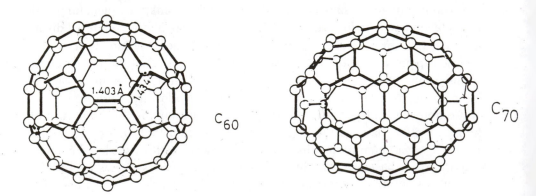

C_{60} C_{70}

Fig. 5 - Structures of the fullerenes, C_{60} and C_{70}

Because of the orientational freedom, plastic crystals usually crystallize in high symmetry (cubic) structures. It is significant that cubic structures are adopted even when the molecular symmetry is incompatible with the cubic crystal symmetry. For example, t-butyl chloride in the plastic crystalline state has a fcc structure even though the isolated molecule has a three-fold rotation axis which is incompatible with the cubic structure. Such apparent discrepancies between the lattice symmetry and molecular symmetry provide clear indications of the rotational disorder in the plastic crystalline state. It should, however, be remarked that molecular rotation in plastic crystals is rarely free; rather it appears that there is more than one minimum potential energy configuration which allows the molecules to tumble rapidly from one orientation to another, the different orientations being random in the plastic crystal.

As the name suggests, plastic crystals are generally soft, frequently flowing under their own weight. The pressure required to produce flow of a plastic crystal, as for instance to extrude through a smàll hole, is considerably less (2-14 times) than that required to extrude a regular crystal of the same substance. t-butyl alcohol, pivalic acid and d-camphor provide common laboratory examples of plastic crystals.

Existence of a high degree of orientational freedom is the most characteristic feature of the plastic crystalline state. We can visualize three types of rotational motions in crystals: free rotation, rotational diffusion and jump reorientation. Free rotation is possible when interactions are weak, and this situation would not be applicable to plastic crystals. In classical rotational diffusion (proposed by Debye to explain dielectric relaxation in liquids), orientational motion of molecules is expected to follow a diffusion equation described by an Einstein-type relation. This type of diffusion is not known to be applicable to plastic crystals. In the case of plastic crystals it would be more appropriate to consider diffusion, interrupted by orientational jumps.

Computer simulation is helpful in understanding the nature of the plastic crystalline state as distinct from the crystalline state. By freezing a plastic crystal one can obtain an orientational glass, as distinct from positionally disordered glasses discussed earlier. We have indeed observed such an orientational glassy state in C_{60}.

The liquid crystalline state

The liquid crystalline state is another intermediate state of matter which several molecular crystals pass through before they become isotropic liquids on melting. The term *mesophase* is often used to describe such a state of matter. Mesophases share the characteristics of both liquids and crystals. In liquid crystals, there is no translational order but there is orientational order. The nature of the mesophase depends on the molecular shape; spherical or globular molecules gain rotational freedom easily and form plastic crystals. Rod-shaped molecules, on the other hand, gain translational freedom more easily than rotational freedom and give rise to liquid crystals. Liquid crystals have become a vast subject of study, and we shall only introduce the various types of transitions exhibited by liquid crystals in this section. Newer varieties of liquid crystals and transitins are being reported all the time.

Liquid crystals can be broadly classified into lyotropic and thermotropic. *Lyotropic* liquid crystals are formed by amphiphilic substances in the presence of a solvent. The majority of liquid crystals of current interest are *thermotropic* in origin, that is the liquid crystal state arises as a result of a temperature change. Both finite molecules as well as polymers form thermotropic liquid crystals. *Nematic, cholesteric* and *smectic* are the three Friedelian classes of liquid crystals formed by rod-like elongated molecules. Since the cholesterics are nothing but nematics formed by chiral molecules, in present-day classification, only nematic and smectic are recognized as distinct classes of liquid crystals.

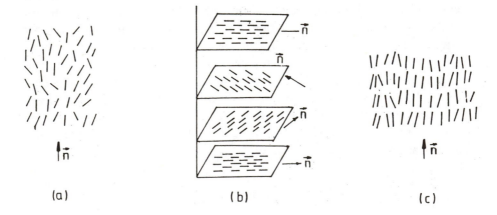

(a)   (b)   (c)

Fig. 6 - Representation of a) nematic, b) chiral nematic and c) smectic liquid crystals

In the nematic phase, molecules are arranged almost parallel to one another in the direction of the long axis (fig. 6), but there is no long-range positional correlation in the direction of the long axis. Cholesteric phases are spontaneously twisted nematics, the twist arising from the optical activity of the molecules. A number of cholesterol derivatives exhibit this phase and hence the name. Smectic phases are, in the Friedelian sense, lamellar phases with no long-range positional correlation in the individual lamellae (fig. 6). Smectics possess, in addition to the orientational order as in nematics, some positional order as well; the centres of the molecules are on the average in equidistant planes. An interesting class of liquid crystals is that formed by disc shaped molecules (discotics).

Quasicrystals

It has long been known to crystallographers that a five-fold axis of symmetry cannot be present in crystals. It is an article of faith that a five-fold symmetry in any properties of crystalline materials, including diffraction patterns, cannot occur except as a result of twinning. The report of Shechtman *et.al.* in 1984 that specimens of a rapidly quenched alloy with approximate composition $MnAl_6$ were found to exhibit icosahedral diffraction symmetry was therefore received with astonishment by crystallographers and others. There was considerable scepticism regarding the proffered explanation that this rapidly quenched material is not crystalline in the conventional sense, but may instead be *quasicrystalline*, representing a three-dimensional analogue of Penrose tiling. This proposed state of matter differs from the crystalline state in possessing no lattice periodicity (i.e. no translation group), and is therefore characterized as *aperiodic*. The crystal lographic community preferred at the time to accept the alternative hypothesis put forward on chemical structure considerations, by the celebrated chemist Linus Pauling that the material was actually an icosa-twin (i.e. a composite of twenty identical crystalline individuals twinned together with icosahedral symmetry). Condensed matter scientists have however preferred the quasi-crystal concept, or the concept of an icosahedral glass or a compromise between the two.

While it has been known for many years that five-fold and icosahedral symmetry are forbidden in the presence of two- or three-dimensional translation groups (represented by two- and three-dimensional Bravais lattices respectively), local icosahedral arrangements of atoms in crystals are possible, as in complex metallic materials. The icosahedral arrangements do not have ideal icosahedral symmetry. Each icosahedral grouping has *local*, nearly icosahedral symmetry, the ideal icosahedral symmetry being broken by the requirements of the crystal lattice and the associated physical forces that distort the icosahedral shape. In some cases it is also coarsely broken by differences in chemical identity among atoms of the group.

With the quasicrystal model, the detailed atomic arrangements in the icosahedral phases (as exemplified by rapidly quenched $MnAl_x$) are not known exactly. Their relationships with icosahedral groupings in known metal alloy structures are therefore still not known with certainty. Icosahedral groupings and linkages seem to be the important features of the structures of these materials. We should note that some icosahedral phase compositions are approximately the same as those of crystalline alloys containing icosahedral groupings and also that there is some correspondence between the two in observed diffraction intensities. In the last few years, a number of alloys displaying icosahedral symmetry has been characterized. Transmission electron microscopy (electron diffraction) has played a central role in deciding on the quasicrystalline nature of these alloys (fig. 7).

Fig. 7 - Electron diffraction pattern of a quasicrystal taken along the 5-fold symmetry axis

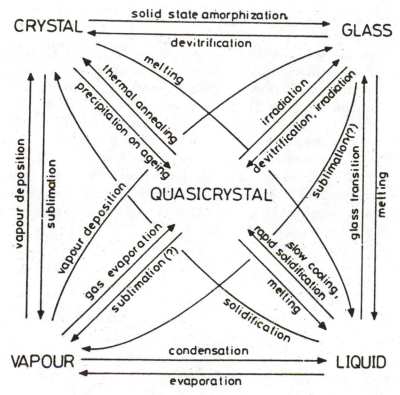

Fig. 8 - Kinetic routes to prepare quasicrystals

The formation of the quasicrystalline phase depends on the competition the material faces from other phases during the process of nucleation and growth, both of which depend on the process parameters. In fig. 8 we show a schematic diagram depicting the different kinetic routes for transformations involving the quasicrystalline phase (QC). The liquid to QC transformation is thermodynamically first-order. Interestingly we can also have a glass-QC transformation and there seems to be a structural similarity between these two phases. The glassy state is a higher energy state with respect to the QC state. Therefore, the QC→ glass transformation requires high input of energy (e.g. electron radiation) and the reverse process requires thermal annealing. Proper annealing can transform a QC to the ordered crystalline phase. It appears that in terms of order, the QC state is between the glassy and the crystalline states.

The Icosahedral Motif

We have just seen how the icosahedral motif is present in the quasicrys-talline state. Crystalline alloys with icosahedral coordination are also known (fig. 9). Buckminsterfullerene, C_{60}, the supermolecule of carbon has the ico-sahedral symmetry, I_h (fig. 5). It is noteworthy that I_h is the highest point group symmetry. Amazingly clusters of noble gases and metals also exhibit icosahedral structure almost as if it were Nature's preferred packing arrange-ment (fig. 10). The number of metal atoms in stable clusters in the gas phase corresponds to magic numbers associated with electron or atomic shells. The atomic shells seem to have preference for the icosahedral structure over the face-centred cubic structure (cubic close packing) although the two arrange-ments have the same number of atoms per shell.

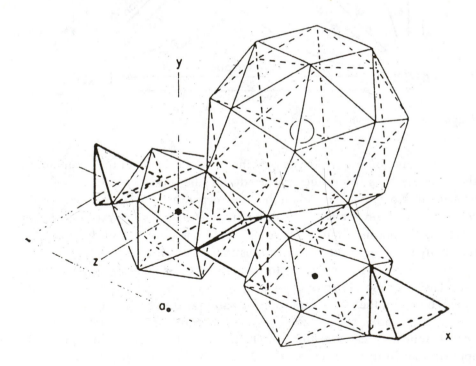

Fig. 9 - Part of the crystal structure of $NaZn_{13}$ where the Zn_{13} icosahedral cluster is shown

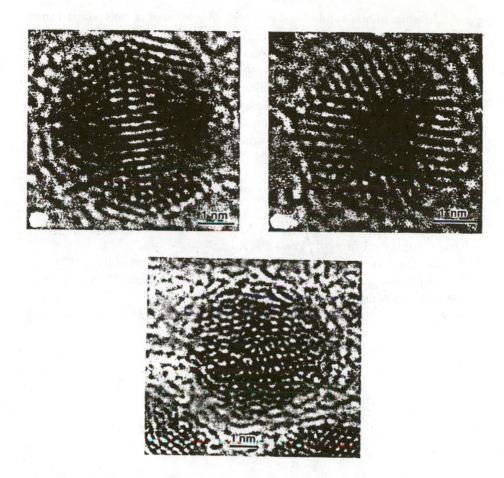

Fig. 10 - Electron micrographs of icosahedral gold clusters

Concluding Remarks

From a study of defects in crystalline solids it would appear that large concentrations of random defects or disorder are not tolerated; an ordering mechanism manifests itself to eliminate disorder in such solids. Many of the complex crystalline solids however show departure from periodicity associated with compositional changes due to occurrence of intergrowths of units of related structure.

All crystalline solids including metals can be rendered amorphous or glassy. In amorphous solids (and glasses) where there is only short-range and intermediate-range order, one can develop models to describe and simulate "structure". Orientationally disordered solids are an interesting class of solids formed by spherical molecules; they give rise to a orientational glassy state on quenching. The quasi-crystalline state of certain alloys is an aperiodic metastable somewhere between glassy and crystalline states. Icosahedral structural units seem to be present in quasi-crystals. Interestingly, even atomic clusters of metals seem to have icosahedral structures.

Bibliography

1. C.N.R. RAO and J. GOPALAKRISHNAN, *New Directions in Solid State Chemistry*, Cambridge University Press 1989.
2. Faraday Discussion 92, The Royal Society of Chemistry, London, 1991.
3. D.P. SHOEMAKER and C.B. SHOEMAKER, in *Quasiperiodic Structures* (ed. M.V. Jaric) Academic Press, New York, 1988.
4. S. RANGANATHAN and K. CHATTOPADHYAY, *Ann. Rev. Mater.* Sci. *21*, 437 (1991).

RELATIVISTIC COMPLEXITY IN PHYSICS

M. MOSHINSKY*

Instituto de Física, UNAM
Apartado Postal 20-364, 01000 México, D.F. México

It is a great pleasure pleasure and honor to speak at this Plenary Session of the Pontifical Academy of Sciences on "Complexity in Mathematics, Physics, Chemistry and Biology".

It is also a difficult task to present the subject of "Relativistic Complexity in Physics" as, in view of the diverse fields of the members of the audience, only very elementary mathematics can be used, and also because the time allotted is very short. Let me plunge immediately into the subject.

Since the XVII century, particularly after Newton's *Principia*, two fundamental concepts have been introduced in physics. One of them was the momentum of the particle which, in one dimension, can be expressed as

$$p = mv \tag{1}$$

where m is the mass of the particle and v its velocity. The other is the kinetic energy of the particle

$$E = mv^2/2 \tag{2}$$

Clearly these two concepts are related, as eliminating the velocity v between them we obtain

$$E = \frac{p^2}{2m} \tag{3}$$

* Member of El Colegio Nacional

This, very simple, and enormously important, relation pervaded all of physics until the seminal work of Einstein in 1905.

Without going into any detail of the concepts that lead to its derivation, we shall simply state the equation that Einstein proposed instead of the Eq. (3) which is[1,2]

$$E^2 = p^2c^2 + m^2c^4 \qquad (4)$$

where c is the velocity of light, which now is assumed to be the same in all frames of reference.

At first sight Ep. (4) is an algebraic relation as simple as (3), but it has turned out to be a revolutionary concept in physics and, in one aspect, to bring about an enormous complexity in describing relativistic phenomena.

Let us first consider the positive square root of Eq. (4) and assume that $(pc) \ll mc^2$, so that using Newton's binomial theorem we have

$$E = (p^2c^2 + m^2c^4)^{1/2} = mc^2 + (p^2/2m) + \dots \qquad (5)$$

We thus recover the non relativist relation (3) between energy and momentum, but with a constant term added, which does not interfere with the equations of motion, but indicates that the simple fact of having a mass m implies an energy mc^2.

Now let us return to the relativistic equation (4) and take its square root with both signs i.e.

$$E = \pm (p^2c^2 + m^2c^4)^{1/2},$$

and compare it in fig. 1 with the non-relativistic expression (5) in which the energy of the rest mass is included.

The part with diagonal lines indicates the region where energy levels are present associated with different values of the momentum p giving rise to a continuous spectrum. For the non-relativistic free particle problem, these levels occur above mc^2, but in the relativistic case they appear both above mc^2 and below $-mc^2$, due to the $\pm$ sign of the square root of (6).

The latter situation did not seem to bother those that used relativistic mechanics for the twenty years after 1905, as they always took the positive sign of the energy.

---

[1] A. EINSTEIN, *Zur Electrodynamik Bewegter Korper* Ann. d. Physik 17, 891, (1905).

[2] A. SOMMERFELD Ed. *The Principle of Relativity*, Dover, New York 1952 pp. 37-96; M. PLANCK, *Zur Dynamik bewegter Systeme*, Ann. d. Physik 26, 1 (1908).

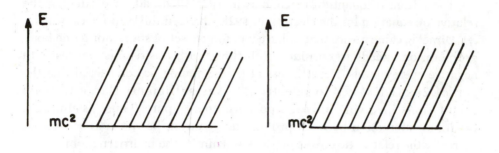

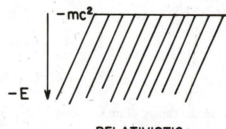

NON-RELATIVISTIC   RELATIVISTIC

Fig. 1

The advent of quantum mechanics in 1925-1926 and, in particular, the relativistic equation for the electron in 1928, changed this attitude radically.

Dirac[3] became aware that to have a complete set of states, for a one body problem, one needs to consider both those with positive as well as with negative energy levels. Nevertheless at the beginning he considered that the gap of $2mc^2$ (1 $M e V$ for electrons), was sufficiently large to trust calculations involving only the positive energy part. Thus, assuming a Coulomb interaction between proton and electron, he was able to predict correctly the relativistic effects on the spectrum of the hydrogen atom. In fig. 2 we illustrate the resulting spectrum, where in the ordinate we have the energy and in the abscissa we indicate the orbital angular momentum.

Because of the ± sign of the square root we have reflected levels in the negative part, and the diagonal lines indicate a continuous spectrum.

. The discrete levels below mc^2 but above $-mc^2$ are greatly exaggerated if the charge of the nucleus is e i.e. a proton, but would be approximately correct if it is $Z e$ with Z of the order of 92. The levels are also slightly split because of the spin orbit coupling, but the figure is not derailed enough to show it.

If $Z = 137$ the two lowest levels join at $E = 0$ and thus we can no longer ignore the negative energies which start to invade the positive part. This may be seen in future experiments where, for example in the collision of Uranium with Uranium, a complex nucleus with $Z > 137$ could be formed, for a small amount of time.

What to do with the negative energy levels? Before entering in their discussion we wish to analyze a two body relativistic problem in the same manner used in the one body case.

Assuming that the masses of the two particles are the same, their energies are given

$$E_1 = \pm(p_1^2c^2 + m^2c^4)^{1/2}, \qquad\qquad (7a)$$

$$E_2 = \pm(p_2^2c^2 + m^2c^4)^{1/2}, \qquad\qquad (7b)$$

and in the center of mass frame, where the total monument of the system vanishes i.e.

---

[3] P.A.M. DIRAC, *The Principles of Quantum Mechanics*, Oxford, at the Clarendon Press, Fourth Edition, revised 1967, Chapter XI.

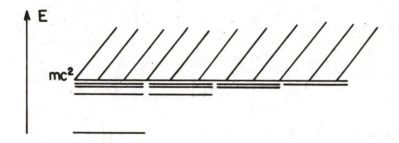

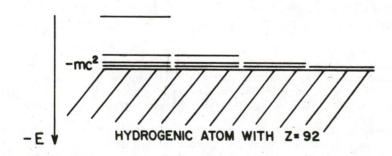

Fig. 2

$$p_1 + p_2 = 0, \quad \text{or} \quad p_1 = -p_2 \equiv p \qquad \text{(8a, b)}$$

the total energy is

$$E = E_1 + E_2 = \begin{cases} E_+ = 2(m^2c^4 + p^2c^2)^{1/2} \\ E_0 = 0 \\ E_- = -2(m^2c^4 + p^2c^2)^{1/2} \end{cases} \qquad (9)$$

which is shown in fig. 3.

When we look at the corresponding problem in quantum mechanics, *i.e.* the Hamiltonian of two *free* Dirac particles[4], we obtain the same type of spectrum, a continuous one above $2mc^2$ or below $-2mc^2$ indicated by the diagonal lines, and a single *infinitely degenerate one* at $E_0 = 0$. Again, we could dismiss the last one together with negative energy eigenvalues but, if an interaction is introduced, the latter become active and states start pouring from below and in particular from the degenerate level E_0. We have denoted the latter as a cockroach nest ("cucarachero" in Spanish) because it looks like an innocuous crack in the wall until you put some food near it i.e. apply an interaction.

We have several examples of the behavior of this cockroach nest. In one case, the Dirac oscillator interaction[5], only part of the levels of definite parity pour out. In another, in which we have a particle-antiparticle system[6] with a Dirac oscillator interaction, all the levels pour out from the original $E_0 = 0$. Finally, for a square well potential, the cockroach nest subdivides into two parts.

The question remains though of how to deal in general with a relativistic quantum mechanical problem[7]. The answer was given by Dirac when he suggested that all the negative energy states are filled with particles that obey Fermi statistics and so by Pauli's exclusion principle a transition to them is forbidden.

---

[4] M. MOSHINSKY, G. LOYOLA, A. SZCZEPANIAK in *J.J. Giambiagi Festschrift*, edited by H. Falomir *et al* World Scientific, Singapore 1990, pp. 324-349.

[5] M. MOSHINSKY, G. LOYOLA, C. VILLEGAS, *Anomalous basis for representation of the Poincaré group*, J. Math. Phys. 32, 373 (1991).

[6] M. MOSHINSKY and G. LOYOLA, *Mass spectra of the particle-antiparticle system with a Dirac oscillator interaction*, Proceedings of the 1992 Conference on The Harmonic Oscillator (NASA publications, in press); M. MOSHINSKY, L. BENET, G. LOYOLA, A. SALINAS, *Comparison of perturbative and variational procedures for a relativistic problem*, Rev. Mex. Fís. 38, 778, (1992).

[7] J.D. BJORKEN and S.D. DRELL, *Relativistic Quantum Mechanics*, McGraw-Hill, New York, 1964.

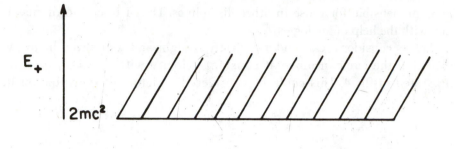

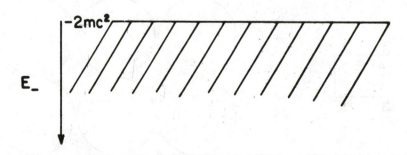

Fig. 3

The above remark sounds trivial for physicists, and possibly incomprehensible for those in other disciplines. Thus I intend to discuss it only with the help of analogies.

Let us consider a car, and a perfectly straight and well paved highway, on which the car can move at constant velocity with no hindrance, as illustrated in fig. 4. This could be considered a picture of the situation in

Fig. 4

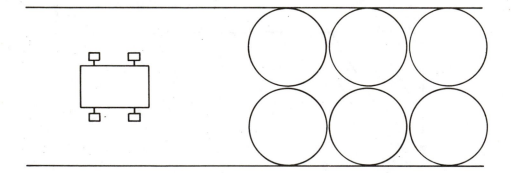

Fig. 5

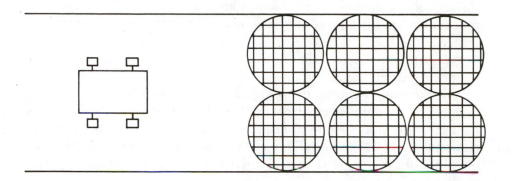

Fig. 6

non-relativistic quantum mechanics. In the relativistic case the highway would be smooth on the left hand side, i.e. positive energies, but full of holes of the dimension of the car on the right hand side, i.e. negative energies, as shown in fig. 5. This in an experience I have lived through on some secondary Mexican highways and I know that the reasonable thing to do is to turn back. Suppose though that a hand (Dirac's) comes from heaven and puts a cover on each of the holes as illustrated in fig. 6. Then you can proceed, but the nature of the problem has changed completely. Before you had a single car on a smooth highway, now there is the car and innumerable covers.

If you proceed at slow speed the motion can still be almost rectilinear and uniform but with the feeling of periodic bumps. If you travel at high speed the covers will fly apart and a veritable chain reaction can be initiated, with hundreds of covers moving around and a good chance that the car will end up in one of the holes.

Thus we see that the existence of negative energy levels tremendously increases the complexity of our problems, as it is impossible, at least at high energy, to speak of a single particle, as one has to consider its interactions with all the others with negative energies (op.cit. n. 3, 7).

How are physicists dealing then with the problem? Again using analogies we could say that there are three ways: Walking over the negative energy sea, swimming in it or diving.

Walking over it implies restricting ourselves to the positive energy states as, for example, Dirac did when dealing with the hydrogen atom. In today's more complex calculation involving relativistic effects in heavier atoms[8], variational calculations are used which involve negative energy states, but procedures are introduced in the computer program to eliminate their contribution.

Walking on the sea, should not be surprising to members of the Pontifical Academy, as Jesus-Christ did it 2,000 years ago.

By swimming we mean the perturbative calculations that are made, for example, in quantum electrodynamics[9], in which the coupling between the electron-positron and electromagnetic fields is small, i.e. the fine structure constant α = (1/137). The validity of these procedures, particularly using Feynman diagrams, is shown by their extraordinary agreement with experiment.

Finally by diving we imply the non-perturbative type of procedure when the coupling between fields is large, as happens in quantum chromo-dynamics[10], which seems to describe systems of particles with strong interactions. For these procedures the largest computers are barely sufficient and give rise to a technique known as lattice theory.

In conclusion we see that the change in the relation between energy and momentum from (3) to (4), while insignificant from the algebraic point of view, has increased the complexity of our physical world tremendously.

---

[8] *Relativistic Many Body Problems*, Proceedings of the Adriatic Conference, Editor I. LINDGREEN, Physica Scripta, Vol. RS7, (1987).

[9] J.D. BJORKEN and S.D. DRELL, *Relativistic Quantum Fields*, McGraw-Hill, New York, 1965.

[10] T. MUTA, *Foundations of Quantum Chromodynamics*, World Scientific, Singapore (1987).

DISCUSSION
(H.R. CROXATTO, chairman)

CROXATTO: Dr. Moshinsky's interesting paper is now open for discussion, questions, comments.

CABIBBO: What definition of complexity do you have in mind when you say that the relativistic picture is more complex?

MOSHINSKY: I would say, for example, that in nonrelativistic physics, either Newtonian or quantum mechanical, you can treat the one body problem by itself. Even in the motion of the planet, when you eliminate the center of mass motion, you have a one body problem. The quantum non-relativistic hydrogen atom is a one body problem. But if you are concerned with problems like those Prof. Rubbia deals with in CERN, or other people in high energy laboratories, they never have the simple situation of one body problems, so for that reason I call these phenomena complex.

CABIBBO: It was nice to see this complicated concept presented in a way which I think makes it transparent. I personally feel that your statement that this small modification to the question adds complexity, I would say it adds richness, in the sense that the vacuum which in common language is considered to be nothing, now turns out to be a very rich thing. Indeed the vacuum shares in its own essence a lot of properties with matter, just because the positive and the negative and zero energy are essentially current and similar properties. Not only does it carry the properties of the electron position field which, as you have mentioned, is the most obvious field, because it is connected with the hydrogen atoms and with things we know about, but it contains all other properties of all other particles, first of all fermions, other particles (you have mentioned quantum chromodynamics), quarks and anti-quarks. They also make atoms, they also satisfy the Dirac equation. Therefore they are also present in one way or the other in the vacuum. In fact if your equations are extended from the Dirac equations to other types of equations because the expression $E = Mc^2$ is true for all particles, they also fill up the vacuum with all kinds of other particles which are different in nature. Then you get a field in which those particles sit, but they interact with each other so strongly that they fill the vacuum with a vacuum expectation value which is non zero. So it is a long journey which starts from the Dirac equation and goes now to the total revision of the vacuum, which is no longer just emptiness but what I would call the lowest energy level, which has less energy but as much richness as matter itself. In fact today, we elementary particle physicists are mostly concentrating on the study of the vacuum. Our most profitable target is the vacuum itself. We excite the vacuum, we hit a state of zero energy which will indicate a new graph of particles. Thus we excite out of the vacuum the states which are indicated by your graph and then the vacuum becomes a real horn of plenty. Those of you who have seen the event produced in the lab have learned that two particles coming together annihilate, vacuum follows, but is vacuum excited? The result of energy in the vacuum starts a spontaneous transformation of this excited vacuum into other particles. In fact, all possible forms of matter come out, not only the ones which have survived 15 billion years instability, but all possible forms of matter independently of whether they exist today, whether thay have been observed by men or not, as long as they exist as forms of real matter they will be coupled through that little zero line that you indicated to be a source, of course I would not call it a snake pit, I would call it a real horn of plenty.

MOSHINSKY: I certainly agree with the word richness as applied to the vacuum. I also would like to point out that the vacuum state appears in the second quantized ordinary Dirac equation and concerns all the negative energy states that are filled with electrons. In the two body problem I discussed, the line at the center is not the vacuum, but a new type of state we called cockroach nest as we are not dealing with second quantization. Atomic physicists that deal with variational procedures for the sum of Dirac Hamiltonians plus Coulomb interactions, get these types of cockroach nests, but pay no attention to them as they only want states in the vicinity of Zmc^2 (where Z is the charge of the nucleus, and m the mass of the electron). They should though take these nests into account because they affect the variational results, and we are discussing this situation with atomic physicists interested in relativistic effects. In relation to whether the word richness is more appropriate than complexity, for the problem I am discussing I would like to go back to the presentation of Lovasz on Information and Complexity, who related the latter term with the length of the program required to get some specific information. Now if you want to get numerical information on, say, the Bohr energy levels in the hydrogen atom, the program is only a few lines long, but anything that is done in relativistic physics with present day accelerators requires very long programs on the most powerful computers. That is why I call relativistic phenomena and, in particular, the vacuum as visualized in relativistic quantum mechanics, a very complex phenomena.

RUBBIA: I believe that the vacuum being the state in which all possible physical phenomena are present, in a virtual way, but still present, will win the record for the highest complexity.

THIRRING: I just wanted to add a word to what Rubbia has said, because he seems to be a vacuum fan and I think there is one thing one can say about the vacuum inspite of its complexity. It is usually the only state which is invariant under Lorentz transformations. You could also say the vacuum fills like a gas or so all of space and contains a lot of stuff inside, but it has a remarkable invariance property, it is not only isotropic and translation invariant but it is even invariant under the Lorentz transformation, where we mix space and time; this is what sets it apart from other substances which we see.

THE NON-LINEAR UNIVERSE:
CREATIVE PROCESSES IN THE UNIVERSE

MICHAEL HELLER

Pontifical Academy of Theology

Faculty of Philosophy Cracow, Poland

and Vatican Observatory V-00120 Vatican City State

1. Introduction: Cosmology as a Theory of the Ensemble of Universes

Cosmology is exposed to a seemingly irrefutable objection: it cannot pretend to be a truly empirical science since it studies the object, the Universe, of which we are given a single copy. What does a single "copy" mean? Are stars, cluster of galaxies, atoms, and quarks given to us in "many copies"? Are they given to us or constructed by us?

Laws of physics are usually expressed in the form of differential equations.[1] Solutions of these equations reconstruct general patterns of the world's behaviour, whereas suitably chosen initial or boundary conditions account for individual characteristics of phenomena. This is where "many copies" enter our theoretical investigations: many copies of atoms, electrons, quarks, of our Galaxy and, if necessary, of the Universe. Let us see how this method works by considering the example of the Universe.

The paradigmatic equations of XXth century cosmology are the field equations of Einstein's general theory of relativity. Having the equations we try to find as many of their solutions as possible. We treat each class of solutions as if it represented a family of potential universes and then we attempt to identify by the initial or boundary conditions the subclass of solutions which best fits the large scale structure of the world we actually observe.

---

[1] But also in the form of symmetries. It seems that there is no serious philosophical analysis of this twofold mathematical expression of physical laws (i.e., laws as differential equations and laws as symmetries).

Although we try to narrow this subclass as much as possible we are always left with a certain observational indeterminacy[2] which does not allow us to single out the unique world model.

From the theoretical point of view we are also interested in universes (solutions of Einstein's equations) that are not very similar to the world we inhabit. The only way to learn which properties of the world are its essential features and which are its accidental ones is by comparing various solutions, i. e., by investigating the "space of solutions". For instance, we study how the properties of having singularities or horizons, of admitting favorable environment for carbon life, etc., etc., are distributed within the space of the solutions of Einstein's equations.

Even the results of our experiments (which ultimately are just real numbers) give us no direct access to the "world in itself". We must construct it with the help of our theoretical machinery, and from the very nature of this construction, when producing one copy, we automatically produce many "nearby" copies. In this way, the possibility of *producing* many copies of a given instance turns out to be the key property of the empirical method. As we have seen, cosmology is not an exception in this respect.

The set (or space) of all solutions of Einstein's field equations is sometimes called the *ensemble* of universes. From the mathematical point of view, it is a very complicated space. In fact, only its rather small subspaces have been investigated. The above considerations justify treating cosmology as a *theory of the ensemble of universes*[3]. Such an approach to cosmology turns out to be especially fruitful, or even unavoidable, when topics like those which will be studied below are concerned.

In section 2, I briefly discuss some features of Einstein's field equations which are rooted in their non-linear character. One of such features is the existence of the deterministic chaos in some (or possibly in the majority) of their solutions. Strangely enough, it would be better to play dice than make computations to predict the evolution of a world modelled by such a solution. This is analysed in section 3. Another problem is that the commonly used criterion of chaos (Lyapunov exponents) depends on time reparametrization and consequently it cannot serve as its invariant characterization.

---

[2] Even without invoking quantum uncertainties, simply because of unavoidable measurement errors which in cosmology are usually bigger than in other branches of physics; usually, but not always, see, for instance, recent COBE measurements of the temperature of the blackbody background radiation.

[3] M. HELLER, *Theoretical Foundations of Cosmology*, World Scientific, Singapore-London 1992.

Recent results in the relativistic (i.e., invariant) theory of chaos are reported in section 4. On the one hand, equations admitting chaotic behaviour are non-integrable. On the other hand, integrable equations are at the core of almost all developments in theoretical physics. The interaction between chaos and integrability is discussed in section 5. It is argued that the phase spaces of physically important chaotic systems are, in a sense, organised by the phase spaces of "nearby" integrable systems. Some examples are quoted and analysed. And finally, in section 6, some doubts are considered concerning the view that the existence of life is a generic property of the ensemble of universes.

2. Non-Linear Strategies in Space-Time

The Einstein field equations are

$$G + \Lambda g = -\kappa T$$

where G is the Einstein tensor, g - the metric tensor of space-time, T - the energy-momentum tensor (responsible for the distribution of all form of energy in space-time besides that of the gravitational field), Λ - the so-called cosmological constant, and κ - the gravitational constant the value of which guarantees the smooth transition to the limit of the Newtonian theory. In this short string of symbols a very rich structure is encoded (if written in components without any simplifications the above equations contain several thousand terms). From the mathematical point of view Einstein's equations constitute the set of ten strongly non-linear second order partial differential equations. For a Lorentz metric g (which is the standard choice) they become hyperbolic, i.e., wave equations.

One often hears that, on the strength of Einstein's equations, it is the distribution of "matter" (T) that determines the space-time geometry (g), but in fact the interaction between T and g is more complicated. One cannot first specify T and then solve the equations for g since all physically sensible formulae for T depend on g. Consequently, any g whatsoever is a "solution" of Einstein's equations if no restrictions are imposed either on T or on g[4]. The important fact is that these equations admit four arbitrary functional degrees of freedom among ten components of g. This fact reflects the free-

---

[4] They can assume the form of space-time symmetry assumptions or restrictions imposed on the algebraic type of the Riemann tensor, or on boundary conditions.

dom of making arbitrary changes of coordinates, and reduces the number of ten equations to an independent six.

The much popularized fact that, according to general relativity, gravity bends space-time is a symptom of the non-linearity of this theory. It is exactly to this property that the theory owes some of its most appealing properties. I shall mention only those relevant to the present paper.

1. The holistic character of the theory. The essential property of non-linearity is that in a non-linear equation, in contradistinction to a linear one, the superposition of two solutions is not a solution. Roughly speaking, a non-linear structure cannot be built up by simply adding or juxtaposing its component parts; it consists of a subtle hierarchy of interactions between all elements of the totality, interactions between interactions included. The degree and strength of this hierarchy may be of various types and is determined by the concrete form of non-linearity. In general relativity we are faced with its strong form.

2. Strictly connected with the above property is another aspect of the non-local character of general relativity. Although solutions of Einstein's equations are determined only locally (as is the case with solutions of all differential equations), non-local topological questions enter the theory from the beginning. In connection with this, two remarks are to be made. First, this kind of non-locality refers not to space alone but to space-time, i. e., also the histories of objects or of observers are part of a global play. Moreover, the concept itself of history strongly depends on the structure of the totality. It turns out that the very existence of the global history of the Universe is not a generic property in the ensemble of universes. To have a global history, a given universe must satisfy rather demanding causality conditions[5]. Second, this non-locality refers not only to space-time, but also to various matter fields which, through the field equations, are coupled to the gravitational field.

3. Non-linearity is the necessary, but not sufficient, condition for the existence of deterministic chaos. The essence of deterministic chaos consists in a complete unpredictability of the system (described by equations which are in principle fully deterministic) originating from an arbitrarily small, but finite, indeterminacy of the initial data. Unfortunately, not much is known about the nature and existence of chaos in partial differential equations.[6] In

---

[5] M. HELLER, *Time and History: the Humanistic Significance of Science*, European Journal of Physics, 11, 203-207 (1990). M. HELLER, *Time and Causality in General Relativity*, The Astronomy Quarterly 7, 65-86 (1990).

[6] The standard approach is to approximate partial differential equations by difference equations, the so-called maps, and investigate chaos in terms of the latter.

principle, because of their strong non-linearity Einstein's equations, even in the absence of any non-gravitational energy, allow for solutions of the chaotic type for the metric field, but in fact only those cases have been studied which can be reduced to a system of ordinary differential equations (this can be done for spatially homogeneous world models, the so-called Bianchi cosmology). Although only several individual chaotic solutions are known, there are good reasons to believe that the existence of the deterministic chaos is a generic property within the ensemble of universes.

4. In the light of the above one should reconsider the determinism issue in relativistic cosmology. In addition to the fact that the hyperbolic character of the metric introduces strong limitations to classical determinism and causality (light cones, domains of dependence, Cauchy horizons...), the unpredictability connected with the deterministic chaos unconditionally ruins all Laplace-like programs in cosmology[7].

5. Non-linear effects make possible the origin and growth of self-organizing structures. It is known to-day that such structures - from clusters of galaxies to living organisms - can persist and evolve in the far-from-equilibrium states owing to various non-linear gravitational, chemical, thermodynamic, etc. strategies. Since such structures consist, in essence, of spacial patterns and temporal rhythms it seems that the necessary precondition to produce them must be for the Universe to admit a certain non-purely-local history. If the existence of the history is not a generic property in the ensemble of universes, the same should be said about the existence of self-organizing structures. Moreover, the question could be asked: is the existence of self-organizing structures a generic property in the subensemble of universes admitting the existence of histories? Time (history) and life seem to be epiphenomena of non-linearity. (Some remarks concerning the generic character of life will be made in section 6.)

6. The non-linear character of equations is sometimes so tedious in solving problems that one would like to have the possibility of linearizing the problem, i. e., to find a set of linear equations which would sensibly approximate the situation. If one regards the ensemble of universes as a geometrical space (the points of which are solutions of Einstein's equations), then the linearized Einstein equations would determine the tangent space to the ensemble. Of course, one can meaningfully speak of the tangent space only if the ensemble

---

[7] M. HELLER, *Laplace's Demon in the Relativistic Universe*, The Astronomy Quarterly, 8, 219-243 (1991).

has the structure of the differenciable manifold (in a neighbourhood of the considered solution). In such a region Einstein's equations are said to be *stable with respect to linearization*. As shown by Fisher et al.[8], this is not the case around the spatially closed solution with sufficient symmetries[9], where the ensemble exhibits an unexpected complexity (conic singularities appear). The fact that in the neighbourhood of such solutions the ensemble is not even a manifold shows its geometrical richness, and also poses serious problems, especially when the method of "small perturbations" is applied to Einstein's equations, e. g. in studying the problem of the origin of galaxies and their agglomerations (in such circumstances spurious perturbations can easily be fabricated, (1)).

7. The smoothing-out problem. This problem is especially important for the RWFL cosmology.[10] World models belonging to this class (which itself is of zero measure within the ensemble) are spatially homogeneous and isotropic and, as testified by many observational data, surprisingly well approximate the actual Universe. The problem is that our Universe is manifestly non-homogeneous over scales smaller than 10 or even 100 Mpc, exhibiting rather complicated texture of the "galactic tapestry". On such scales the space-time geometry, as determined by strongly non-linear Einstein's equations, is expected to be extremely complex. How could it happen that it is so well approximated by the spaces of constant curvature corresponding to the isotropic and homogeneous distribution of matter? For non-linear equations, an "averaged solution" is, in general, not a solution. Are we faced with an exceptionally benevolent type of non-linearity that makes the life of cosmologists easier than allowable in principle? It seems that this is really the case. Carfora and Marzuoli[11] have performed a smoothing-out procedure for models which may be considered to be near (in the ensemble of universes) to the closed RWFL cosmological models. The procedure has been made in the full theory, having nothing in common with the linearizing of the Einstein equations[12]. It has been demonstrated, for the considered class of models,

---

[8] A.E. FISCHER, J.E. MARSDEN & V. MONCRIEF, *Symmetry Breaking in General Relativity, Essays in General Relativity — A Festschrift for Abraham Taub*, ed. by F.J. Tipler, Academic Press, 1980.

[9] Although it has been formally shown for Einstein's equations with vanishing energy-momentum tensor, the result could be extended to classes of matter fields, such as gauge theories coupled to gravity.

[10] RWFL stands for Robertson-Walker-Friedman-Lemaître.

[11] M. CARFORA & A. MARZUOLI, *Smoothing out Spatially Closed Cosmologies*, Physical Review Letters, 53, 2445-2448 (1984).

[12] The idea is to select initial data, the Cauchy development of which leads to the model to

that the smoothing-out procedure, for decades used in cosmology without an explicit explanation, is indeed a legitimate method.

I have listed above some spectacular features and problems generated by the non-linear character of general relativity. However, one should remember that in the "every-day life" of this theory in contemporary physics its non-linearities seldom assume more pronounced forms. In analysing many phenomena the linearized field equations are enough and, as we have seen, the smoothing-out procedure in cosmology effectively screens out more dangerous non-linear effects. The strongly non-linear character of general relativity shows up in such exotic states (exotic from our point of view) as the very early moments of the Universe and violent phenomena in astrophysics. They also play decisive roles on the conceptual level, for instance in all attempts to quantize gravity.

3. Does God play dice?

It is an interesting property of our mind that it takes for granted events of high probability, but it asks for explanations if confronted with improbable events. It looks as if the high probability of an event were for our mind a kind of "sufficient reason" to justify the occurrence of the event. A random outcome when playing dice does not evoke our surprise, but if the outcome is ten times in succession the same number we begin to inquire "why?". This, when applied to our understanding of the world, leads to what has been called by Ellis the *probability principle*[13]; it states that "the universe model should be one that is a probable model within the set of all universe models". To make this principle a workable tool one should define an appropriate probability measure on the ensemble of universes. Till now no such measure is known, and instead one usually employs the other principle, called the *stability principle* stating that the world model, when slightly perturbed, should retain its original shape. Rigorously speaking, this requires at least a reaso-

---

be averaged out. These data are then smoothly deformed into the initial data for a closed RWFL world model. The deformation is defined in such a way that the four constraint equations are automatically satisfied. The flow of deformed initial data generates a one parameter family of solutions which interpolate between the space-time to be averaged out and the closed RWFL solution which now can be considered as a smoothed out version of the original space-time.

[13] G.F.R. ELLIS, *Relativistic Cosmology: Its Nature, Aims and Problems, General Relativity and Gravitation* — Invited Papers and Discussion Reports of the 10th International Conference, Padua, July 1983, ed. B. BERTOTTI, F. DE FELICE & A. PASCOLINI, Raidel, Dordrecht-Boston 1984.

nable topology on the ensemble of universes to make meaningful such terms as "slightly perturbed" and "to retain the original shape"; in practice, however, people perturb a given quantity describing the Universe (e.g., its Hamiltonian) by adding to the equations a term with a small coefficient, and look how this procedure affects the considered world's property.

Another rather technical point is that one seldom considers merely the set of all possible universes, i.e., the set of solutions of Einstein's equations, but rather the set of all possible initial conditions which lead to these solutions. The probability principle in such a formulation would imply that our Universe should be modelled by the solution evolved from the initial conditions chosen at random. Many people believe that in fact this is not the case. In their opinion, the Universe evolved from random initial conditions would have been very different from the one we inhabit and investigate. Some properties of our Universe, such as the high degree of its spatial isotropy and homogeneity, its precise expansion rate, physical and chemical conditions enabling the biological evolution to start and continue, make it very "improbable" within the ensemble of universes; in fact, they make it belong to its zero-measure subset. We can picture the Creator, equipped with a pin which he throws at random onto the space of initial conditions. The initial data shown by the pin are to be created to produce the Universe. If the Creator plays his game honestly, the chances for a Universe like ours to be created are null.

It seems that the way out of this "improbability dilemma" is the so-called mixmaster program in cosmology. According to this program, the initial data of our Universe were completely chaotic, i.e., chosen at random from among all possible initial data, but they came to appear very special because some dissipative (mixing) effects ironed out all original irregularities. Particle creation, neutrino viscosity or hadron collisions were suspected to mixmaster the early Universe, giving it its present shape.

Of course, this program "explains" the special character of the actual Universe only if the above mechanisms work equally well in all sufficiently typical initial conditions. To see whether this is really the case we must turn to those solutions of Einstein's equations in which mixmaster processes can occur. For this to happen the geometry and the evolution must be properly synchronized. For instance, if the expansion is too fast it can prevent mixing to overwhelm the entire universe. Only very few mixmaster solutions are known. The Bianchi IX mixmaster world model serves as a case-study.

In 1921 Kasner[14] found a seemingly simple solution of Einstein's equations which was first studied as a cosmological generalization of the RWFL model by Schücking and Heckmann[15]. The model is flat and has ellipsoidal symmetry at any instant of time; it expands along two perpendicular directions, and contracts along the third perpendicular direction. Its rate of evolution is such that the total space volume expands proportionally to time, but implosion in one direction causes the model to evolve towards an ever-flattening (but expanding in two directions) pancake.

Historically, the Kasner model (in fact, it is a one-parameter family of models) was interesting as the first discovered spatially homogeneous but anisotropic solution of Einstein's equations (to-day classified as a Bianchi I solution); now its importance stems from the fact that it serves as a building-block of the Bianchi IX mixmaster world model, originally investigated by Misner[16] to whom we owe the mixmaster idea in cosmology[17]. This world model is spatially closed, and has the initial singularity of the Big Bang type and (probably) the final singularity of the Big Crunch type. Its evolution backwards in time (towards the initial singularity) consists in the sequence of Kasner-like epochs separating cycles during which two axes perform a certain number of "small oscillations" accompanied by the monotonic decrease of the third axis. Then, suddenly, the Kasner epoch occurs: the axes change their roles in an unpredictable manner, two of them collapse, whereas the third one is stretched in a violent tidal deformation. The evolution jumps to another cycle, the amplitudes of all oscillations become bigger, and the process leads to a continuous squeezing of volume. The jumps succeed one after the other. An infinite number of Kasner epochs will occur before the initial singularity is reached (see fig. 1).

One can suspect that such a special evolutionary pattern would require a special tuning of the initial conditions. Indeed this happens to be the case. Although rather simple algorithms exist to compute the number of small oscillations in every successive cycle (provided we choose this number in the first cycle), the oscillation amplitude and the periods of oscillations, the system is entirely unpredictable. To compute the state of the system at a cer-

[14] E. KASNER, *Geometrical Theorems on Einstein's Cosmological Equations*, American Journal of Mathematics, 43, 217-221 (1921).
[15] E. SCHÜCKING & O. HECKMANN, *World Models, La structure et l'evolution de l'univers*, Onzième Conseil de Physique (Solvay), Stoops, Bruxelles 1958.
[16] C.W. MISNER, *Mixmaster Universe*, Physical Review Letter 22, 1071-1074 (1969).
[17] C.W. MISNER, *The Isotropy of the Universe*, Astrophysical Journal 151, 431-457 (1968).

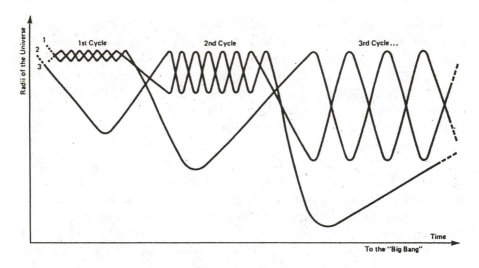

Fig. 1 - The mixmaster evolution of the Universe.

tain moment one should know the initial data with infinite precision. A tiny error in our knowledge of these data, for instance, at the tenth decimal place, after only eight cycles would completely ruin the possibility of predicting the oscillation number in subsequent cycles. Such behaviour is technically called *deterministic chaos*. If not usual measurement errors (we could suppose the Creator knows things precisely), the quantum uncertainties (which, after all, the Creator himself has invented!) would make such a world a completely unpredictable system. Therefore, we could suppose that even for God it would be better to play dice (or toss a coin if He prefers zero-one outcome) in order to compute the evolution of such a universe rather than make computations.

Unexpectedly, the last sentence turns out to be almost literally true. As shown by Chernoff and Barrow[18], the dynamics of the mixmaster model can be reasonably well approximated by the so-called Bernoulli process (an iteration map generated by a simple difference equation[19]) which in turn is isomorphic to the ideal coin tossing or dice playing[20].

---

[18] D.F. CHERNOFF & J. BARROW, *Chaos in the Mixmaster Universe*, Physical Review Letters, 50, 134-137 (1983).

[19] J. FORD, *How Random is a Coin Toss?* Physics Today, April, 40-47 (1983).

[20] In fact, the Bianchi IX mixmaster dynamics is mixing, ergodic, Bernoullian and has positive Kolmogorov entropy.

4. Towards the Relativistic Theory of Chaos

Standard indicators of the chaos existence in dynamical systems are the so-called Lyapunov exponents measuring an exponential separation rate of nearby trajectories which, in turn, causes effective "forgetting" the initial conditions and, consequently, the onset of chaos. To be more precise, let us consider the phase space of a dynamical system and two trajectories in it evolved from the initial conditions separated by a small distance Δ_0. At some later time t, under the action of dynamics, this distance becomes Δ_t. If the ratio Δ_t/Δ_0 grows exponentially with time, the quantity $(1/t)\ln|\Delta_t/\Delta_0|$ exists and is finite. The limit λ of this quantity when Δ_0 goes to zero and t goes to infinity is called the (*principal*) *Lyapunov exponent*[21]. Chaos exists in the system if $\lambda > 0$.

The criterion of chaos existence defined above explicitly depends on time reparametrization, and this makes its application to general relativity misleading. For instance, in current literature three "times" are used to analyse the mixmaster Bianchi IX model: 1) the cosmological time t in which the Big Bang occurs at t = 0; 2) the so-called Λ-time (the "standard time" for Bianchi cosmologies) in which the initial singularity takes place at $\Lambda \rightarrow \infty$; and 3) the Misner-Chitre time T in which one approaches the initial singularity as $T \rightarrow \infty$. All three times are interrelated by trasformations of an exponential type. For the mixmaster Bianchi IX world model the Lyapunov exponent is equal to zero in the Λ-time, is positive in the Misner-Chitre time and is inapplicable to this model in cosmological time since the region of chaos is not near t = ∞ but near t = 0; however, by using other methods it can be demonstrated that in the cosmological time t trajectories diverge even faster than exponentially[22]. No wonder that when Lyapunov exponents are used for numerical computations one obtains indecisive results[23].

It has been shown by Chitre[24] that the mixmaster Bianchi IX model can asymptotically be represented by the geodesic flow on a curved space, and the fact that the curvature of this space is negative testifies to the presence of

---

[21] Usually, one Lyapunov exponent is defined for each dimension of the phase space (the separation Δ_t between trajectories is regarded as a vector in the phase space, and projected along the unit vectors of the coordinate system). The principal Lyapunov exponent, defined in the text, coincides with the largest of these exponents.

[22] J. PULLIN, *Time and Chaos in General Relativity*, reprint Syracuse University, 1990.

[23] J. PULLIN, *art. cit.*, and S. RUGH, *Chaos in the Einstein Relativity*, reprint of the Niels Bohr Insitute, Copenhagen, 1991.

[24] D.M. CHITRE, Ph. D. Thesis, University of Maryland.

chaos. The idea has been fully elaborated by Szydlowski and his co-workers[25]. Its main features are the following.

It can be shown, by using the Maupertius least action principle, that any Hamiltonian dynamical system can be reduced to the problem of the geodesic flow on a certain (pseudo)riemannian space with the so-called Jacobi metric (essentially, determined by the potential). The trajectories of the system are now represented by geodesics on this (pseudo)riemannian space, and from their behaviour one can make inferences about the existence of chaos (with the help of the geodesic deviation equation).

To present the Bianchi IX cosmology in the form of a Hamiltonian system is a standard thing, and it turns out that the rest of the procedure works equally well.

However, the deviation equation is too complex to be solved directly, but there is a simple procedure[26] which gives us the sign of the Ricci scalar (with respect to the Jacobi metric). If this sign is negative, the geodesics of this metric are unstable against small changes of the initial conditions, i.e., deterministic chaos appears (in compact areas of the phase space)[27]. The chaotic behaviour itself can be identified with the negative sectional curvature averaged over all two-directions[28].

As we can see, the above construction is purely geometric and can serve as an invariant indicator of deterministic chaos. It liberates us from all troubles connected with the reparametrization of time.

5. Chaos and Integrability

In the light of the prevalent view, a certain degree of deterministic chaos seems to be a necessary condition for the origin and evolution of organized

---

[25] M. SZYDLOWSKI & LAPETA, *Pseudo-Riemannian Manifold of Mixmaster Dynamical Systems*, Physics Letters, 1484, 239-245 (1990), M. SZYDLOWSKI, J. SZCZESNY & M. BIESIADA, *A Criterion for Local Instability of a Geodesic Flow from the Jacobi Equation in Fermi Basis*, Chaos, Solitons and Fractals, 1, 233-242 (1991); M. BIESIADA & M. SZYDLOWSKI, *Mixmaster Cosmological Models as Disturbed Toda Lattices*, Physics Letters, A 160, 123-130 (1991); M. SZYDLOWSKI & A. LAPETA, *The Local Instability of Mixmaster Dynamical Systems*, General Relativity and Gravitation, 23, 151-167.

[26] The potential of the system depends on two vectors, the tangent to the geodesic and the normal one, which are chosen at random from all possibilities.

[27] One should notice that the chaotic behaviour, being a qualitative property, can be destroyed by the averaging procedure, but it cannot be created by it. Therefore, the negative sign of the Ricci scalar is only a sufficient condition of chaos.

[28] The relaxation time of the system can also be defined in terms of the sectional curvature; it turns out to coincide with the Kolmogorov entropy.

structures (biological life included). On the other hand, theoretical physicists would be inclined to say that everything that exists in the world (the world itself included) is modelled by a solution of a certain differential equation[29]. Even if this statement might seem exaggerated, one should admit that the world our physics reconstructs would be unthinkable without solving differential equations. One should suspect that one owes the existence of structures (oneself included) to an interaction between chaos and integrability.

As far as chaos is concerned there seem to be no major problems: as we noticed earlier, the existence of chaos is most probably the generic property within the ensemble. But there are problems with integrability: our experience with the Hamiltonian dynamical systems tells us that any small perturbation of the form of an integrable equation almost always destroys its integrability. In view of these, all successes of modern physics, based on modelling phenomena in terms of integrable equations, seem to be an edifice constructed on a razor's edge. It was René Thom who first stressed the contradiction between (to use his wording) "structural stability" and "calculability". He wrote: "Today's physics has sacrificed structural stability to calculability; I

Fig. 2 - The invariant tori of the Kolmogorov-Arnold-Moser theorem.

---

[13] This statement need not be incompatible with the fact that certain laws of physics assume the form of symmetries; see footnote 1.

would like to believe that physics would not have to regret its choice"[30]*. It seems, however, that this need not necessarily be the case. The disjoint either-or could in fact be replaced by a nice consonance.

The problem of integrability is a difficult problem, and to go beyond intuitions one would have to limit one's considerations to the Hamiltonian dynamical systems. The integrability of this class of systems was intensively studied in connection with the famous problem of the stability of the planetary system. This is not to say that dissipative systems play a lesser role in science. On the contrary, the very mechanisms of life are inseparably linked with dissipative structures. We only consider the Hamiltonian systems as an indication of how tricky things could be when we change to more general situations.

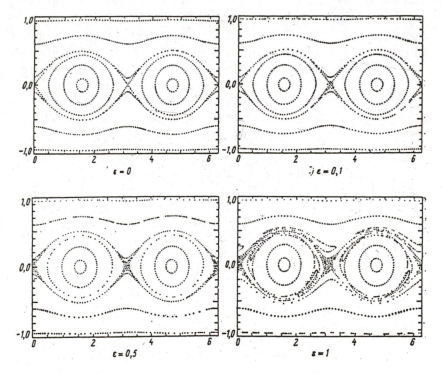

Fig. 3 - Stochastic layers in the space of a fast asymmetric top.

[30] R. THOM, *Stabilité structurelle et morphogénèse, Interedition*, Paris, 1977.
* Strictly speaking, there is a difference between "calculability" and "integrability", but I do not think Thom's idea differs very much from that developed above.

First of all, we must remember that there is no single concept of integra-bility; as Poincaré said "a system can be only more or less integrable". However, for the sake of present — of necessity very sketchy — reflections it would be enough to understand integrability in its broadest sense as the pos-sibility to solve a system in a finite number of steps consisting of algebraic operations and computing integrals of known functions (quadratures)[31]. The general result is that any perturbed integrable Hamiltonian system is non-integrable. Moreover, if a Hamiltonian dynamical system depends on a para-meter (i.e., if we have a one-parameter family of Hamiltonian systems), then integrable systems correspond, in a general situation, to exceptional isolated values of the parameter.

If in phase space the regions of the different behaviour of trajectories are separated from each other by invariant stable or unstable trajectories, called *separatrices*, they play an important role in determining the topological structure of the phase space of the considered dynamical system. The exi-stence of separatrices is a characteristic feature of integrable systems (the separatrix itself is a solution of the system). It turns out that although inte-grability is an unstable property of dynamical systems, some structural ele-ments of integrable systems survive the action of small perturbations. An outstanding example is the Kolmogorov-Arnold-Moser theorem concerning the stability of the solar system. The structure of the phase space of the cor-responding non-perturbed Hamiltonian system is such that its trajectories wind up around invariant surfaces forming a family of concentric tori (fig. 2). The theorem states that the majority of invariant tori survives any sufficien-tly small perturbation but undergoes deformations. This means that for the majority of initial conditions, for the dynamical systems remaining close to the integrable ones, the motion continues to be quasi-periodic. These pheno-mena are accompanied by a dissolution of separatrices in the phase space of the perturbed system[32].

In some examples (for instance in the motion of a fast asymmetric top), when an integrable system is perturbed its separatrices disappear and in their places stochastic layers are created (fig. 3). In this way, the structure of the phase space of nearby models is somehow organized by the structure of the integrable model or, in other words, an integrable trajectory (separatrix)

---

[31] W.W. Kozlow, *Integrability and Non-Integrability in the Hamiltonian Mechanics*, Usp. Mat. Nauk, 38, 3-67, (1983).

[32] W.I. Arnold, *Mathematical Methods of the Classical Mechanics*, Polish Scientific Publishers, Warsaw, 1981 (Polish translation from Russian).

is approximated by chaotic ones. Bogoyavlenskii[33] has even elaborated a method allowing him to reconstruct complex dynamical regimes having some stochastic properties by investigating the behaviour of separatrices (which approximate, the topology of trajectories)[34].

Let us turn to another important example. An interesting type of dynamics has been discovered by Toda[35] when investigating the problem of the wave motion in non-linear media. It has been found by him that solitons in a one-dimensional non-linear lattice are conserved and behave like particles. A simple model of this kind, in which particles of unit mass with an exponential potential are considered, is integrable, but some of its "disturbances" exhibit stochastic properties. The model has led to a number of theoretical works[36]. It turned out that the mixmaster Bianchi XI cosmological model belongs to the class of disturbed three-particle Toda lattices. Already in the third order of approximation such a Toda system has clear stochastic properties and exhibits all qualitative features of the mixmaster Bianchi IX world model[37]. We are entitled to say that the phase space structure of the non-integrable and chaotic mixmaster model is organized by the phase space structure of the integrable Toda lattice to which the mixmaster model is an approximation.

Examples of this kind offer an interesting possibility of reconciling the discrepancy (alluded to by Thom) between integrability and structural stability: the wealth of evolving forms in the real Universe is possible because of a subtle interaction between these two properties. To have an organized structure one needs chaos, but not any chaos; it must be of such a special sort that would lead, in the limit as the perturbation goes to zero, to an integrable system.

This would also explain another puzzling property of the world, namely its capability of being approximated by relatively simple mathematical models. To this property science owes its very existence, and so far it has see-

[33] O.I. BOGOYAVLENSKII, *Methods in the Qualitative Theory of Dynamical Systems* in *Astrophysics and Gas Dynamics*, Nauka, Moscow, 1980 (in Russian), pp. 31-34.

[34] G.M. ZASLAVSKII & R.Z. SAGDEEV, *Introduction to Non-Linear Physics*, Nauka, Moscow, 1988 (in Russian), pp. 129-139.

[35] M. TODA, *Waves in Nonlinear Lattice*, Prog. Theor. Phys. Suppl., 45, 174-200 (1970).

[36] O.I. BOGOYAVLENSKII, *op. cit.*, pp. 300-313. G. CONTOPOULOS & C. POLYMILIS, *Approximations of the 3-Particle Toda Lattice*, Physica, 24 D, 328-342 (1987); S. UDRY & L. MARTINET, *Orbital Behaviour Transition from Hennon-Heiles to the Three-Particle Toda Lattice Hamiltonian*, Physica, 44D, 61-74 (1990).

[37] M. BIESIADA & M. SZYDLOWSKI, *Mixmaster Cosmological Models as Disturbed Toda Lattices*, Physics Letters, A 160, 123-130 (1991).

med to be one of the greatest mysteries of natural philosophy. If the above considerations are correct, they throw new light on this question. Forms in the real world are by no means simple, they are non-integrable and chaotic[38], but they inherit at least some of their structural properties from integrable mathematical structures. In our investigations of the world we are hunting for simple *integrable* structures. If we find a correct one (we do surprisingly often), the world responds to it through various experimental data. These data, being always "contaminated" by unavoidable measurement errors, do not fit exactly to a unique integrable model, but rather to a class of nearby chaotic and non-integrable ones. In a sense, it is not our simple (integrable) structures that approximate the complex (non-integrable) structures of the world, but rather vice versa: the wealth of structures in the real world approaches, in a sort of a limiting process, our integrable models.

To sum up, we have these three: structural stability of chaos, integrability of equations, and the world's property of admitting approximations by our simple models. Neither of these three is independent of the other two. They are all aspects of the strong non-linearity of Einstein's equations.

I should point this out clearly, that the ideas developed in this section are but hypotheses based on examples taken from the class of Hamiltonian systems. However, these hypotheses persuasively suggest that deterministic chaos and integrability need not be antagonistic properties of the world but rather mutually cooperating aspects of its non-linear structure. We should expect that a non-Hamiltonian dynamical system could offer us yet richer examples of this cooperation.

6. Our Place in the Ensemble

The measurements of the blackbody background radiation give us the ratio of the number of photons to the number of baryons (the so-called *entropy per baryon*) which is of the order of 10^8. This number imposes strict constraints on the amount of chaos admissible in the early history of the Universe. All chaos had to be effectively smoothed out before the baryon

---

[38] The Reader is asked to forgive me for abusing the language. I should have said "forms in the real world are modelled by non-integrable and chaotic equations", etc., but such pedantry would make the text non-readable. I hope this concession on behalf of the smoothness of language, met from time to time throughout the present paper, will not lead to misunderstanding.

number first imprints upon the world's structure, and this occurs when the Universe is 10^{-35} second old. If this happened later, the heat production due to chaos dissipation would have been so great that life would have had no chances to appear.

This again raises the question of how special are we in the ensemble of universes. As we have seen, according to the philosophy implemented in the probability principle (see section 3 above) our existence is justified provided the life admitting conditions are generic in the ensemble of universes. A modification of the same philosophy guides many works connected with the so-called *inflationary scenario*. The inflation phase is the very early period in the world's evolution when, owing to the splitting of the strong nuclear force from the electroweak force, the quantity $\rho + 3p$ becomes negative (ρ is here the energy density and p the pressure of the fluid filling the Universe), causing the exponential expansion which is superimposed on the ordinary Friedman-like evolution[39]. Because of this dramatic expansion the Universe effectively forgets its pre-inflationary state. In this way, the present properties of the world could evolve from any initial conditions, and we owe our existence not to the fine tuning of the initial conditions, but rather to the inflation mechanism itself. It goes without saying that this philosophy works only if the set of initial conditions admitting the inflationary phase is "large enough". This, however, is an open question.

The philosophy underlying the probability principle, although very popular nowadays, is not the only option. Roger Penrose, has adopted just the opposite view. In his opinion, to explain the second principle of thermodynamics and to solve the time arrow problem one must assume that the initial singularity (in contrast to the final one) had to have a very smooth structure (its "gravitational entropy" had to be extremely low) which, of course, requires carefully chosen initial conditions[40]. Penrose believes that the fundamental asymmetry between the structures of the initial and final singularities (or equivalently, the very special choice of the initial conditions leading to this asymmetry) will be incorporated into the future quantum theory of gravity.

---

[39] *Inflationary Cosmology*, eds. L.F. ABBOT, SO YOUNG PI, World Scientific, Singapore, 1986; S.K. BLAU & A.H. GUTH, *Inflationary Cosmology, 300 Years of Gravitation*, eds. S.W. HAWKING, W. ISRAEL, Cambridge University Press, 1987; A.D. LINDE, *Physics of Elementary Particles and Inflationary Cosmology*, Nauka, Moscow, 1990 (in Russian).

[40] R. PENROSE, *Singularities and Time-Asymmetry in General Relativity*, An Einstein Centenary Survey, eds. S.W. Hawking, W. Israel, Cambridge University Press, Cambridge, 1979; R. PENROSE, *The Emperor's New Mind*, Oxford University Press, Oxford, 1989.

In other words, he suspects the existence of a new law of physics governing the fine tuning of the initial conditions.

With Penrose's philosophy there is nothing to do but wait till we have a quantum theory of gravity and see whether it is right or wrong. With respect to the philosophy underlying the probability principle I would like to make the following remark. In my view, this philosophy is premature as long as a suitable measure is not available on the ensemble of universes (or equivalently on the space of corresponding initial conditions), and this does not seem likely to occur in the near future. The ideology of perturbing a given equation with the help of a "small parameter" might easily lead to spurious results since by using this method we arbitrarily isolate a single property (wanting to know whether this property is generic or not), whereas different properties of the world are certainly mutually interdependent, and we should expect that they will together contribute to determining the measure in terms of which the stability or genericity is to be defined. In this manner, a set of properties could turn out to be "typical" even though some of the properties belonging to this set, when considered alone, seem to be non-generic[41]. The necessary conditions for biological evolution might seem non-generic, but the problem could look very different if they are considered as structurally linked with other properties of the world. Without a suitable measure of the ensemble of universes such philosophies can only have the value of vague speculations. Sometimes a philosophy may seem highly attractive, but attractiveness alone cannot be regarded as a rational justification.

---

[41] To put some flesh into these intuitions let us remember the Banach metric on the space of dynamical systems in terms of which the structural stability of these systems is destined. All terms of the corresponding equation (not just a small perturbing parameter) play a role in defining this metric. S. SMALE, *The Mathematics of Time*, Springer, New York - Heidleberg, 1980.

DISCUSSION
(COYNE, chairman)

ARECCHI: Some time ago there was a seminar in which Lee Smolin dealt with a vision of quantum cosmology which looked like a bifurcation tree. However, the control parameter is not external to the system but it is due to feedback within the same bifurcation tree. In other words it is a self-consistent structure, as if in your case the perturbation parameter, by some kind of feedback process, were stabilized at a given value. Are you familiar with Smolin's approach? Is there any correlation with yours?

HELLER: Smolin's ideas concern quantum gravity, whereas in my talk I dealt with purely classical cosmology. I think, however, that the problem of time is essential in both approaches. I have alluded to the fact that the usual criterion for the occurrence of chaotic behaviour, i.e., that the Lyapunov exponents should be positive, does not work in the relativistic context. Lyapunov exponents are not invariant: after time rescaling the positive Lyapunov exponents can vanish or become negative. Here we have the problem of time. We try to avoid it by replacing the Lyapunov exponents by some other, more geometric, criteria, nevertheless the problem of time enters into the game in this or in another way. Already in the seventies there was some work, the goal of which was to introduce an "internal time" to cosmology. This time was assumed to be measured by the complexity of what happens in the Universe rather than by an external parameter. When we come to the quantum gravity regime, things can drastically change, time can entirely disappear. But, of course, the problem in classical and quantum regimes are strictly linked by the mechanism of the emergence of time in changing from one regime to another.

PULLMAN: D'abord je veux vous féliciter pour une conférence qui est vraiment fascinante et suscite des questions intéressantes. Je vais vous en poser une moi-même. En relation avec le problème des conditions initiales, nécessaires pour l'apparition de la matière de l'univers, dont vous avez parlé, certains le présentent sous forme d'une nécessité d'une sélection extrêment rigoureuse, extrêmement précise, des constantes universelles qui nous régissent, condition absolument nécessaire pour que le monde soit tel que nous puissions y exister et l'observer. Je voudrais vous demander quelle est votre position, quelle est votre opinion, si vous en avez une, par rapport au principe anthropique de Brandon Carter.

HELLER: I have revealed a little of my attitude to the anthropic principle in the last paragraph of my talk. Generally speaking, I think it is premature to draw strong cosmological conclusions from the existence of life. We have no well-defined probability measure on the set of all initial conditions leading to various universes, and consequently we are unable, in a precise manner, to say which property in this set is typical and which is non-typical. Moreover, the origin and the evolution of life is a very complex phenomenon which must be treated in connection with many other phenomena. And, as I said at the end of my talk, it might easily be that the phenomenon of life seems to be non-typical when it is considered in isolation from other properties of the world's evolution, but if a class of properties, including life, is considered together, as a certain entirety, it might turn out that it is less non-typical or even generic. I think that the main advantage of the anthropic principle is that it forces people to think about certain questions which so far have never been asked in cosmology.

JAKI: It seems to me that Einstein, who filled the pivotal role in the very foundations of your presentation, may object to it in the same graphic way in which he objected to the Copenhagen interpretation of quantum mechanics. In 1948 he and Pauli were walking in Princeton and just when they reached Einstein's house he became rather impatient with Pauli's insistence on the Copenhagen interpretation, he turned to him and said — the full moon was visible: — "My dear friend do you think that the moon is there only when you observe it?" To your presentation Einstein might object: "Do you think that the universe is there only when scientific cosmologists write equations about it?" This is the first point I would like to make. The second point is this: It seems that a set of inferences may follow from your paper. The first step is this: Until we know the exact measure of our possible solution to Einstein's equations we cannot know that there is a universe which encompasses all material existence. The second step: Assume that we learn that measure. In that case the inference bifurcates, one direction is the Platonic direction, namely, that the geometrical or mathematical formulas turn themselves into material reality. But I do not want to be concerned with Platonism at this point. The second direction is in the Gödelian direction. Let us assume that we learn that measure. Then in terms of Gödel's theorem we cannot have certainty about that measure. Then comes the third step. Immanuel Kant then could argue that he was in substance right with the first two of his four antinomies, namely, the two cosmological antinomies, that the universe is an unreliable notion and therefore the cosmological argument is impossible.

HELLER: I would be very happy if I could discuss with Einstein, and if I could be shown to be wrong by him. The "philosophy" underlying my talk is the following. There are some mathematical structures, and there is the Universe; these mathematical structures model the Universe; and nothing more. Therefore, in my talk there was no real philosophy, I would even say that there were some accents warning against a too simplistic philosophy, as for instance the anthropic principle in its exaggerated forms. If I speak about the set of universes, or the set of initial conditions, it is a purely mathematical concept. The probability measure on such sets is necessary in order to say something sensible about the probability of an occurrence within these sets. And nothing more. I think what I have presented is perfectly neutral with respect to your questions.

RICHTER: When talking about general relativity there are always these two levels to distinguish, one is the description of the motion of particles within the given metric as when you apply Einstein's theory to the motion of planets and derive the perihelion motion of Mercury, etc., but that is not the context of your talk. It is a context, however, in which I would immediately accept that the relevant equations that one derives from general relativity should be of a Hamiltonian type. Now you were talking at a level where the metric itself was derived from Einstein's equations, which are in a sense similar to Maxwell's equations, except that they don't have boundary conditions and therefore are usually only solved on manifolds without boundaries. Doing so you obviously have freedom, because you have so many equations and you have to introduce specifications before you can reduce your system to a system of a few degrees of freedom as you did. A priori I think there is nothing that necessitates a Hamiltonian structure in the system that you derived here as a chaos. For example if you take the Robertson Walker metric as a starting point you wouldn't get Hamiltonian equations there. So what is the reasoning behind reducing these equations to equations with only a few degrees of freedom rather than to continuous models like a Navier-Stokes equation, where you also would have chaos, of course, but of a different kind?

HELLER: The main reason is that we want to have as simple equations as possible. There are also deeper physical reasons behind it. Observations show that our Universe is spatially homogeneous and isotropic to a very high degree. If you drop the assumption of isotropy, you are left with a homogeneous space, and in such a case all cosmological solutions are divided into nine classes, the so-called Bianchi classes. All these Bianchi cosmologies can be reduced to Hamiltonian systems. The Mixmaster cosmology I spoke about belongs to the Bianchi class IX. However, modern cosmology also deals with dissipative phenomena. For instance, particle creation in a strong gravitational field can phenomenologically be described as the bulk viscosity effect.

GERMAIN: I want to make a similar remark. I am not competent in Relativity and Cosmology. But why do you assume a Hamiltonian system? Why do you neglect dissipation? We have a lot of examples in ordinary continuous mechanics in which you find singularities, chaos, shocks and similar difficulties. You have a different picture if you deal with non-dissipative models and a very different picture if dissipation is taken into account. Will your conclusions be changed by the introduction of dissipation?

HELLER: A lot of work has been done on dissipative processes in cosmology. However, the present Universe is very well approximated by the Hamiltonian world models. As I said, the Universe is spatially homogeneous and isotropic, that is to say, it enjoys spherical symmetry. If you have an isotropically expanding sphere there is no shear viscosity. There might be some bulk viscosity, but it turns out that it could be at work only at the very early stages of the Universe. The strict observational constraint to the quantity of dissipation present in the Universe comes from the observed ratio of the number of photons to the number of barions in the Universe which is equal to 10^8. This implies that all dissipative effects should be over before the Universe was 10^{-35} of a second old.

LICHNEROWICZ: Je voudrais avant tout remercier notre conférencier. Ensuite, je voudrais prendre un ou deux points sur les modèles d'univers, comme vous avez très bien dit. Les conditions initiales possibles sont beaucoup moins nombreuses qu'elles n'ont l'air à priori, parce que, à priori, nous devons choisir notre petit modèle d'univers de poche. Il faut se choisir une variété, une topologie déterminée, et avoir quelque part des conditions initiales de portée globale, ce qui n'était pas toujours dit clairement au point de départ. Et le phénomène fondamental de la gravitation, c'est l'interdépendance des mouvements des masses matérielles. Les champs de gravitation ou les métriques intérieures à tous les corps et toutes les distributions d'énergie se raccordent les uns sur les autres, de manière techniquement différenciable. C'est cela qui assure la liaison qui fait des champs de gravitation. Ces conditions initiales capables de secréter des univers globaux posent actuellement des problèmes mathématiques très importants et très difficiles. Il y a quelques théorèmes d'existence globale des conditions initiales locales sur une épaisseur courte de temps, beaucoup de théorèmes qui remontent à moi-même et à Evan Sjaker; mais la plupart des conditions initiales vont donner des univers dans lesquels il va y avoir des singularitées et des chocs, des troubles, tout ce que vous voudrez. Enfin je signale que les physiciens et nous-mêmes, nous avons des goûts pour les topologies simples. Nous pensons toujours que l'univers est fait, comme nous disons, avec un plat, ou bien avec une sphère multipliée par une droite temporelle. Il y a beaucoup d'autres modèles pris par des physiciens, et nous avons beaucoup de choix, mais, comme j'ai dit, pour le moment ce n'est pas Dieu, c'est nous qui avons chacun notre modèle d'univers de poche.

LEJEUNE: Monsieur le Président, j'ai honte de vous demander la parole, car je voulais faire non pas une réflexion philosophique, mais une réflexion théologique. J'ai été un petit peu surpris, que vous finissiez votre calcul, à propos de Dieu, joue-t-il aux dés, en disant que puisque le calcul est si difficile, il devrait jouer aux dés. Alors je me suis posé la question: comment pouvez-vous jouer aux dés quand vous n'avez pas encore créé l'univers?

DEL RE: My question really is perhaps very complicated. I was aware ot two meanings of the word philosophy. One is the traditional European meaning; thinking about truth, being, the theory of knowledge and so on. The other is the meaning of more pragmatistic societies where by philosophy is meant a set of rules guiding your practical choices. Now, I don't see exactly which definition you adopt in your use of the word philosophy. You speak for instance of Penrose's philosophy. Could you specify what you mean there by philosophy?

HELLER: Perhaps some day I shall have an opportunity to deliver a talk on my views on philosophy and its relationship to science. There is nothing to hide in this respect. However, in the present talk I used the word "philosophy" in the colloquial sense, often used by scientists when they speak to each other about their work. You could easily eliminate this word by replacing it by "interpretation", "image of the world", "Weltanschauung"... I have discussed scientific questions, perhaps open questions but purely scientific ones.

MALU: Je voudrais répondre à la question de Lejeune de façon peut-être facétieuse en disant, si j'étais Dieu, j'aurais laissé la chance jouer son rôle, et plusieurs univers se créer au hasard, et alors j'aurais joué aux dés pour pouvoir sélectionner un de ces univers.

LICHNEROWICZ: J'aurais envie de renforcer en disant que la seule chose que Dieu ne puisse faire c'est jouer aux dés.

HELLER: At the end I have a question for everybody: How large is the set of all initial conditions leading to a Universe containing a cup of coffee?

THE EPOCH OF GALAXY FORMATION

MARTIN J. REES

University of Cambridge

Introduction

The topic of my talk is the epoch of galaxy formation. Two months ago the Pontifical Academy convened a working group on this theme under my chairmanship. It was attended by about twenty specialists, of whom one, Dr. Coyne, is here today. I shall try to mention some of its conclusions, and to set them in a broader context that relates to the overall theme of the present plenary session. After all, the emergence of galaxies and clusters is the first stage, indeed a prerequisite stage, for the formation of our Earth and the complex phenomena which have evolved on it.

Our discussions focussed on the *empirical*, rather than the *'philosophical'*, aspects of cosmology. Pre-eminent among the puzzles we addressed is the origin of cosmic structure. Our universe seems to have evolved, in between 10 and 20 billion years, from a dense amorphous fireball to its present state, where its dominant features are galaxies, distributed in clusters. We aimed to address the basic question "how did the observed structures emerge and evolve?" We are forced to suppose that the early universe was not completely smooth and uniform. If it had been, it would still, even now, consist of nothing but smoothly distributed cold diffuse gas. There must, even at very early epochs, have been some irregularities in the density or in the expansion rate. Initially these were of small amplitudes. However, any region that was slightly overdense would have lagged more and more behind the overall cosmic expansion, eventually condensing into a gravitationally bound system: a galaxy or cluster, depending on its scale.

The fluctuations must have originated at an ultra-early epoch in cosmic history, when the relevant physics is highly uncertain and speculative. The working group did not address the actual *origin* of the fluctuations, although there have recently been important new insights into how they might be the outcome of quantum effects during a so-called "inflationary" phase. However, we can fortunately infer a good deal about the nature of the fluctuations from direct observations of galaxy clustering, distant objects, and the background radiation. All these latter aspects were discussed in some detail.

The cosmologist can learn about the past, as does a geologist or paleontologist, by inferring from the history of our Sun, and other stars in our own Milky Way. But we have an advantage over those other scientists in that we can probe the past directly by looking at distant objects, which may be so far away that their light set out when the universe was significantly younger. The most important objects in this context are the quasars, bright hyperactive central regions of galaxies. These can now be observed out to distances so great that their light set out towards us when the universe was 200 times denser than it is now, and probably rather less than a tenth of its present age. The very existence of these quasars, identified by their very high redshifts, implies that some galactic-scale structures had already formed when the universe was 1 billion years old.

Quasars are perhaps rather exceptional objects, and it would be even better if astronomers could detect more typical galaxies at similarly great distances. Modern photon detectors can in fact register objects down to such low light levels that several hundred thousand galaxies are detected in every square degree of sky. However, the distance of these galaxies is uncertain, because they are too faint to have proper spectra taken. They are clearly very remote, and may still be in the process of formation, but they are probably not quite as far away as the most distant quasars, which may involve an atypical subset of galaxies that formed specially early, from initial fluctuations of unusually high amplitude.

The general context: the "big bang" cosmology

In the perspective of the cosmologist, even entire galaxies are just "test particles" which indicate how the material content of the universe is distributed, and how it moves. Galaxies are clustered, sometimes in small groups like our own Local Group of which the Milky Way and Andromeda are the dominant members, sometimes in big clusters with hundreds of members. Moreover, there has recently been great progress in mapping out the apparent large-scale filamentary or 'sheetlike' structures in the universe, on scales much larger than individual clusters. But nevertheless on the *very* largest scales the universe genuinely seems simpler and smoother. If one imagined a box whose sides were a few hundred million light years, dimensions still small compared to the observable universe, its contents would be about the same wherever we placed it. There is, in other words, a well-defined sense in which the universe is roughly homogeneous. It seems that all parts of the universe evolved in the same way with the same history.

Our observable universe is simpler, in its overall structure and symmetry, than we have any right to expect. It is of course sensible methodology to start off by making simplifying assumptions about homogeneity, uniformity of natural laws, etc., and cosmologists did this. But what is surprising is the degree to which these models remain relevant. The kinematics are simple. The universe expands in the same way in all directions in accordance with Hubble's law. Individual galaxies in the universe evolve as they age. But does the entire universe evolve? Did everything really start 10 or 20 billion years ago?

The quasars allow us to probe more than 90% of cosmic history. But the first evidence concerning the very early phases of the so-called "big bang" came in 1965 when the microwave background radiation was discovered. The discoverers did not themselves know how to interpret this radiation. But others quickly realised the momentous implications. Intergalactic space wasn't completely cold, but at about 3 degrees above absolute zero. This may not sound much, but it implies about a billion quanta of radiation for every atom in the universe. This radiation is now known to have a very precisely thermal black body spectrum. It is a direct relic of an epoch when the entire universe was hot, dense and opaque.

According to the so-called 'hot big bang' theory, the universe would have been hotter than 10 billion degrees for its first second of existence. After about half a million years, it would have cooled to about 400 degrees. At that stage, the primordial protons and electrons recombine into hydrogen atoms, and the universe becomes transparent. The microwave background radiation has propagated uninterruptedly since that time, and is therefore a direct probe of an early phase before any galaxies or clusters existed. During the ensuing 10 or 20 billion years, the radiation would be cooled and diluted by the expansion, but would still be around. It fills the universe and has nowhere else to go.

Corroborative evidence for the "big bang" theory comes from consi-derations of the origin of the chemical elements. One of the triumphs of astrophysics over the last 40 years has been the realisation that chemical elements such as carbon, nitrogen, and oxygen could be synthetised in the centres of stars, which subsequently explode, their matter being recycled into later generations of stars, and into our Solar System. Our Milky Way galaxy is a kind of ecosystem in which gas is recycled and progressively enriched into elements further up the periodic table. All the heavy elements on Earth, and we ourselves, are the ashes of dead stars. This process of stellar nucleosynthesis could never, however, account for the large amount of *helium* in the universe and why it was so uniformly distributed. It was therefore encouraging when the first calculations of nuclear reactions in the primordial fireball, carried out in the 1960s, showed that the simplest assumptions led to the conclusion that the material emerging from the fireball would contain about 24% of helium, agreeing almost precisely with what is now observed in the oldest objects.

The so-called 'hot big bang' theory formed the basis for all the discussions of our working group. How justified are we in accepting this framework? The great Soviet cosmologist Zel'dovich once claimed in a lecture that it was "as certain as that the Earth goes round the sun". Perhaps he was unaware of his compatriot Landau's dictum that "cosmologists are often in error but never in doubt!" We must never forget that cosmology is still a science where facts are scarce, and where we depend on observations made at the limits of the sensitivity of our instruments. But over the last 25 years, the empirical case for a big bang has greatly strengthened in several specific respects. The spectrum of the background radiation seems to be that of a black body to a precision better than one part in 1000; the abundance of cosmic helium and deuterium and lithium accord very well with the predictions of the model. Moreover, there are several discoveries that *might* have been made, which would have invalidated the hypothesis, and which have *not* been made. Astronomers might have found objects with *zero* helium abundance; there might have been objects whose inferred age was inconsistently high; physicists might have found that neutrinos had a mass of, say, 1 keV, in which case primordial neurinos, if present in the numbers predicted by the hot big bang would contribute too much mass in the present universe. I personally believe we are justified in placing at least 90% confidence in an extrapolation back to $t = 1$ second in general accordance with the "big bang" theory. However, when we consider an *even earlier* phase, as I shall mention later, everything is much more speculative and we can be correspondingly less confident.

Fluctuations and the growth of cosmic structure

Two contrasting views have been adopted for the origin of structure. The first is that structure evolved in a "top-down" fashion, with clusters forming first, and subsequently fragmenting into galaxies. The second alternative is that structure develops in a hierarchical 'bottom-up' fashion, with galaxies forming, first, and then agglomerating into clusters, and later still into superclusters. There was a consensus among the working group that a "bottom-up" scenario is strongly favoured. A few years ago, this was still an open question, and the current agreement is based on a combination of several lines of evidence which have emerged in recent years.

There was less consensus, however, on the question of the *cosmic dark matter*. It is embarrassing to astronomers that 90% or more of the mass of the universe is unaccounted for, and could be in a variety of possible forms, ranging from elementary particles up to black holes. The existence of dark matter should not surprise us — indeed there are all too many forms that dark matter may take. The range of option is now being narrowed down by observation and experiment. Moreover, we may get clues to the dark matter from a better understanding of how galaxies form. After all, since the dark matter is gravitationally dominant, the luminous part of galaxies is essentially just a tracer for how the dark matter is distributed. We may also benefit from advances in fundamental physics, which tell us more about what kind of particles might have survived from the early universe. Some participants in the working group presented the results of elaborate computer simulations of how galaxy clustering proceeds. It turns out that the present appearance of the universe depends rather sensitively on what the dark matter is made of, and better simulations may therefore in themselves help to settle this basic question.

It has been a general article of faith that, whatever the matter consists of, the present structures form by a process of *gravitational instability*. If the early universe is slightly irregular in its density or expansion rate, then the density contrast grows as the universe expands, until eventually the overdense regions stop expanding and condense into bound systems. But one can calculate how rapidly these density contrasts grow. It turns out that, to account for the present structures, the early universe must have possessed irregularities which perturb the gravitational field by about 1 part in 10^5. Because of these fluctuations, we would expect that the microwave background temperature would not be exactly the same all over the sky. The radiation effectively comes to us directly from an epoch when the universe was about half a million years old. Photons coming from an incipient cluster would lose an extra bit of energy in climbing out of the associated gravitational potential well, and would therefore be cooler than average, but only by about one part in 10^5. For 20 years, experimenters have been searching, with gradually improving sensitivity, for these very small predicted irregularities in the background temperature. Early in 1992 they were for the first time detected, by the COBE satellite. If these fluctuations had not been present, it would have been an embarrassment for supporters of the consensus view that clusters and superclusters evolved by pulling themselves together gravitationally from small amplitude initial irregularities. COBE provided the first evidence that incipient or embryonic structures actually existed at the expected level in the early (pregalactic) universe.

Within the next year, we can confidently expect further information on those fluctuations, on a range of angular scales. As we discussed at the working group, the exact amplitude of the initial fluctuations is an important datum which can be compared directly with the sizes of present-day structure in the universe (and the depth of the associated gravitational potential wells), thereby allowing us to confront the alternative models more quantitatively with relevant data.

As already mentioned, I believe our inferences about cosmic history back to the time when the universe had been expanding for 1 second should be taken seriously — perhaps, indeed, as seriously as ideas about the early history of our Earth, which are also based on rather indirect inferences, fossils, etc. But those who believe that there was a "big bang" must accept that few of its detailed consequences and implications for later stages are yet understood. For instance, the fluctuations are undoubtedly a cosmogonically crucial feature of our universe but their precise origin is still a mystery. To extend the geophysical analogy, we know that the Earth is round and something about its global properties, but the features on its surface, the continents and oceans are harder to understand: the details are still controversial, and some topographical features will never be understood as more than accidents. But our bafflement about geophysical problems does not lead us to doubt that the world is essentially round. Likewise, in the cosmological context the perturbations to the overall smooth curvature of the universe induced by the gravity of even the largest superclusters are no more than 1 part in 10^5. To understand these details we must refine, rather than abandon, the various models already being explored.

The complications of cosmic structure and how it emerged, the themes of our working group, are those of complexity and nonlinearity. The basic physics that is relevant at late cosmic epochs is well understood: Newtonian gravity, gas dynamics, etc. But when, after some millions of years, the first bound systems condense out (perhaps to form the first stars), a range of complex nonlinear feedback processes set in, which make simulations of cosmic evolution at least as hard as weather prediction here on Earth.

The problems of the ultra-early universe

A quite different type of difficulty arises when we consider the *very* early stages of cosmic expansion, the first millisecond. Everything was then probably almost "linear" and structureless, but was squeezed to densities exceeding that of an atomic nucleus; as one extrapolates still further back one has *less and less confidence in the adequacy or applicability of known physics*. But explanations of why cosmic fluctuations exist, why the universe contains the observed mixture of matter and radiation, and why it is expanding in the symmetrical and simple way which makes progress possible, must await a better understanding of the earliest phases of the big bang.

We have no quantitative explanation for the fluctuations. Indeed, the basic force of gravity is still not properly unified with the forces of microphysics.

It is *gravity* that holds together individual stars and entire galaxies; without it no cosmic structures could exist. Regions in the expanding universe that were even *slightly* overdense to start with would have suffered extra deceleration, and lagged behind more and more, until their expansion eventually stopped. It was via this *gravitational instability* that galaxies and clusters condensed (after a billion years or so) from the expanding universe.

And gravitating objects have the peculiar property that when they *lose* energy they get *hotter* — as Professor Thirring has explained, their specific heat is *negative*. For instance, if the sun's radiative losses *weren't* compensated by nuclear fusion, it would contract and deflate, but would end up with a *hotter* centre than before: to establish a new and more compact equilibrium where pressure can ballance a (now *stronger*) gravitational force, the central temperature must *rise*.

From the initial big bang to our present Solar System, this 'antithermodynamic' behaviour has been amplifying density contrasts (in the manner drammatically shown in the N-body simulations) and creating temperature gradients within galaxies and within stars ("self-gravitating" fusion reactors). These processes are prerequisites for the emergence of any complexity.

A second key feature of gravity is its *weakness*. In a single hydrogen molecule, the force of gravity is about 36 powers of 10 weaker than the electrical binding forces. In any large self-gravitating object the positive and negative *electric* charge almost cancel, but everything has the same sign of *gravitational* "charge". Gravity "gains" relative to microscopic forces roughly as the two thirds power of the mass involved. (Gravitational binding energy per atom goes as M/R. For objects of the same density, R goes as M 1/3, yielding a net dependence going as M 2/3). A body must therefore contain about 10^{54} atoms before gravity starts to crush it. This corresponds to about the mass of the planet Jupiter, and anything much larger than this would be compressed and would in effect become a star. It is because gravity is so feeble that stars are so big. If gravity were somewhat stronger, for instance 26 rather than 36 powers of 10 weaker than microphysical forces, a small-scale speeded up universe could exist, in which stars (gravitationally-bound fusion reactors) had 10^{15} of the Sun's mass, and lived for less than a year. This might not allow enough time for complex systems to evolve: there would be fewer powers of 10 between astrophysical timescales and the basic microscopic timescales for chemical reactions. Moreover, no organisms could get very large without themselves being crushed by gravity. Our universe is vast and diffuse, contains structures on so many scales, and evolves so slowly, *because* gravity is so weak.

So a force like gravity is essential if structures are to emerge from amorphous beginnings. But (paradoxically) the *weaker* it is, the grander and more complex are its consequences.

Three hundred years ago, Newton showed how his law of gravitation explained why the planets traced out ellipses. But it was a mystery to him why they were "set up" in almost coplanar orbits. In his *"Optics"* he writes: "blind faith could never make all the planets move one and the same way in orbits concentrick... Such a wonderful uniformity in the planetary system must be allowed the effect of choice". In the present century, this coplanarity is recognised to be a natural consequence of the Solar System's origin from a spinning gaseous disc. Cosmologists are now offering physical explanations (albeit tentative ones) for features of the cosmos that previously seemed as inexplicable (or even providential) as the structure of the Solar System seemed to Newton.

Three features of the present Universe which seem prerequisites for the existence of observers are the excess of matter over antimatter, the existence of fluctuations, and the closeness of the expansion rate to its 'initial' value. Twenty years ago, these were all regarded as part of the 'initial conditions', perhaps to be understood only in 'anthropic' terms. But most cosmologists would now attribute the first of these, the cosmic baryon-to-photon ratio, to potentially calculable processes related to small asymmetries in the strong interactions. And the idea of 'inflation', although its details depend on still-uncertain physics, now offers an interpretation for the scale and 'flatness' of our Universe that seems compellingly attractive. The demarcation between cosmogonic phenomena which we can address scientifically and those we cannot has shifted, since Newton's time, from the Solar System to the first microsecond of the "big bang" which triggered the expansion of the observable part of the universe.

Many speculative ideas about the ultra-early universe figure prominently in popular books, and they are indeed fascinating. However, I am slightly uneasy about presentations which use the same tone of voice to describe ideas on the first 10^{-36} (or even 10^{-43}) seconds as are used in describing the Hubble law, the microwave background, etc. On the one hand, readers may treat speculations more credulously than is merited. But on the other, they may be justifiably sceptical of these 'far out' ideas, and be tempted to dismiss the rest of cosmology as based on equally shaky foundations. I believe a very sharp distinction must be drawn between the first microsecond of cosmic expansion, when the relevant basic physics is completely uncertain, and the later stages, after the first second has elapsed, when cosmologists are building up at least the outlines of a self-consistent picture involving primordial helium, background radiation, young galaxies and quasars, on the basis of well-established laws of physics. It was the impressive recent observational progress in these latter areas which we focussed on during our working group. We all came away encouraged to believe that it may be only a few years before the outlines of the correct cosmogonic scheme start to come into clear focus.

DISCUSSION
(COYNE, chairman)

ARECCHI: Can you please put the last transparency again? I have a question related to it. Presumably, until the end of the early era, the system was phase coherent, which means that beyond it if I had to use the common techniques used to characterize an extended structured system, I would measure the power spectrum versus the wave number. But now, I know that if I have a single lengthscale, I get a sharp exponential cut off. Otherwise in case of fractal clustering I get a 1/f power spectrum. Now, the main question, I heard without reading because I'm not a cosmologist, about a power law spectrum measured, for instance, by Pietro Nero (Rome Univesity) and other people. My question is the following: a laboratory experiment is a real spatial spectrum, whereas here the reconstruction is over space and time because you are looking backward in the universe. So, I wonder, what is the meaning of such a space-time spectrum, and such a space-time structure?

REES: The assumption that is normally made is that, at epochs when the fluctuations are still linear, they can be treated as gaussian fluctuations with random phases, and with a power spectrum which we hope to be able to calculate on the basis of some physical ideas about the early universe. The origin of the fluctuations may lie in quantum effects at ultra-early phases of cosmic expansion. In terms of a given initial spectrum, we can in principle calculate the clustering properties of the universe now, and thereby confront our theories observationally. We can also, incidentally, check whether the fluctuations are indeed gaussian both by the statistics of present-day clusters and by studying the fluctuations in the microwave background. I might just add that there are some theories, particularly those based on topological defects in the early universe, which would predict non-gaussian fluctuations.

ARECCHI: Excuse me, my question was not about models of theories but about observational data. What is known about that?

REES: The fluctuations in the present distribution of galaxies are analysed by determining correlation functions, percolation methods, etc. Over some range of scales the properties resemble those of a fractal. However, it is important to realise that above a certain scale the fluctuation amplitude diminishes and the universe becomes genuinely smooth. We are not in a universe consisting of clusters of clusters of clusters *ad infinitum*. In fact we need help from mathematicians in devising better tests for the significance of linear structures and filaments. The eye has evolved to be almost too good at picking out patterns. It was presumably better for our ancestors that they should see many tigers that weren't there than that they should miss the one that was! In comparing data on clustering with our theories we also have to remember that the luminous galaxies may not be clustered the same way as the gravitationally-dominant dark matter. To infer where the dark matter is concentrated we need to study the *motions* of the galaxies, which will be gravitationally pulled towards places where the dark matter is densest, but that is a much more challenging job observationally.

PULLMAN: I have two short comments. You just said that we know more about the center of the sun than we do about the center of the earth. Could we rationalize this observation by saying that to some extent complexity is a low temperature effect, provided the temperature is not too low? Thus when we look back to the beginning of theworld, to the Big Bang, the four forces, the four fundamental forces were not separated yet. So, we can imagine, that the universe was governed then by one law. Now we have a multiplicity of laws, which is thus a low temperature effect. But of course, the temperature should not drop to absolute zero.

REES: Yes, I completely agree with you. At high temperatures everything is going to be broken down into its simplest constituents, and therefore no complex structures can survive. Moreover, according to some ideas in unified field theories, the 4 basic forces of nature acquired their distinctive identity as a result of phase transitions as the universe cooled from a very hot initial state. If these theories are correct, then your statement is true in even a more fundamental sense, because at the highest temperatures even the basic forces themselves do not operate in the same different and distinctive ways as they do at low temperatures. It is because the early universe could have been "simple" in at least one sense of the word that I do not think we are necessarily being over-presumptuous in trying to understand it. Indeed it may well offer less of a challenge than the understanding of the smallest living organism. It is the biologists, not the cosmologists, who have set themselves the most ambitious goal.

PULLMAN: Now my second comment. You have put a number of questions in the form "why". Now, in sciences we generally ask "how". If you want to continue with "why", what about the queen of questions, the hundred thousand dollar question, which I think was asked first by Leibniz: why is there anything rather than nothing?

REES: We may be able to push back our chain of inference to the initial instants of the big bang, and perhaps even to understand why the universe is expanding the way it is, and why it contains the mixture of ingredients that we observe. However, there is still, as you say, a sharp demarcation between *how* questions and *why* questions. Indeed, it is perhaps apposite to recall where Newton would have drawn the dividing line. He believed that his laws could explain the orbits of the planets, but, as I mentioned, he could not understand why they had almost coplanar orbits. We now understand enough about stars and protostars to explain that. But although the demarcation line has moved back to the ultra-early stages of the big bang, it exists now as surely as it did for Newton.

MÖSSBAUER: There seems to be overwhelming experimental evidence that there is a hierarchy in the formation of matter, in clustering first of galaxies and then later clusters of galaxies. Is that experiment sure, then why do you still keep neutrinos in your list of potential candidates for dark matter?

REES: Part of my answer is simply conservatism. Neutrinos have the virtue of being known to exist, unlike the heavy neutral supersymmetric particles which some cosmologists speculate about. But it is true, as you imply, that the simplest cosmogonic model dominated by neutrinos runs into trouble, in that it predicts all small-scale perturbations are erased, and that structures would then develop in a "top-down" fashion, contrary to observations which support a hierarchical 'bottom-up' picture.
But there are alternative models, slightly more complicated but not entirely unrealistic, which could maintain a role for neutrinos, but invoke other kinds of dark matter as well, or different kinds of initial

perturbations. I pesonally think it would be a big boost for cosmology if you discovered that the heaviest type of stable neutrino had a mass somewhere in the range between 10 and 30 electron Volts.

ODA: Prof. Rees has coordinated many pieces of observational knowledge and theoretical consideration. We note the healthy progress of physics by successive alternations of theories and experiments; experiments are led by theoretical predictions and theories are led by new unexpected discoveries. Here, however, I would like to stress that improvements and breakthroughs in sensitivity or accuracy of experimental techniques without being guided by theoretical requirements are often followed by unexpected discoveries. I call this the "artisan's approach" and I may reword the theoretician-experimentalist-alternation as "prophet-artisan-alternation". The COBE experiment may be this case.

REES: Compared to most theorists I would class myself as a cautious empiricist. I do that because, as you say, what has made cosmology a science has been the input from experimenters and observers over the last 25 years. Turning to the COBE experiments in particular, these were of course the culmination of a 20-year quest, by many different groups, to detect the expected cosmic fluctuations. In a sense, the fluctuations were expected. It would have been a big embarrassment if the early universe had been smooth even at the level of 10^{-5}. The discovery of the fluctuations reassures us that we are on the right lines in basing our models on the assumption that large-scale structures grow via gravitational clustering, rather than requiring a more mysterious and more efficient process. The sensitivity and accuracy of these experiments is amazing, and in the next few years we can expect a great wealth of data on the microwave background, as well as on high redshift objects, etc. I think cosmology is overall a subject where theorists have a more modest and subsidiary role than they do in, for instance, particle physics.

ODA: I, as an experimentalist or an artisan, strongly believe that for us the most important thing is to sharpen our knife in physics more and more.

RICHTER: I have two questions. First, given that gravity interaction itself does not have any typical length, do you nevertheless have arguments for a typical size of a galaxy, and maybe also for typical angular momentum?

The second question concerns the reliability of what is known about the distribution of the galaxies. Is it not true that our look into outer space is hindered by the presence of our own galaxy and therefore we have information only about a part of that universe?

REES: To deal with your second question first, it is true that we cannot readily see outside our Galaxy, in the optical waveband, in directions close to the plane of the Milky Way. I do not think that is a serious handicap. We can observe *most* of the extragalactic sky optically. Moreover, complementary *infrared* and *radio* observations, to which the Galaxy is trasparent, reassure us that there is no great difference between that part of the universe in directions lying close to the galactic plane and the rest.

Turning to your first question: you are quite right that I didn't say anything in my talk about why galaxies exist — what is special about the dimensions of these entities, which are the most conspicuous features of the large-scale universe. Even if there is no obvious preferred scale introduced by gravity, the *gas* that eventually forms the luminous content of galaxies *is* influenced by other effects, particularly dissipation and radiation cooling. There are in fact some arguments, which I helped to develop about 15 years ago, which suggest that these dissipative processes can be important up to a certain maximum mass and maximum scale. Their scales agree roughly with those of galaxies. Arguments of this kind may therefore explain why there is an upper limit to the size of galaxies and why larger-scale gravitationally-bound structures manifest themselves as clusters of individual galaxies, rather than as single amorphous supergalaxies. Regarding the angular momentum, it is true that this cannot have been stored in the dense early phases of the universe. The favoured view is that galaxies acquire their spin at the stage when they are just starting to collapse. A typical protogalaxy would not be exactly spherical and, because it had a quadruple moment, it would exert a gravitational torque on its similar neighbours. Each protogalaxy can thereby, it is believed, acquire enough angular momentum to explain the observed rate of spin and the observed extent of the discs or galaxies like our own.

SELA: In view of the fact that you evaluated the chance of the Big Bang theory being ninety per cent correct my question is what was there before the Big Bang?

REES: At least you did not ask "why didn't the big bang happen sooner?", which is an even more impossible one to answer! All I can say in answer to the question you did ask is that, as we extrapolate back closer to the initial instant, conditions in the big bang become more and more extreme, and we have to jettison more and more of our commonsense concepts. Right back at the beginning we have to worry about quantum effects on the scale of the entire universe, and the idea that there are three dimensions of space and one of time may have to be abandoned, along perhaps with the concept of an "arrow of time". For this reason, even if, as some people speculate, our observable universe is just one member of an ensemble, we cannot put the elements of this ensemble in any kind of chronological sequence.

DALLA PORTA: I don't know if my question is related to complexity. What is your opinion on the so-called experimental evidence on anomalous red shifts?

REES: Let me first explain the background to this question. For over 20 years some astronomers have claimed that the high redshifts of quasars need not necessarily indicate that those objects are at great distances, and that some as-yet-unknown physics, and not the Hubble expansion, accounts for the large redshifts. I personally do not feel the case for anomalous redshifts is strong. Firstly, there is growing evidence that quasars are indeed at great distances (from, for instance, the observations of gravitational lensing, and absorption lines, due to intervening galaxies which are themselves at great distances). Second, the case made by Arp and his colleagues over the last 20 years seems to me to have rather little cumulative weight. Naturally, as observational programmes continue, more and more anomalous phenomena are discovered, but in this instance these do not hang together in any coherent alternative picture. I would like to add a psychological point. Arp and his associates would naturally be delighted if the anomalies were real. On the other hand there are some astronomers who would be very upset if unsuspected complications meant that they were further than they previously believed from correctly delineating the big picture. I am myself in a middle position. I think it would be *wonderful* if astronomers could really discover something fundamentally new such as an unsuspected redshift mechanism. My predisposition is therefore to look positively at Arp's claims, and I am therefore disappointed that his case still seems, after more than 20 years, to be depressingly feeble.

PULLMAN: I wish to remark that an answer to Dr. Sela's question was proposed by Saint Augustin, who taught that time was created together with the material universe. Thus time did not exist before the Universe and the question becomes meaningless. But God did and what was he doing there? Saint Augustin had a sense of humour and he recalled an anonymous answer to this query: He was imagining tortures to be inflicted on those who ask this type of question.

ARECCHI: May I comment on the question raised by Richter why a uniform gravitational force provides a peculiar scale for galaxies. If instead of being a gravitational problem it were a Maxwellian, an optical problem, which is more familiar to me, I would invoke some kind of non-linear refraction index which depends upon the same radiative intensity, and which modulates the gravitational waves in such a way that there are some natural clusters. There is a movie by William Firth of Strathclyde University in Glasgow, he starts from a homogeneous situation in a Kerr-like medium, that is, which has a non-linear refraction index, and then in a very natural way he arrives at a pattern formation, to cluster formation, if you like, on a natural scale. So I wonder whether there is anything which is like a refraction index for gravitational ways and a non-linear refraction index as well.

REES: I am not sure about the particular process you mentioned, but there are, in addition to the one I mentioned in my answer to Professor Richter, a number of other dissipative processes which can imprint preferred scales during the expansion of the universe. These may be connected with, for instance, photon viscosity, or the streaming distance of light neutrinos. There may also be late phase transitions which give astronomically interesting scales. But I'm sorry I don't know enough to comment sensibly on the particular work you referred to.

COYNE: It is obvious that we could continue for a long time but we have to close. There are a few announcements to make very quickly. First, that we worked you so hard this afternoon that so you can sleep in late tomorrow. Actually tomorrow's session will begin at 9.30 and the reason for that is not because you have been good today, it is because several of the Council members must be with the Secretary of the State to plan the Papal audience. They are also critically needed for the kind of discussion we are going to have tomorrow. So, the meeting will begin at 9.30, therefore the buses will depart at 9.10.

The other announcement I would like to make concerning that meeting is to repeat that it is limited to the Academicians themselves. The reason for that is not to exclude others. It's simply that it is Academy business; it's family business that we want to talk about, so thanks to our other guests for understanding that the meeting tomorrow morning is limited to the Academicians themselves. The President, Prof. Marini Bettòlo, has responded to our message and I would like to read that to you. It's just a few lines: "To all participants of the Plenary Session, I've received with great emotion your kind telegram, and I'm very grateful to all of you. I apologize for not being with you these days, but I hope to continue the collaboration after my recovery which is rather rapid. Thank you again. Best wishes and regards. G.B. Marini Bettòlo".

WHY IS THE EARTH'S ENVIRONMENT SO STABLE?

J. L. LIONS
Collège de France, Paris

Introduction

In this lecture we are going to consider the climate system of the planet earth *from the view point of control theory.* Very roughly speaking, the goal of "control theory" is to find *algorithms* so as to be able to *act* on a given system in such a manner that *the system behaves according to our wishes.* This theory, which is the heir of the Calculus of Variations, has been introduced, and used, for *human built systems,* for research, technology, engineering. It has also been considered, in particular by Norbert WIENER, in the framework of living systems. Without entering at all into any discussion about the GAIA hypothesis, we want, in this lecture, to show how ideas from control theory *could be* used for the Planet Earth System.

This "planet earth system" is certainly "complex", whatever the definition of "complexity" may be.

The simple statement of "controlling climate" can, rightly, raise eyebrows! But it is not a new question and, more importantly, *it is already in use,* even if connections with control theory are implicit rather than implicit.

The first mention that human activity could possibly modify average temperatures on the Earth seems to go back to J. FOURIER[1]. Let us quote: "The organization and progress of human societies, the action of natural forces can significantly change even in vast areas the state of the soil, the water distribution and the main flows of the air. Such effects can change, in

---

[1] J. FOURIER, *Rapport sur la température du Globe Terrestre et sur les espaces planétaires. Mémoires Acad. Royale des Sciences de l'Institut de France,* (VII) 590-604 (1824).

the course of centuries, the average temperature...". In other words, human actions can play the role of a "control function": *changing this function could change the state of the system,* and this is precisely the definition of a control function.

The next step is to use control functions in order to have the system "behaving according to our wishes".

This idea was introduced (for the first time?) by John Von Neumann in 1955[2]. "It is of course beyond human power to modify the amount of solar energy. But what really matters is not the amount that hits the Earth, but the fraction retained by the Earth, since that reflected back into space is no more useful than if it had never arrived. Now, the amount absorbed by the solid earth, the sea, or the atmosphere seems to be subject to delicate influences. True, none of these has so far been substantially controlled by human will, but there are strong indications of control possibilities". We "could" act - *but to achieve what?* This is the question of *criteria*. We have to make precise *in a quantitative manner*[3] what our "wishes" are. In the context of the planet earth, as indicated by Von NEWMANN himself, this is highly debatable "What could be done, of course, is no clue to what should be done; to make a new tropical, "interglacial" age in order to please everybody, is not necessarily a rational program."

Nevertheless *some kind of active control on a world basis had begun.* Rules for chemical emissions, for instance, are *exactly* a control decision, where the goal to achieve, the "wish" to fulfil is to stop or, at least, to slow down, the supposed (verified, proven?) global heating.

It is with these considerations in mind that we present here, and with great caution in the possible conclusions, some *remarks on the use of control ideas and techniques* in the framework of the Planet Earth System.

We shall follow the following plan.

1. Remarks on feedbacks.
2. Controllability.
3. Control of chaos.
4. Criteria and complexity.

---

[2] J. VON NEWMANN, *Can we survive Technology?* Nature (1955).
[3] J.L. LIONS, Lecture at the French Academy of Sciences, 10th CADAS Anniversary *'Le Temps du Contrôle'*.

1. *Remarks on feedback*

Firstly let us briefly examine the *classical situation in system sciences*. A "feedback" is an action on the system under study which is implemented in "real time" and whose "amount" depends on the available information. In other words the *feedback depends on the observed state of the system, and it is computed in "real time"*.

This action is supposed to have the system behaving "as it should".

Let us make these things a little bit less imprecise.

The system is supposed to be represented by its *"state"*.

Example 1.1.

The "state" of a space probe is its position and its velocity.

The state changes according to evolution equations. It is a dynamical system.

Example 1.1. (follows).

For a space probe, the Newton laws give the state equations.

Example 1.2.

Suppose we want to "control" the melting of ice in a box containing solid and liquid water. The state can be the temperature, governed by heat (diffusion) equation with a free boundary (the interface ice/water).

The state equations contain "control variables", which can be used (or not) according to our decision.

Example 1.2. (follows).

We can modify the course of melting by changing the heating applied on the boundary of the box.

The "action" has to be implemented in "real time", depending on the information available (stochastic elements generally come in to the picture at this point).

The "action" means (roughly) that the action should be implemented at a time where it is still "optimal". This notion is *not* intrinsic. It depends on the "scale" of the time evolution of the system.

Example 1.3.

Feedback in "instantaneous" time, based on the Kalman filter, was essential in the Apollo mission to the moon.

Example 1.4.

The control of an industrial plant such as in steel or aluminium processing should also be implemented in "real time", and the control of hydro electrical plants as well. The notion of "real time" is not the same in these various systems.

The notion of "feedback", in the above sense, goes back to Huyghens, Leibniz (it may appear before, in the framework of Automata, to mimic "nature"). It contains two aspects:

1) as described above, the rule to act,

2) the moment we use a feedback, we *connect* two (or more) parts of the "system" *which would otherwise be "unrelated subsystems".* In other words *we make a system out of a subsystem.*

Example 1.1. (follows).

In a space probe, the probe itself and the engines can be thought of as two somewhat "unrelated" subsystems until the moment when the engines are used.

It is this second aspect which is used in the notion of feedback used in "natural" systems, such as the Planet Earth System.

One of the best known feedback processes in the latter system is certainly the one introduced by J.G. Charney[4]: "The Desert 'feeds back' its dryness". Three of the commonly discussed feedback mechanisms are[5] the water vapour feedback, the snow-ice albedo feedback, the clouds feedback. Depending on the situations, a feedback can accelerate or slow down a phenomenon (such as global warming). It is then called a positive (resp. negative) feedback.

There are also feedbacks between vegetation and, say, humidity. We emphasize two points.

1) all these feedbacks are an important part of the complexity: finding them and expressing them a quantitative way is a far from trivial part of the analysis of the earth system;

---

[4] J.G. CHARNEY, *Dynamics of Deserts and Drought in the Sahel. Quart. J. Roy. Meteor. Soc.* 181, 193-202 (1975).

[5] J.T. HOUGHTON, G.J. JENKINS & J.J. EPHRAUMS, *The IPCC Scientific Assessment.* WMO/U.N. Environment Programs, C.U.P. 1990.

2) the feedbacks, as alluded to above, are different from the classical feedbacks of, say, engineering disciplines. They are not computed and implemented in order to achieve a given goal. They are "natural". But

a) They *could* achieve the stabilization of the system automatically;

b) World decisions (recommendations) on greenhouse gas emission norms *are* of the feedback type — with the goal of "not changing the present situation too much";

c) Decisions on the car traffic in megacities subject to pollution are definitely of the feedback type, this time in the classical sense of engineering sciences.

The question we want to address now is: what could be achieved by using control action on the Planet Earth System?

2. Controllability

The obvious completely general question is as follow: is it possible, given the constraints of physical and economical nature, to have the system behave according to our wishes?

Example 2.1.

Is it *possible*, in the framework of the Apollo mission, to send humans to the moon and back? (We now know the answer is "yes"!)

This type of question is within the framework of *controllability*. Let us represent the "state space" in Fig. 1 schematically. (In the case of the Planet

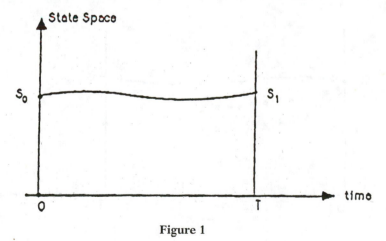

Figure 1

Earth it is an infinite dimensional space, of course approximated in any "practical" computation by a finite dimensional space).

The time $t = 0$ represents the "origin" of our computations (we are oversimplifying the presentation) and the time $t = T$ in the time horizon. It is chosen according to the problem. To fix ideas, T can be one year if we wish to improve the state of pollution over a city, a few years if we wish to "clean up" a river basin, 10 years or more for greenhouse effects, etc.

Let S_0 represent the present state in Figure 1.

If there was no human action (in a way which is not so simple to define), the system would describe a trajectory (represented in Figure 1) and would reach S_1 at time T.

Remark 2.1.

We will have to take into account the very high sensitivity to initial conditions! Suffice it to say, for the time being, that we are dealing here with *averaged* quantities.

Suppose now we are not pleased with the idea of reaching S_1 at time T. We start acting on the system, in order to reach S_2 (Figure 2) which is considered as "better".

Instead of following its "natural" path, the system will then describe "curves" indicated in Figure 2. Can one choose control functions so as to reach S_2 (the system is then said to be *controllable*) *or to get as close as one wishes to* S_2 (the system is then said to be *approximately controllable*)?

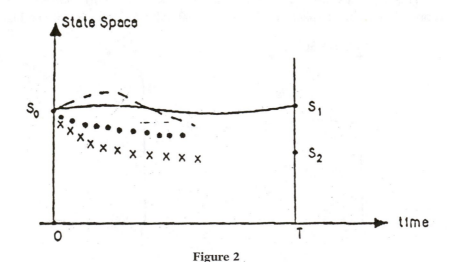

Figure 2

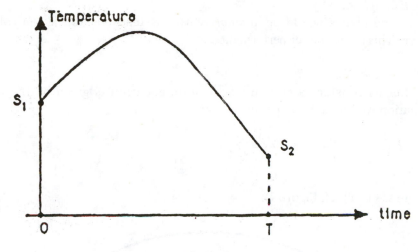

Figure 3

One should add constraints:

1) Physical constraints on the state.

Suppose the state is the globally averaged temperature. It is certainly not advisable to come exactly to S_2 at time horizon T after reaching, in mid course, very high average temperatures.

2) Physical and economical constraints on the control variables.

For instance, one cannot stop energy production!

Therefore "controllability" has *not* an absolute and intrinsic meaning. A system on which we can act "everywhere all the time" will certainly be more easily "controllable" than a system on which we can act "just a little from time to time". Controllability has to take into account all constraints.

This being said we now want to give "pieces of evidence" in support of a conjecture[6][7] which can be (in vague terms) stated as follows:

Conjecture 2.1.

The more unstable a system is, the "easier" and the "cheaper" it is to control.

---

[6] J.L. LIONS, *Are there connections between turbulence and controllability?* Proc. IX Int. Conf. INRIA 'Systems Optimization'. Springer Lecture Notes in Control and Information Sciences 144 (1990).

[7] J.L. LIONS, *WMY 2,000 and Computer Sciences.* Remarks Lecture at the 25th Anniversary of INRIA.

Example 2.2.

Some planes would be *unstable* without steady feedback control on them. This allows better performances.

Example 2.3.

Let us consider a system whose state equation is given by the heat equation with a "stability parameter" λ:

(2.1) $$\frac{\partial y}{\partial \tau} - \Delta y + \lambda y = v \chi_\sigma$$

in $\Omega \times (0,T)$ (cf. Figure 4),

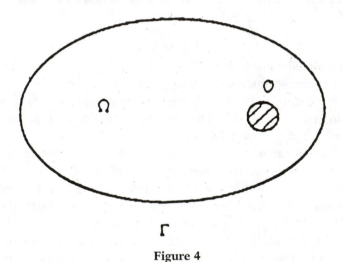

Figure 4

where $v(x, t)$ is the control function applied in a small region $\sigma \subset \Omega$. We assume that (by proper scaling)

(2.2) $y = 0$ on the boundary Γ of Ω, and $t > 0$,

(2.3) $y(x,0) = 0$.

Let $y^1(x)$ be a given temperature we wish to get as close as possible to at time horizon T. Let B be the unit ball of, say, the space of square integrable functions on Ω. We wish to find v in such a way that

(2.4) $y(x, T) \in y' + \beta B$

where β is given (and small).

Moreover, we want to achieve (2.4) in *the cheapest way*, i.e. we want to find

(2.5) $\inf. \iint_{\sigma \times (0,T)} v^2 \, dx \, dt,$

for all v's such that (2.4) holds true.

The result of (2.5) is a function of λ, say $\varphi(\lambda)$.

For $\lambda > 0$, it can be shown that $\varphi(\lambda)$ decreases as λ decreases. It is cheaper to control system (2.1) where λ is smaller, i.e. when (2.4) becomes (loosely speaking) "less stable".

Example 2.4.

A very stable and "viscous" system can be *non*controllable. Let us consider the state equation given by

(2.6) $\dfrac{\partial y}{\partial \tau} - \Delta y + y^3 = v\chi_\sigma$

with (2.2) and (2.3) unchanged. We have "only" replaced λy by y^3 in (2.1).

But it completely changes the situation. In the present case, $y(x,T)$ describes a *"small"* subset of square integrable functions in Ω. This fact is due to the very "viscous" term added to classical "energy estimates" contributed by y^3. One indeed adds to the energy the term

$$\int_\Omega y^4 dx.$$

Example 2.5.

Along somewhat similar lines, some models arising in multiphase flows seem to be noncontrollable under any "reasonable" control action (cf. [8]). In this context it seems plausible to think that interfaces between ice caps and liquid ocean could be one of the most difficult subsystems on Planet Earth to "control"[9].

---

[8] I. Díaz, *Positive and Negative results in exact controllability for a general class of non linear parabolic equations.* C.R.A.S. Paris 1992.

[9] J.L. Lions, *El Planeta Tierra*, Instituto de España. Espasa Calpe, Madrid 1990.

Remark 2.2.

A system not "reasonably" controllable could undergo *irreversible changes*. One more reason, this time of a mathematical origin, for monitoring the ice caps as closely as possible. Very interesting results will no doubt follow from a close analysis of data given by the ERS1 Satellite (ESA satellite launched by Ariane 4 in July 1991)[10].

Example 2.6.

Let us now consider the model of perfect viscous fluid flows given by Navier-Stokes equations. We conjecture that this system is, under reasonable controls, approximately controllable. Numerical approximations, now under way, point towards a positive answer.

Let us mention here a more technical conjecture, which could be numerically verified.

We consider simultaneously the Stokes and the Navier-Stokes systems, given respectively by

(2.7)
$$\frac{\partial y}{\partial \tau} - \Delta y = v \chi_\sigma - \nabla p, \quad v = \{v^1, v^2, v^3\}$$

$$\text{div } y = 0, \quad \text{in } \Omega \times (0,T) \quad (\Omega \subset \mathcal{R}^3)$$

$$y = 0 \quad on \quad \Gamma \times (0,T),$$

$$y(x,0) = 0$$

and

(2.8)
$$\frac{\partial y}{\partial \tau} - (z\nabla)z - \Delta z = v \chi_\sigma - \nabla q,$$

$$\text{div } z = 0$$

$$z = 0 \quad on \quad \Gamma \times (0,T),$$

$$z(x,0) = 0$$

---

[10] J.K. RIDLEY, S. LAXON, C.G. RAPLEY, D. MANTRIPP, *Topography of Antarctic Ice Sheet mapped with the ERSI Radar Altimeter. Earth Observation Quarterly*, ESA pub. May-June, 14-15 (1992).

It is proven[11] that, even with one of the v components identically 0, (2.7) is approximately controllable. If B denotes the unit ball of square integrable vectors on Ω, one can therefore define (as in (2.5))

$$(2.9) \qquad \inf. \iint_{\sigma \times (0,T)} v^2 dx\, dt = M_1, \qquad y(x,T) \in y^1 + \beta B,$$

where y^1 is given.

It is *conjectured* that we still have approximated controllability for (2.8), so that we can define

$$(2.10) \qquad \inf. \iint_{\sigma \times (0,T)} v^2 dx\, dt = M_2, \qquad z(x,T) \in y^1 + \beta B,$$

We conjecture that

$$(2.11) \qquad M_2 \le M_1.$$

If true this would be a further indication that "turbulence" *helps*, at least from a theoretical view point, as far as "control" is concerned. Indeed, it is the non linear term $(z\nabla)z$ which contributes, by "mixing the scales", to turbulence.

This remark leads us to the question of high sensitivity to initial conditions.

3. Control of Chaos

High sensitivity of the weather to initial data is well known. Let us recall a classical example.

Example 3.1. - President's Day Snowstorm.

A very strong snowstorm struck the U.S. mid-Atlantic states in February 1979. This legendary storm, often referred to as the "President's Day Snowstorm" was notable not only for its severity but also because of the very poor forecast[12]. Why? Investigators discovered that the predicted evolution of the President's Day snowstorm in the western Atlantic was extremely

---

[11] J.L. LIONS, *Exact Controllability for Distributed Systems. Some trends and some problems*, *Applied and Industrial Mathematics*, ed. R. Spigler, Kluver, 1991.

[12] R. DALEY, *Atmospheric Data Analysis*, Cambridge Atmospheric and Space Science Series, 1991.

sensitive to small errors in the initial analysis in the Northwestern Pacific four days earlier.

Example 3.2. - Lorenz equations.

These equations, classical by now, are the "prototype" of chaotic situations. It is a simple looking system of 3 ordinary differential equations obtained by projection on the space generated by 3 basis functions (the so called Galerkin's method) of the classical model of *thermo-hydrodynamics* (which couples by convection and thermal expansion the Navier Stokes and the heat equations).

These equations are taken as examples of the very high sensitivity of the solution to small initial errors, in the framework of weather prediction.

Remark 3.1.

It should be clear that for very complex systems such as the planet Earth system, pointwise controllability is *impossible*, both from a theoretical and from a practical viewpoint. What we are interested in are *averages* in space and in time (for a season, say). And then questions of controllability can make sense.

Remark 3.2.

Indeed very interesting results have been reported recently on the *control of chaos*[13] and on the synchronization of two chaotic systems.

Remark 3.3.

One can state similar conjectures to those of Remark 2 above for the coupled system of thermo-hydrodynamics. Numerical methods are in progress. How far can one go along these lines? Is it possible to "control" Bernard Rolls?

Remark 3.4.

Along the lines of Remark 3.3., it is worthwhile to mention here that many results can be achieved concerning the control of thermo-hydrodynamical systems by using highly vibrating boundaries.

---

[13] E. OTT, C. GREBOGI, J.A. YORKE, *Controlling Chaos*, Phys. Rev. Letters, 64 (11) 1196-1199 (1990). W.L. DITTO, S.N. RAUSEO, M.L. SPANO, *Experimental Control of Chaos*, Phys. Rev. Letters, 65 (26) 3211-3214 (1990).

The preceding Remark points towards the role of periodic control, which leads to a highly oscillating boundary control exerted by the Sun for a time horizon of several years.

All these remarks, which are clearly of an introductory nature, seem to bring some elements of justification to the remark made by J. von Neumann "The climate is simpler to control than to predict".

We can *summarize* where we stand - emphasizing once more that everything which has been presented so far is of a very preliminary and conjectural state of affairs.

i) The "natural" feedbacks, if thought of as natural control functions, stabilize the planet Earth system. The turbulent and chaotic sub-systems of the global system make the role of these controls "easier" and *more efficient*, rather than less efficient.

The irreversible (physical) changes could rather be related to changes (if any) in the most "viscous" parts of the system, such as the ice caps.

ii) But of course the argument of (i) can — and *should* — be reversed. If the system can be "easily" controlled, it could mean it is going to be, at a time scale not at all easy to estimate[14], very sensitive to human actions.

We could act significantly on the system, and actually, an "inadvertent global experiment"[15] has begun.

How to proceed?

4. Criteria and complexity.

For applying control techniques and methods, one needs three basic elements:

i) *State equations*, depending on the time horizon we are interested in, and on the geographical region we want to study with particular care, and also on the "subsystem" which is our main interest, etc.

ii) Control variables. We have indicated some of them. Without any attempt at being exhaustive, one could add:

— Volcanoes, as "natural" and "stochastic" controls,

— Economic incentives,

— Demography (a particularly delicate "control").

---

[14] B. DATKO, J. LAGNESE, M.P. POLIS, *An example of the effect of time delays in boundary feedback stabilization of wave equations.* Siam J. Control Optim. 24 152-156 (1986).

[15] V. RAMANATHAN, *The Green house theory of climatic change: a test by an inadvertent Global Experiment. Science* 240 293-299 (1988).

"Active" controls have been mentioned already by J. von Neumann. More recently the sinking of large masses of steel in cold oceans has been presented.

iii) Criteria. — i.e. functionals we want to "optimize" so that the system behaves "according to our wishes".

The choice of criteria is a very fundamental topic, which is beyond the scope of the present lecture. We are confining ourselves to the following remarks of an introductory nature.

Remark 4.1.

For any problem, there are a large number of reasonable or natural criteria one would like to optimize. Simultaneous optimization *all of them* is impossible: no solution exists in general. This is a classical situation in Economics. One replaces "optimization" by the search for *equilibria* of some sort, such as Pareto equilibria.

Remark 4.2.

The number of criteria adds to the *"complexity"*.

Faced with such complexity when one considers the Planet Earth System or any of its subsystems, one could be tempted to do nothing and to proceed with "business as usual", a very dangerous procedure if what is summarized in the conclusion of section 3 is true.

We would rather advocate action decided on a "no regret basis"[16]: we are looking for controls to improve situations and (hopefully) not to make things worse than "business as usual".

In any case, if feedback control can stabilize a system whose many components are of an unstable nature, that stabilization could explain what can be considered as stability of climate and, at the same time, it indicates that great attention should be paid to see that this "stability" is maintained.

But the complexity of the system, the intricacies of feedbacks, the variety of possible control variables, the number of criteria where conflicts of interest may appear, even for apparently "local" problems, can make any kind of coordinated decision impossible in the absence of a reference to philosophical and moral values.

---

[16] J.L. LIONS, *Contrôles à moindres regrets*, C.R.A.S. Paris 1992.

DAY THREE
29 OCTOBER 1992

BIOLOGICAL COMPLEXITY

PETER H. RAVEN

Botanical Garden, St. Louis, Mo., USA

Biological complexity, as several speakers today have noted, is enormous, difficult to understand, and very difficult to reduce to the kinds of formulas that were so beautifully discussed in earlier papers on physics, cosmology, and astronomy. We view biological diversity in a variety of different ways. To give you one example, we view it though the eyes of taxonomy or classification. It is interesting that, before there was a formal classification recorded in books, we had throughout the world a series of systems of classification, which were local in distribution and which would now be called folk taxonomies[1]. If we look at the ways that the Eskimos in the north, the indigenous peoples in the central Amazon, or the rural peasants in southern France — in short, any people who do not depend primarily on written language for describing the world of nature — we find a series of regularities in the way that they deal with the world. Those taxonomies, for example, always have to deal with a finite number of specific identified items, usually in the order of several hundred. They are not deep, in the sense of hierarchies; they do not have families or orders or phyla in them; and every speaker of the language — everyone who is familiar with that system — will recognize all of the items, because every item has to be recognized in order for there to be communication, so that there is no need for a hierarchy and there can be only a finite number of items.

This is regardless, as I have implied, of the biological richness of the area concerned. It also bears mentioning, briefly and in passing, that the human

---

[1] P.H. RAVEN, B. BERLIN & D.E. BREEDLOVE, *The origins of taxonomy.* Science 174, 1210-1213 (197).

mind itself, and our capacity for dealing with the outside world, must have a lot to do with our original ability to, and necessity for dealing with biological items — because, from the very beginning, human survival has depended on that ability, and our minds must be continued somewhat by the kinds of classification that were inherent in doing that. As human progress moved forward into written manuscripts, people began to try to codify more and more about the plants and animals around them, and of course we have the famous early efforts of men such as Theophrastus, Dioscorides, and Aristotle, who tried to deal with the biological universe. The system really began to move forward, however, with the introduction of movable type and, therefore, of widely distributed books in Europe, which eventually led to the age of the encyclopedists when, probably towards the close of the 17th century, people really began to take a serious interest in classifying biological reality in its entirety. In those systems, a deeper and deeper hierarchical classification came to be and the reason for a hierarchical classification, that is, grouping of individual kinds of organisms into genera or families or orders or phyla, was that the characteristics, the conditions under which the folk or verbal taxonomies could operate were no longer met. There were so many organisms being classified that no one could hold them all in mind. It became useful to group those organisms into categories of ascending complexity or inclusiveness, and the characteristics of those increasingly inclusive taxa, or kinds of organisms, came to be very important as a way of communicating information about them. Even at the time of Carl Linnaeus, in the middle of the 18th century, the biological hierarchy, or the hierarchical classification, as we now know it, was not yet completed, and several levels have been added over the past 200 years.

The obvious way to complete this brief exposition of this particular way of looking at the diversity of life on earth is to point out that, with the invention of computers, one is freed, in a logical sense, from the necessity of a hierarchical classification, although actually people have maintained it. With computers and relational databases, it is perfectly possible, in principle, to retain all of the original information upon which the classification and the items being discussed are logically based and, in an operational sense, to retrieve all of that information, when needed, and still to use categories of increasing inclusiveness, as they may be needed.

Now, another way of looking at the complexity of life on earth is to view it in an evolutionary perspective. Our planet is something like four and a half billion years old, with the oldest rocks that remain being something like 3.9

billion years old. Fossils in South Africa and Australia go back to about 3.5 billion years ago, and that is when we know that there were prokaryotic organisms and a simple level of organization in existence, including the cyanobacteria, which were engaging in photosynthesis more than three billion years ago. That process of photosynthesis gradually led to the accumulation of sufficient oxygen to make the characteristics of the Earth's atmosphere more like what they are now, with about 20 percent oxygen. Very importantly for an environmental consideration of the earth, that increase in oxygen concentration in turn led inevitably to the production of an ozone layer in the stratosphere sufficiently thick to protect the organisms on Earth from the ultraviolet B radiation that continually bombards the Earth from space, and which is intrinsically extraordinarily damaging to biological molecules. Specific examples of that damage are skin cancer and melanomas, and other types of injuries caused to biological molecules are well known. At least two billion years after the first evolution of life on Earth, one detects the first eukaryotic cells; and eukaryotic cells, larger in volume, are also the product of serial symbiosis, the cooperative or mutualistic organization of prokaryotic cells within one another.

Very early in the evolution of eukaryotic cells, one had, through the organization of a particular group of bacteria into those cells, the evolution of the almost universal organelles called mitochondria; and by a subsequent incorporation, probably of cyanobacteria into some of those organisms, one also had the origin of chloroplast, the photosynthetic organelles that are characteristic of the relatively few eukaryotes that are capable of photosynthesis. For the first 750 million years of the evolution of eukaryotes, although the cells sometimes adhered, they were all unicellular, and it was not until something like 730 million years ago that one first saw the evolution of multicellular eukaryotes, more characteristic of the kinds of organisms that we have now. The emergence of the first organisms onto land occurred something like 430 million years ago. This is an important point: during 90 percent of the history of Earth there were no organisms on land whatsoever. Starting only about 430 million years ago, the ancestors of the modern fungi, then eventually the ancestors of the arthropods, and of the terrestrial vertebrates, tetrapods, came onto land and began to form communities of increasing complexity. By 300 million years ago, which is, in geological terms, just yesterday, forest had appeared on the land, and you had development of the kinds of complex communities that exist now, and whose integrity and diversity are so threatened by the human population at the present time. Starting about 300 million years ago, one began to see these complex co-evolved communities whose appearance signalled a profound change in the relevant physical characteristics of the Earth, which, depending on how mystical your thoughts are, may be regarded as the creation of a kind of a Gaia Earth, functioning as a physical unit.

The last of several major extinction events on Earth was at the end of the Cretaceous Period, approximately 65 million years ago. The ancestors of the mammal were already in existence, but our ancestors at that time were very small, probably mostly arboreal creatures, bearing only a generalized resemblance to the primates and the other mammals that exist at the present time. Following the great extinction at the end of the Cretaceous Period, which most scientists now believe to be coupled with the landing of a giant asteroid on Earth, perhaps, as it now seems, off Yucatán in Mexico, we had an increasing evolution of numbers of species on Earth, and an increasing evolution of the complexity of biological communities. They have been going on and on and on, forming more and more species of greater and greater complexity, for the past 65 million years.

With the opening of the great passage between southern South America and Antarctica about 27 million years ago, the conditions were set for the development of the circum-Antarctic current that connected the southern oceans into a single system and which seems, in turn, to have led to the formation of continental glaciation in Antarctica, starting about 16 million years ago. This coincides with a period of worldwide increased aridity, which began at about that time and which greatly diversified and changed the communities of organisms all over the Earth. As the temperature gradient from the Equator to the Poles gradually became more extreme, tropical lowland forests became as extreme a habitat as they are. They must in no sense be thought of as an aboriginal kind of habitat from which other kinds of life on earth emerged. The generally mild climate habitats that occupied much of the Earth before the middle of the Miocene, about 15 million years ago, gave rise gradually to the savannahs. These changes and the increasing aridity accentuated the extent of the deserts, and gave rise to the Mediterranean-type vegetation, the Arctic tundra/taiga vegetation, and other kinds of more specialized communities that we see at the present day, all of which include components of these more moderate communities, occurring in equable climates, from which they evolved.

Our first ancestors in the genus *Australopithecus* appeared on Earth four to five million years ago, the exact time being uncertain. The genus *Homo* appeared about two million years ago, and fossil records of individuals similar to *Homo sapiens*, people of modern aspect, appeared about 500,000 years ago, and the rest of the course of evolution is well-known to you. For about 10,000 years, starting in three diverse centers, people have been cultivating plants and animals for their agricultural purposes. So what began about 10,000 years ago as a few million people scattered over the face of the Earth, has grown into the five and a half billion people explosively gathering speed at the present time, and we will return to that problem in a minute. There were about 130 million people at the time of Christ, one billion people in about 1830, two billion in about 1930, two and a half billion in 1950, and an additional three billion have been added to the world population since 1950, so I will say, for the first time, but not for the last time, that anyone who thinks that the world today is like any world that existed in the past, in terms of the pressure that human beings exert on the capability of the world to support itself, is simply in error[2].

Let us turn now to a more general discussion of some of the aspects of biodiversity. How many species of organisms are there on Earth? Before we begin making general and very theoretical statements about life on Earth, it would seem to be logical to begin to try to understand something about the dimension of that life. The facts are that we understand very little. We do know that there are about 20,000 species of butterflies, diurnal Lepidoptera, on Earth and, we understand their patterns of distribution very well. We do know that there are about 45,000 species of vertebrates on Earth, of which we know most, with the rather major exception of the fresh water fish faunas of Latin America[3,4]. We know that there are about 250,000 kinds of plants on Earth, counting only vascular plants and bryophytes as plants. But we do not know very much at all about other groups of organisms. In fact, our state of knowledge is abysmal.

There are about 1.4 million kinds of organisms that have been given names on Earth since the time when Linneaus began with plants and animals in the 1750s[5]. There are, as I said, about 250,000 species of higher

[2] ARROYO, M.T.K., P.H. RAVEN & J. SARUKHAN, *Biodiversity. In: An Agenda of Science for Environment and Development into the 21st Century* (J.C.I. Dooge et al., eds.) Cambridge: Cambridge University Press 1992, pp. 205-219.

[3] LOWE-MCCONNELL, R.H., *Ecological Studies in Tropical Fish Communities.* Cambridge: Cambridge University Press 1987.

[4] WOODWELL, G.M. (ed.). *The Earth in Transition. Patterns and Processes of Biotic Impoverishment.* Cambridge: Cambridge University Press 1990.

[5] E.O. WILSON, *The current state of biological diversity.* In: *Biodiversity* (E.O. WILSON, ed.). Washington D.C.: National Academy Press 1988, pp. 3-18.

plants; about 45,000 species of vertebrates; and about 750,000 named species of insects. But if you go on to think about how many there really are and how the ecosystems of the world are put together, you immediately are plunged into severe difficulties. We used to think there might be about three or four million kinds of organisms on Earth until a scientist named Terry Erwin, who works in the Natural History Museum at the Smithsonian Institution in Washington, began doing experiments about ten years ago that included fogging insecticide up into the canopies of tropical trees and then getting comprehensive samples of all the kinds of insects found in particular kinds of trees[6]. He particularly studied arboreal caryatid beetles, very specialized organisms, and attempted to estimate how many were distinctive on each kind of tree; what kind of overlap there would be from one kind of tree to another; what kind of overlap there would be 100 kilometers away, and so forth. He came up with an estimate that there might be 30 million kinds of insects living in the canopy of Latin American trees alone.

Erwin's calculations, like some early estimates of the diversity of marine organisms[7], seem to have been faulty for various technical, mainly statistical reasons. However, the current conservative estimate that is most reasonable, as nearly as my colleagues and I could work it out, is that there are eight to ten million kinds of organisms on earth[8] [9]; this is, as I say, a good conservative estimate, which needs to be qualified in several ways. First of all, there is no way to include bacteria in those totals, although bacteria are sometimes included in such estimates. Bacteria are classified by their metabolic characteristics, and there are only 3000 species of bacteria that have been properly classified and named. On the other hand, using modern molecular methods like DNA probes to determine relatively different sequences in different bacteria, one gram of soil under a Norwegian beech forest has been estimated recently to contain 5000 different kinds of bacteria. Since there are only 3000 bacteria named in the whole world, and since it is unlikely that a Norwegian beech forest in the richest possible habitat for bacteria, it seems almost certain that there are many, many, many more kinds of bacteria.

---

[6] ERWIN, T.L. *Beetles and other insects of tropical forest canopies at Manaus, Brazil, sampled by insecticidal fogging*. In: *Tropical Rain Forest: Ecology and Management* (S.L. Sutton et al., eds.). Edinburgh: Blackwell Scientific Publications 1983, pp. 59-75.

[7] GESAMP, *The State of the Marine Environment, UNEP Regional Seas Reports and Studies No. 115*. Nairobi: United Nations Environment Programme 1990.

[8] MAY R.M., *How many species are there on Earth?* Science 241, 1441-1449 (1988).

[9] GASTON, K.J., *The magnitude of global insect species richness*. Conservation Biology 5, 283-296 (1991).

An even more serious problem than getting a vague idea of the number of bacteria on Earth is that we do know how to define, even approximately, the kinds of units that we would want to call species. We can define species of bacteria in a functional sense, in a sense of what they do in nature, but we cannot really define them in terms of species that compare to species of eukaryotic organisms. For viruses, of course, the case in even clearer, viruses being essentially pieces of the genomes of prokaryotic or eukaryotic organisms that have broken loose in some way and are functioning in other cells, by their very origin and mode of operation. You not only have the difficulties involved in deciding how many you want to recognize in any but the most functional way, but it becomes quite impossible to make any estimates. Therefore, when I say eight to ten million, I am talking about only eukaryotic organisms. For many groups of eukaryotes, some of considerable economic importance, and I mention particularly mites, nematodes, and fungi, fewer than 50,000 species have actually been described and named, but general estimates are that there may be about a million species in each of these groups (Wilson *op. cit.*)[10].

The whole point of this diversion has been to indicate that we really are abysmally ignorant of life on Earth at the very time when are destroying it with unprecedented speed[11]. We really do not have the basic tools that we need for managing the physical biosphere in any but the crudest and roughest way. If we know so little about individual kinds of organisms, how can we understand the linkages between organisms in natural communities, which are basically and ultimately responsible for the Earth's ability to capture a certain amount of sunlight, which bombards the Earth continually, and expend that sunlight in a regulated way? We do know that leaf-cutter ants in Central America gather leaves and grow fungi on them, which they use as their primary food. We know that a male euglossine bee takes a scent from the petals of a particular orchid, and then uses it to mark its mating site in the jungle; and that in turn that bee carries a pollinium with pollen of that orchid as it goes from flower to flower. We have come to understand, during the last decade, that certain kinds of fruit bats are absolutely indispensable in the spread and re-establishment of forests in West Africa,

---

[10] HAWKSWORTH, D.L. *The fungal dimension of biodiversity: magnitude, significance, and conservation.* Mycological Research 95, 641-655 (1991).

[11] EHRLICH, P.R. & E.O. WILSON, *Biodiversity studies: science and policy.* Science 253, 758-762 (1991).

and if you kill off the bats, there is no way to re-establish the forests. But those are just vignettes that show how profoundly ignorant we are[12].

We operate on a kind of blind faith that the ecosystems and communities of the world are really infinitely interchangable or substitutable[13]. But while we are assuming that, we have lost an estimated 20 percent of the topsoil on the world's agricultural lands in the last 40 years and are losing it right now at a rate of 24 billion tons per year [14] [15]. In other words, the world each year is losing, from its agricultural lands, the amount of top soil equal to all the topsoil on all the wheatlands of Australia, and we are really not doing anything about it. We do not know how to manage or manipulate communities nor do we acknowledge our enormous dependence on organisms, which are the only sustainable elements that we have in the world system for our survival[16]. For those of us who live and practise agriculture in the temperate parts of the world, we have to acknowledge that the genetic diversity that underlies our ability to use those crops lies mostly in foreign countries and mainly in developing countries. For example, about twelve years ago on a hillside in Mexico, botanists discovered a new species of corn[17], which is a perennial and occupies the area about the size of a football field, which easily could have become extinct without ever being discovered , and yet which has in it genes that make the bearer resistant to seven of the nine major viral diseases that lower the yield of corn as it is grown throughout the world. Of course, you know that maize, or corn, together with rice and wheat, provide an estimated 60 percent of all the calories consumed by human beings, directly or indirectly[18], so this is not a trivial discovery, and the fact that an organism like this could have been lost easily, without ever having been seen, is not a trivial observation. We do not, as a global community, invest in the development of tropical crops. We do

---

[12] WILSON, E.O., (ed.) *Biodiversity*. Washington, D.C.: National Academy Press 1988a.

[13] SOLBRIG, O.T., *From Genes to Ecosystems: A Research Agenda for Biodiversity*. Report of a IUBS-SCOPE-UNESCO Workshop. Paris: International Union of Biological Sciences 1991.

[14] BROWN, L.R. *et al.* (9 additional authors). *State of the World* 1990. New York: W.W. Norton and Company 1990.

[15] BROWN, L.R. (12 additional authors). *State of the World* 1992. New York: W.W. Norton and Company 1992.

[16] DI CASTRI, F & T. YOUNES, *Ecosystem function of biological diversity*. Biology International, Special Issue 22. Paris: IUBS 1990.

[17] ILTIS, H.H., J.F. DOEBLEY, R. GUZMAN, M.B. PAZY, *Zea diploperennis* (Gramineae): *a new teosinte from Mexico*. Science 203, 186-188 (1979).

[18] *Prescott-Allen*, R. & C. *Prescott-Allen, How many plants feed the world?* Conservation Biology 4, 365-374 (1990).

not try to get crops in places that will feed people in the tropics or subtropics. Worldwide investment in crop agriculture is almost exclusively done in temperate, industrialized countries, and that is, of course, where we are paying for these enormous surpluses that have us fighting continually with one another as we try to work out rational global trading situations. But, in the tropics, as I will explain in a few minutes, where people are desperately in need of the development of new crops for domestic consumption, for export, and for economic support, very, very little is being invested on a scale adequate to come up with new commodities that we really could use and to create markets for them[19].

Of course, the use of plants as medicines is something that I need refer to only briefly[20]. Indigenous people in the Amazonian region of South America still hunt with arrows, darts really, dipped in an extract from a plant; this extract in known to us as curare. In the hands of modern medicine, of course, curare and medicines related to curare are used in all operations involving open heart surgery, or any other surgery requiring thoracic relaxation, but we would not even have known that it existed if we had not learned about it from the indigenous people who used, and still use, it for hunting. Every single one of the 20 top-selling pharmaceutical drugs worldwide is either taken from a natural product or modelled on one. The only drug that is effective against all forms of malaria in the world at the present time, especially the resistant strains of plasmodium that have appeared recently, is an extract from a Chinese herb called *Artemisia annua*, and the efficacious molecule in that plant is utterly unpredictable from any of the other molecules that have ever been used to treat malaria. One final example is *Catharanthus roseus*, a native of Madagascar, which is the source of vinblazstine and vincristine, drugs whose use has grown into a $200 million a year industry for Eli Lilly, put on the market in 1971, and which have raised the chance of survival from childhood leukemia from 1 in 20 to about 9 in 10, when used in combination with other drugs.

Plants are essentially an endless store of commodities that we might use to support ourselves, and when you then go on to genetic engineering, which is the ability to transfer useful genetic material from one organism to another, thereby improving the characteristics of the recipient, the potential

---

[19] SAGASTI, F. *Co-operation in a fractured global order.* New Scientist 127, 18 (1990).
[20] OLDFIELD, M.L., *The Values of Conserving Genetic Resources.* Washington, D.C.: US Department of the Interior, National Park Service 1984.

uses of plants multiplies further[21]. I remind you that the first example of successful genetic engineering was performed only in 1974. When you consider the inherent potential of genetic engineering, then you realize that each individual organism being lost is not like losing an article from the shelves of a grocery store; it is like losing a whole bag of useful commodities. Being just at the beginning of our genome sequencing projects, we do not even know how many distinctive genes organisms have, how they differ one from the other, nor how they function; and losing those organisms at the present time is certainly losing a great deal more, but of an unknown quantity, than simply losing them for their own sake[22].

Now, unfortunately, there are a number of factors by virtue of which we are losing this very, very poorly known biological diversity, and I repeat that we have recognized and named only about 15 percent of the kinds of organisms on Earth. We are virtually totally ignorant of the connections between them, the ways in which they form efficient communities the degree of substitutability between them, or indeed of anything that really can be characterized as sustainable use of the Earth. The fundamental factor that is driving to biological extinction, which is also the fundamental factor driving global warming, and the fundamental factor destroying the ozone layer, and the fundamental factor altering regional climates , and the fundamental factor creating a human demographic situation in which the HIV virus could grow into a worldwide pandemic situation, is human population growth. From two and a half billion people in 1950, the population has increased by an additional three billion in the last 42 years.

Almost every developing country believes that it is desirable to slow down the growth of its population, and most have implemented policies that reflect that belief, but because the populations of developing countries characteristically include 35 to 45 percent of people less than 15 years of age, and although an increasing proportion of those people choose to have smaller families, there are so many people having families at all that the population continues to grow rapidly. In fact, it continues to grow so rapidly that we are adding approximately 100 million people a year at present and should add about 100 million people a year for the next 25 to 30 years, even if we have sustained attention to the limitation of populations. I need not

---

[21] COLWELL, R.R. *The potential of biotechnology for developed and developing countries.* MIRCEN Journal 2,5-17 (1986).

[22] EISNER, T. *Prospecting for nature's chemical riches.* Issues in Science and Technology 6, 31-34 (1990).

remind this audience that the idea of allowing people to make responsible choices about the number of children that they have, and of making information available to them within the limits of what they consider morally acceptable, is a perfectly acceptable doctrine to the Roman Catholic Church — that is, choosing the size of one's family — and that is precisely what people are doing over much of the world. That being the case, though, there are two important observations that I need to add. One is that, if the global population is eventually to level out, and the earliest it could do so would be about 100 years from now, these next few decades are not some kind of reprieve, during which we can gain the ability to take action. These next few decades are, in fact, the most destructive decades that we will ever have to face. The population is growing rapidly now, we have fewer ways of dealing with it, we do not understand it, our social systems are disordered, and this is the very time when the biological fabric of our common home, the planet Earth, is being torn to shreds. The only time that we can act effectively is now.

The second point I would make, if there is anyone in the audience who does not believe that population is a problem, is that human beings right now, at a level of 5,5 billion, are estimated to be consuming, diverting, or wasting forty percent — 40 percent — of the total photosynthetic productivity on land[23]. Since the most optimistic assumptions of human population stability project a leveling-off at about nine billion at least, one must ask can we ever improve the lot of the human race in situations like that, when we are already appropriating, for our one species, one out of the ten million on this planet, forty percent of everything that the Earth produces?

And what kind of life do we get for most people on Earth while appropriating this 40 percent? Our human society is as unjust a system as could possibly be. The distribution of wealth, the distribution of talent[24] and the distribution of energy throughout the world to manage the world, is about as irrational as it could conceivably be, if we were trying to manage our planet so that there might be some kind of respectable future. Of the 4.3 million people who live in developing countries, 1.5 billion of them live in what the World Bank defines as absolute poverty. Absolute poverty is a

---

[23] VITOUSEK, P.M., P.R. EHRLICH, A.H. EHRLICH, & P.M. MATSON, *Human appropriation of the products of photosynthesis*. BioScience 36, 368-373 (1986).

[24] SALAM, A. *Science, Technology and Science Education in the Development of the South.* Trieste, Italy: The Third World Academy of Sciences 1991.

condition in which an individual cannot expect reliably to find food, shelter, or clothing on a daily basis. Half of those people, or one out of ten people in the World, are suffering from malnutrition, in other words, receiving less than 80 percent of the UN-recommended minimum calorie intake. Their bodies are literally wasting away, and when they are children, their brains cannot develop properly, because they are not taking in enough calories.

One and a half billion people in the world do not have access to dependable supplies of fresh water. That means, put very simply, that the women and children of those families spend their entire lives walking back and forth to get supplies of fresh water. They have absolutely no hope of being incorporated into their societies or exercising any kind of a reasonable input to their societies, but instead are condemned to a life of misery just going and getting water. That's the kind of a world we have now. The 1.2 billion of us who live in industrial countries use about 80 percent of the industrial energy in the world, have about 85 percent of the money in the world, as measured by gross national products; consume anywhere between 75 and 95 percent of anything that you can measure; and the next time you think about population, think about the other side of the equation, which is gluttonous overconsumption in the industrial world, far beyond anything that we could conceivably need. The relationship between these two factors is what is critical, and I can illustrate it very simply. In the face of the relationship that I have just outlined for you, if the 4.3 billion people in the developing world disappeared from the face of the globe tomorrow — disappeared — we would not be even remotely near managing the world in a sustainable way. The 1.2 billion of us who are left are already far from using the world in a sustainable way, never mind what the expectations might be for people in the developing world. That is basically why the Brundtland report is, to me, such a sad distraction from the real business of the world, with its vaguely implied pretence that everybody in the world is going to come up to the standard of the industrial world by some kind of mysterious process.

Ladies and gentlemen, those of you who live in the industrial world, we are going to have to seriously limit our consumption, intelligently, in ways in which science, technology, and engineering must play a key role, because our planet cannot give any more than it is giving. To pretend that we can just go on with our lives and that somehow everything else will be okay is, I think, just to ignore the real problems. What we need to do is think seriously about implementing agricultural systems, like agroforestry, that will work in various part of the developing world, as well as the effective production of crops in sufficient amounts to feed people where thay really are. What we need is to find ways that will respect those precious soils of the tropics — often only a few centimeters thick — and allow cultivation to go on there, either allow them to be used, or simply leave them alone. This is not a problem for the Brazilians or the Indonesians; this is a problem for every single one of us. The typical sequence of logging followed by burning for shifting cultivation does not work once the population density gets to a certain level, which it has reached in most developing countries. Shifting cultivation works fine as a temporary enrichment of the soil, as long as there are few enough people that the forests can recover between uses. But with the one and a half billion people who depend on firewood as their source of fuel, and with population levels reaching the levels that they have, shifting cultivation simply no longer works. Every single one of us has a common investment in trying to develop and encourage sustainable systems whereby people can lead lives of dignity in places like the tropics, and until we face up to that fact, the human race will simply not be able to accomplish what it would otherwise, and we will not be able to manage this planet sustainably, regardless of how comfortable any one of us might be individually.

Within the last 40 years, we have seen the loss of about half of all the tropical forests[25], a potentially inexhaustible storehouse of riches for the human race. What is happening, in essence, is a denial of the charge that the Lord gave to Noah thousands of years ago to save the organisms on Earth for the human future. Rather than saving them, or even learning about them, we are simply squandering them as a result of our lack of attention to the problem and our simple lack of charity towards one another. Despite all the beautiful images of our Earth that we have had from space, we are failing that, in the final analysis, we inhabit one single planet, that is all we have, and we have to cooperate in the management of that planet. Why we continue to fall into fanciful dreams of economic development, given that realization, I have no idea.

Most tropical organisms are very widely scattered in their communities, and when the tropical forest is cut into small patches, as you can imagine, you are apt to catch very few of those individuals and consequently such activities put those species immediately at the risk of extinction. One or two individuals will not perpetuate a species. In fact, a tenfold decrease in a given area — whether a single forest or the entire extent of existing tropical

[25] MYERS, N., *Tropical forests and their species. Going, going...?* In: *Biodiversity* (E.O. Wilson, ed.). Washington, D.C.: National Academy Press 1988. Pp. 28-35.

rainforest, for example — will, in general, lead to a 50 percent reduction in the number of species in the remaining area[26]. With the inevitable growth in human population that will occur in the next 30 years, it is also inevitable that a very large percentage of the species on Earth, possibly 20 percent, meaning 50,000 of the quarter of a million species of plants, for example, is likely to become extinct during that period of time[27, 28].

Given the fact that 50,000 to 60,000 species of plants are already known to be economically valuable, it is basically a crime against ourselves that the human race has not found a way to enter into a scheme that would lead to their preservation. It is even more of a crime that we have not used the United Nations, following the Earth Summit meetings in Rio, as a mechanism for drawing us together, putting some executive, legislative, and juridical powers into play, whereby we could negotiate with one another to guard our common ecological future. It seems that we are all just too selfish. Futurists do not agree on much, but they do agree on one thing — they agree that the future will not be like the past.

One final example to illustrate just one single place, on the island of Madagascar, an island about the size of California some 400 kilometers off the east coast of Africa, between 20° and 30° south latitude, there about two to three percent of the world's species of organisms, and most of those live only there, including, for example, all of the lemurs, which comprise more than 40 of the 230-odd species of primates, our closest relatives. Every single one of those species of lemurs is a threatened or endangered species at present and they live, of course, amidst some of the most fantastic vegetation anywhere on Earth, and some of it still in reasonably good condition. Maps generated by remote satellite sensing show an extremely rapid decrease in the forests of Madagascar between the early 1950's and the mid 1980's, coinciding with the explosive growth of the human populations, and those forests continue to be destroyed on at least the same scale as we speak[29]. The reason? Population growth, the cultivation of upland rice, the gradual

---

[26] MacArthur, R.M. & E.O. Wilson. *The Theory of Island Biogeography*. Monographs in Population Biology. Princeton: Princeton University Press 1967.

[27] Raven, P.H. *Biological resources and global stability*. In: *Evolution and Coadaption in Biotic Communities*. (S. Kawano et al., eds.). Tokyo: University of Tokyo Press 1987. Pp. 3-27.

[28] Reid, W.V. *How many species will there be?* In: *Tropical Deforestation and Extinction of Species*. (T. Whitmore & J. Sayer, eds.). *The IUCN Forest Conservation Programme*. London: Chapman & Hall 1992. Pp. 55-74.

[29] World Resources Institute, *World Resources 1990-1991*. Oxford: Oxford University 1 1990.

conversion of Madagascar, which once included large tracts of wet forests as well as the highly adaptive and interesting dry forests, into badly eroded grasslands. That is in fact what about two-thirds of Madagascar consists of at present. The unique and marvelous plants and animals of this island disappear with each new rice-field or pasture, replaced by weedy introductions from Africa, cattle pastures, or even by simple desolation as the land erodes and disappears.

As the world becomes biologically impoverished I think that, as an Academy of Sciences, we have the very same responsibility as Noah had in his time. Biological diversity is the only element on the planet Earth that is capable of sustaining life. Individual biological organisms are the only things that we have on Earth that will produce products that we can use. Communities of organisms, whether original or reconstructed, are the only devices by which we can capture the energy of the sun in an orderly way and protect the soils, the climates, and the atmosphere. Although we are living in an age of enormous extinction, the responsibility is ours at least to make the effort, working together, to try to avoid that extinction and save those organisms that we see as being of the most value. It seems to me that an organization like the Pontifical Academy of Sciences, which has by its very nature enormous moral authority, credibility, and international stature, can play a crucial role in addressing the crisis of biological extinction. Whether we act out of a sense of social justice, or out of a realization of our profound obligation for the stewardship of life on earth, act we must, and without delay.

DISCUSSION
(B. PULLMAN, chairman)

PULLMAN: You have shown us the beauty of nature, the usefulness of nature and also called for a very serious reflection on what's going to happen to this beautiful and useful nature if we take no heed about its future. Moreover, you have raised one of the basic problems which we have to face, the problem of population growth. This paper is now open for discussion.

RUNCORN: You don't think the one consequence of uncontrolled growth of population which is never mentioned in these discussions and yet maybe the most serious one is widespread civil disturbance and war?

RAVEN: One thinks so intuitively, but in fact it is difficult to demonstrate. Certainly, when you look at things like water running out in Africa as it will do possibly over the next 10-20-30 years, and you reflect on the fact that rivers go through many countries, you know that it's going in that direction, but it's difficult to demonstrate scientifically.

SELA: One point that you mentioned but which I think I would like to expand is solar energy. Because this is one of the ways to find a partial solution, especially concerning pollution. I mentioned this because coming from a warm country and having at the Weizmann Institute one of the biggest heliostat fields in which we are trying just to assert the fact that only one third of energy is needed for electricity, one third for transportation, but one third is directly needed as industrial heat. Any methods of creating a "heat-pipe" catching the Sun's energy directly and trasporting it is a very worthwhile topic. Generally, we could think about this approach as we all hope that a large part of the developing world will start living at a much higher level, and will start using cars. If China had as many cars as the United States, there would be really no oxygen. Then we will have to think about alternative methods of running transportation. Thank you.

JAKI: I am glad you mentioned cars. I have already referred to this problem this morning, namely, the 35 million new automobiles being put on the market this year. And the number apparently is increasing. I came across this problem for the first time, when I read a book by Barry Commoner which you probably know, an almost 25 year old book. He pointed out that therein lies an absolutely crucial factor together with road building. Can we in the Western World retain the privilege of having so many cars and deny it to the large majority of the world's population? Do you have any knowledge of any further studies of this problem, namely, the energy consumption of automobiles, of their destruction of oxygen and so forth?

RAVEN: There are plenty of people, including all the major car manufacturers who are developing battery driven electrical cars and other things. But I think automobiles, roads and parking lots are a perfect illustration of what I meant when I said we must not assume that the future will be like the past. People are probably going to have to get a lot closer together, a lot more functionally. They are going to be able on the other hand to work more dispersed because of computer networks and things like that. And I think to assume that cars are going to spread throughout the world on the level that we have them in industrialized countries, is one the most apocalyptic visions that I could dream of. I hope that we can be intelligent enough and come together enough to find alternatives long before that happens. After all we have been using automobiles only since about 1902-1904, there is no reason that the world has to be condemned to an ever increasing expansion of that kind of system. And we really do have to find a new way of thinking. That illustrates my basic point.

DOBEREINER: Let me begin by answering the question of the cars because I think Brazil is the country which so far has found the best solution to this problem, and I never tire repeating the subject because surprisingly enough most people don't think of this. If you burn fossil fuels of course you enrich the atmosphere with CO_2 and cause the greenhouse effect. Most people say, you in Brazil make alcohol but it doesn't help at all, because you burn the sugar cane leaves and you burn the alcohol in cars and it comes out the same. But most people have not thought of the fact at all. Usually the quantity of CO_2 which is put back into the atmosphere when fuel is burned was really taken out of the atmosphere by the sugar cane, when it grows it takes out more CO_2 from the atmosphere than ever is returned even if the leaves are burned. Recently I was very preplexed when I heard that in the United States gasoline is the cheapest in the whole world. Would it not be wise to put a tax on fossil fuels and with this you could afford to use bioenergy? Well, but this wouldn't give the government votes. I wanted to call attention to the responsibility we have as an international scientific Academy, which promotes science for human welfare, to support in whatever way we can, by publications, by lectures, by training, (I would say that it is exactly in the field where I am working) the finding of alternative agricultural systems to produce more food, more energy with less harm to the environment. We have to start right away and if we don't then really it will be too late.

MALU: Monsieur le Président il ne faudrait pas trop vous préoccuper des automobiles parce que nous sommes tellement pauvres que nous ne pouvons pas en acheter. Concernant l'énergie solaire, le gros problème c'est le stockage de l'énergie. Si le coût par exemple du photovoltaïque décroît assez vite, malheureusement le coût du stockage est pratique- ment stationnaire; c'est donc sur ce point-là qu'il faudrait concentrer les efforts de recherche. Je vous remercie.

CHAUVIN: Je m'excuse d'intervenir dans un débat où je ne suis pas du tout spécialiste, mais il y a une chose dont j'ai très rarement entendu parler. Vous savez très probablement tous, que depuis 4 années si ce n'est pas 5, des taxis à hydrogène fonctionnent dans Berlin, je ne veux pas dire des taxis avec de l'hydrogène dans des bouteilles, je veux dire des taxis à l'hydrure de fer. Vous savez que les combinaisons de l'hydrogène avec un métal on été étudiées depuis très longtemps, et que il s'agit d'un hydrure de fer dopé au *tungstène*. Je ne sais pas comment on le dope, c'est ça tout le problème. Les modifications à faire à la voiture sont très légères et portent pratiquement sur le carburateur, et alors, l'hydrogène qui se combine à l'oxygène n'a jamais donné a ma connaissance que de l'eau. Evidemment les voitures sont de 4 à 500 kilos plus lourdes, seulement les personnalités qui ont étudié cela travaillent actuellement sur l'hydrure d'aluminium. Enfin je vous dirai que les brevets qui couvrent l'hydrure de fer sont des brevets Mercedez-Benz, et les idées de rentabilité des Mercedes-Benz sont particulièrement claires à ma connaissance.

RAVEN: We are always properly preoccupied with the importance of more and more energy, but free energy would not solve the problems of the world at all. The world would be destroyed if we had abundant free energy. It is one of the many interlocking aspects of what I tried very hard to present as an overall problem. We tend to get preoccupied with some other way of running an automobile or some improvement that we can make in energy conservation. But I want to emphasise again that if you look at all the factors that I talked about, in fact, abundant energy costing nothing would result in the destruction of the Earth as fast as you could possibly imagine it, we need only to look for new ways of dealing with energy as part of a very, very complex overall readjustment if we want to solve this problem.

PAVAN: When I raised the problem of population this morning I think I was misunderstood as if I wanted to criticize what the Academy was doing about the problem of population; this is not the case. But I think the problem of population growth is so important for a viable future of the human race that I could not come here without telling you my impression about it. It would not be fair for the 8 million children without schools in Brazil, added to a greater number of the ones that don't finish the first grade and the 13,7 million children from 0 to 5 years of age that are very undernourished. Your lecture was scientific, humanistic and really extraordinary in its content I think you put emphasis on the right problems. I thank you very much in the name of other citizens of the Third World.

SINGER: I was equally impressed and depressed by your statements. I was particularly shocked by your conclusion that even if we disregard the contribution to the environmental stress of the 4.5 billion people who already live in misery the magnitude of the problem would be reduced by only 15%. The implication is that mainly we, the rich, have to drastically change our behaviour. This identifies a specific problem, well-known to behavioural therapists and psychotherapists. It is extremely difficult for humans to convert rationally acquired insights into appropriate actions if the latter require a change of behaviour and the former are not backed by direct emotional experience. This is what makes the job of psychoanalysts so frustrating. Thus, if we really want to change something we have to think seriously about mechanisms by which we can change the behaviour of this dangerously inconsequent animal. We have to invest in ways to remedy its inability to translate rationally acquired conclusions into change of behaviour. I guess, we all, even our politicians know what you have just told us, — maybe not in this impressive dimension — and still we seem to be unable to make this a maxim of our behaviour. I don't have a solution to this problem but I am convinced that one key is to try to understand why we are unable to change behaviour merely upon rational insight. Probably it will be necessary to change the way in which we educate our children.

RAVEN: I think I'd like to make a couple of points about that. First of all people do have a very great capacity to change their behaviour if they are properly inspired. If you consider Francis of Assisi of whom Karl Marx said "If I had twelve people like Francis of Assisi I could take over the whole world in 10 years". Or if you consider Ghandi in this century you find inspired leaders whom people respect have an enormous ability to change behaviour. People do not have to go on doing what they are doing. But we do need to demand of our leaders, here in this group, and in all groups, the kind of intelligent leadership that will help us to understand what the choices are, and then to take up those choices. Beyond that, I mainly argue that we need to practise as diligently as we can for ourselves and our children, a real kind of internationalism, not just lip service or not with just neighboring countries that are similar, but a genuine kind of internationalism, and further that we all really do need to look for ways to lead simpler lives, to conserve, to get those around us to conserve, and simply not to be as much of a burden on the world. Example and leadership have wonderful properties and human beings can change. They do change, there are many historical examples.

RAO: I think you have pointed out the kind of areas this Academy should be interested in. Areas of global interest, areas of international importance, where all countries, in principle, should be equally interested, not only developing countries. But, coming from a developing country I must particularly point out, associated with what was said, we must also know how to take a holistic view of energy. Unfortunately people will say: "oh, well, it's just a matter of using shale or some other form of petrolium or solar energy". I don't think there is a way to solve it by a throw of the dice. I myself am interested in the use of amorphosylicone but that is not going to solve the problem. Eventually it will be biomass also. Very few people support, or talk about the use of biomass in energy conservation or other related aspects. I do not know whether you would like to say anything at all about biomass and how, at least in countries like ours, India, biomass is very important.

RAVEN: The only thing I'd like to say is we all need to take as general an overall view as possible. That is one of the many things that we have got to explore.

REES: You emphasized that the first world would need to cut back on consumption. Would you not agree that it is important to distinguish between consumption of resources and economic growth? Certain kinds of economic growth and technical progress can actually be environmentally benign. So the message to the first world is surely not to stop economic growth, not to stop technical progress, but to redirect them towards such directions as telecommunications, miniaturisation and other kinds of sophistication, which can actually reduce pressure on resources.

RAVEN: Exactly. Only consumption is what we need to cut back on. We need to use what we have to continually advance not only for our own sake, for everybody's sake. I couldn't agree more.

ARECCHI: Being an outsider, as a physicist, I was extremely impressed by your data. Let me raise the following question. Can one evaluate what is the amount of reduction that the 1.2 billion people from developed countries can make in their consumption of energy and still keep having a decent life?

RAVEN: There are very important works on that which I can't quote to you in detail now, but many people have worked with systems: John Holdren in Berkeley who is a physicist, Amory Lovins in Boulder, Colorado, who is well-known energy consultant. I can't tell you now but there are certainly many elegant works on that very subject. We waste an incredible amount of energy needless to say, particularly in a place like the USA.

GERMAIN: Oui, je voudrais commenter ce que vient de dire notre collègue Martin Rees. Je suis très frappé par tout ce que nous a dit notre conférencier et depuis longtemps. Mais est-ce que le problème n'est pas, dans nos contrées développées, un problème économique? Quand nous disons, dans chacun de nos pays, que nous sommes engagés dans une guerre économique, cela veut dire qu'il s'agit, en France, d'être meilleurs que les Anglais, ou les Allemands. Une guerre, c'est une guerre dans laquelle il y a des victimes. Les victimes sont aussi chez nous, nous avons 10% de chômeurs en France. Sur le tiers-monde, on écrit de bons articles dans les journaux. Mais on ne voit pas actuellement, que ce soit dans les élections américaines ou dans les élections françaises, la possibilité d'arriver à convaincre nos concitoyens qu'ils doivent changer de moeurs économiques, non pas diminuer leur capacité de production, mais changer leur production. Je comprends que les scientifiques comme nous doivent se manifester et je suis prêt naturellement à soutenir tout ce qui pourrait être fait. Nous allons lancer un message. Mais s'il ne touche pas les responsables, nous aurons fait de notre mieux, et la situation restera la même. Je ne sais si les personnes qui comme vous ont réfléchi à ces problèmes ont une idée sur les moyens de faire changer les mentalités, nos mentalités.

RAVEN: Obviously a tremendously important key question and the only thing I can say at all is that Martin Rees has laid the foundation for the answer. You need not assume that you will suffer or be uncomfortable simply because you are making progress. It is not an image of going into poverty in order to do better. And we find, I think, people all around the world are finding that it is acceptable to try to live a more conservative life not only to recycle, but to use less, to try to live simpler lives, to do better. I think that image is spreading very rapidly right now. Another thing obviously that we can do as scientists, is simply keep pointing out to people that, no matter what we do, we end up having only the one same planet. We do have to live on that one, and we don't really have the luxury of avoiding certain physical limits. In that sense, we need to do all we can to dispel the convenient myth that science and technology will come along and save everybody in spite of everything.

PULLMAN: Arrivé à ce point de la discussion, je voudrais ajouter un témoi-gnage qui se rapproche de ce que vient de dire Germain. Je répète que c'est un témoignage, ce n'est pas nécessairement ce que je pense moi-même. Il y a très peu de temps j'ai eu une discussion sur le sujet de l'énergie, de la population et du tiers-monde, surtout du tiers-monde, comment faire pour l'aider, avec un diplomate français de très haut rang et qui a une grande influence. Nous étions d'accord sur le fait que le bilan des résul-tats de notre aide aux pays du tiers-monde était dans l'ensemble négatif. Je lui demandais alors que faire dans ces conditions, que peut-on faire? Il m'a répondu, grosso modo: "Actuellement il n'y a aucun espoir qu'on puisse faire quoi que ce soit. Aussi longtemps que le gouvernement et les dirigeants des pays en voie de développement ne changeront pas d'attitude, ne se modifieront pas, ne se perfectionneront pas, tout l'effort que nous faisons sera complètement raté. Evidemment, nous arrivons à sauver un millier d'enfants de la mort, mais enfin c'est une goutte d'eau par rapport à ce qui devrait être fait". Alors je lui ai demandé: "mais alors dans ce cas-là qu'est-ce qu'on peut faire tout de même"? Sa réponse était qu'il faudrait faire une "recolonisation éclairée". Il n'y a qu'un diplomate qui peut utiliser des phrases pareilles. Une recolonisation éclairée, qu'est-ce que ça veut dire? Voilà, m'expliqua-t-il: "il faudrait que des pays avancés comme la France, l'Angleterre etc. se voient confier le soin de s'occuper tout spécialement, avec des responsabilités, de certain pays en voie de développement: essayer de leur apprendre la manière de vivre, la manière d'avoir une réaction, une contribution, un effort de leur côté qu'il faut absolument obtenir si nous voulons arriver effectivement à faire quelque chose de positif".

D'une manière plus élégante un effort de responsabilisation des gouvernants des pays du tiers monde a été fait par le Président Mitterand, lors de la conférence des pays européens à la Baule. Le Président a dit très nettement: "l'aide que le gouvernement français est prêt à accorder aux pays africains sera fonction de la démocratisation de ces pays". Le résultat a été un très faible, malheureusement trop faible, effort de démocratisation dans quatre ou cinq pays. Ce n'est pas la solution. Mais je crois qu'il nous faut tenir compte de l'opinion des sphères dirigeantes de notre pays. Nous pouvons dire tout ce qu'il nous plaît ici et faire toutes les propositions que nous voulons, ça ne servirà pas à grand-chose, si ces propositions n'arrivent pas à avoir l'appui de nos dirigeants. Il est donc intéressant et important, je crois, d'avoir présent à l'esprit la manière dont raisonnent nos hommes d'état, nos diplomates, qui finalement vont décider des choses. Et à ce point de vue effectivement, le feed-back de la part des gouvernants des pays en voie de développement est une chose primordiale. Il faudrait que ces pays fassent l'effort nécessaire, d'une manière ou d'une autre. Sinon tout revient à remplir des tonneaux des Danaïdes.

MALU: Monsieur le président, nous avons eu tellement de dictateurs éclairés que je me méfie de votre recolonisation éclairée. Mais ceci étant dit, je crois savoir ou j'ai cru comprendre que peut-être le professeur Arecchi ou quelqu'un d'autre a posé la question de savoir quel était le minimum d'énergie pour assurer une vie décente, si j'ai bien compris. Cette question a été étudiée par plusieurs auteurs et j'ai moi-même sorti un livre sur la question en ce qui concerne l'Afrique aux sud du Sahara. Donc ce sont des questions qui sont connues.

COLOMBO: In the Study Week on "Resources and Population" many things expounded by Prof. Raven were taken into consideration. Not bio-diversity. We missed that point and several others. But I wish to make reference to the issue raised by Prof. Germain in his intervention: someone has to speak. Well, the Holy Father has already spoken in the address he gave to the participants at the end of that Study Week. If you read point 6 of his speech, you see underlined that we all face "new conditions" due to the decline of mortality. Conditions that we are called to meet with recourse to all available intellectual and spiritual energies, rediscovering the moral significance of putting limits on ourselves. And respecting solidarity, because it's easy to speak of population problems of developing countries, but who pays for population control? The poor, not the rich. Is this justice? First try to realize distributive justice and then speak of interventions through demographic policies. This might seem an extreme position. What I wanted to say is that we already have an indication of where to go with our studies, with our reflections. May I make references also to what Professor Rees said about improving technology. There are several problems, economic, legal, political and so on, so that technology stays where discoveries are made and does not go where it is most needed. Thank you.

GERACI: I would like to ask a question. Don't you think that it would be desiderable, while the big enterprises are leading people to change habits and way of life, to organize something that could be of help in the meantime? For example trying to improve the germ line banks, to save, as soon as possible, all the species that are endangered. Because it may well be, as you said and I completely agree, that it will soon be too late to save some of them. I think that it would probably help to organize a commission, or a panel, that identifies the species that are particularly endangered to make special recommendations for their preservation. This is not a solution, I am aware of that, but at least it would be an immediate help that would permit us to wait longer for the already existing organization for plant preservation.

RAVEN: Sure, that was exactly my meaning. In order to preserve biodiversity you need to get a stable world, which is why population, poverty, social injustice, the distribution of wealth, the production of energy, and everything else comes in to it. But obviously we need to do everything we can now. If all of those matters develop, then we have to do things such as internationally agreed upon preservation of selected natural areas, paid for internationally. Secondly seed banks, botanical gardens and so forth. Thirdly, intelligent approaches to the problems, such as the one you were talking about. The most fundamental approach though, I think, to the solution of the preservation of biodiversity, and the creation of a stable world condition is the one that I referred to this morning. With only 6% of the scientists and engineers in developing countries we really will benefit enormously if we can put appropriate institutions in these countries, support them, and provide places for people to work and address the problems in a way that can really be a help to the people of each country. For me this is a fundamental step in the preservation of selected parts of biodiversity on some rational basis which is internationally funded. The worst crime that we can commit is to let everything go through our fingers without doing anything.

WHITE: It seems to me as I listened to Professor Raven and the comments on his talk that we should think of a mechanism that was utilized by the Academy some years ago because of its great concern and that of the Holy Father about the nuclear holocaust. Then, the Academy was empowered in His Holiness' name and in its own name to visit some of the outstanding and important leaders in the major industrial countries, particularly in the then Soviet Union and the United States, and I wondered if... it would be appropriate, because of the great concern expressed here, that in some way the Academy itself be empowered to raise the issue in the same way as happened some years ago with the expression of the world's concern about the nuclear holocaust?

PAVAN: I would buy totally Professor White's suggestion: I think it is a very good one, and the problem is as important as the nuclear bomb holocaust fear was at the time when we discussed it.

DOBEREINER: I will say one word in support of this idea, and in fact I think we do have the responsibility to do this, and if possible we should try somehow to start to talk about this next Saturday, when we are with the Holy Father.

GERMAIN: C'est une très bonne idée. Mais dans le cas de la guerre nucléaire il faut dire que l'affaire avait été préparée par au moins un an de rencontres organisées par cette Académie. Il y en a eu à Vienne, il y en a eu à Londres, avant la réunion des Académicies à Rome. Cette dernière fut un événement qui a été suffisant pour que les Délégations fassent ce qu'a dit le Professeur White, c'est-à-dire aillent trouver leurs autorités pour leur transmettre les voeux. Je suis prêt à soutenir un voeu que nous prendrions ici, mais j'ai peur que nous ayons peu d'écho. Je ne me vois pas revenir en France en disant au Président Mitterand: *"je vais vous apporter un voeu de l'Académie Pontificale des Sciences"*. Je n'aurais pas beaucoup de succès. Donc, je crois que si on veut prendre des décisions dans la ligne de ce qui nous est proposé, il faut effectivement lancer des études, des groupes de travail, rassembler des documents, et agir lorsque nous aurons quelque chose de positif et d'unique en son genre car fortement justifié par une instance compétente et indépendante. En effet d'habitude qui parle de ces choses? Ce sont des groupes qui sont motivés par des idéologies souvent très généreuses, mais qui n'ont pas la qualité de sérieux que l'on peut attendre de notre Académie. Alors à ce moment là je crois que la démarche aurait tout son poids. Mais si elle n'est pas préparée je crois que nous n'aurons pas le succès attendu.

PAVAN: I would also like to agree entirely with you. I think it may take one, two, three of four years but it is worthwhile. Perfect. I agree. Thank you.

PULLMAN: Je crois effectivement que la remarque de Mr. Germain est astucieuse et parfaitement réaliste. Si nous voulons vraiment mener une action efficace, à grand retentissement mondial, il est évident qu'il faut que le Saint-Père en prenne la direction principale, en assume le patronage. Un problème comme celui-là ne peut pas toutefois se régler dans la réunion de samedi, après la réception, malgré la meilleure volonté possible: ça exige beaucoup de préparation, l'élaboration d'un texte très bien fait, très bien documenté. Je sais que les idées qui ont été exprimées ici travailleront beaucoup, beaucoup dans l'esprit de beaucoup de personnes qui sont présentes ici. Nous sommes devant un problème crucial, il faut prendre des mesures mais il faut les prendre d'une façon réfléchie et organisée et surtout lancer une action.

ORGANIZATION, INFORMATION, AUTOPOIESIS: FROM MOLECULES TO LIFE

GIUSEPPE DEL RE

This paper is devoted to the presentation of certain key concepts of the theory of complexity such as they are applied in chemistry, biochemistry and biology, and to a brief discussion of the place of autopoiesis of complex systems, with special reference to that phase of the origin of life in which the transition from aminoacids and purine bases to the first virus-like systems took place.

1. *Key concepts of complexity: organization and information*

The description of complex objects and of their behaviour requires a number of concepts: order, coherence, unity, structure, organization, memory, information, meaning, context, emergence, autopoiesis. It seems obvious that a satisfactory analysis of complexity should be based on a careful definition and discussion of each of them. Organization, information, and autopoiesis (spontaneous appearance and increase of order and organization within a system) come primarily into play in the specific problem this paper will refer to.

From the standpoint of biology, organization is perhaps the concept around which all the others revolve. Its theoretical formulation is provided by the theory of systems, which studies objects formed of several parts, but behaving as stable or stationary wholes. Their behaviour (or activity) essentially consists in processing input signals to yield output signals, in virtue of an internal structure capable of a dynamical activity.

There is a tendency to consider organization close to order, i.e. just a preferential configuration of particles or other elementary components of a system, e.g. neurons in the brain. In fact, the word "structure" is often used as a synonym for it. A more elaborate concept is proposed when order is associated with coherence, but that is still far from corresponding to organization in biological systems. As far as I can tell, in control systems and more so in living systems the two notions of structure (even if coherent) and of organization are distinct and play different roles in the description, or, if you prefer, the explanation of facts.

Since both words are extensively used in everyday language, it may be useful to show in what they differ with an example taken from our ordinary life. Consider an airline, which we call XX-AIR. It may start with one plane and two pilots, accepting passengers as they come at a given airport. So far, it has neither structure nor organization, because one plane with its crew is not all airline systems. But gradually XX-AIR becomes one of the leading airlines in the world, and grows to a hundred planes, crews including stewards, ground personnel, employees. That makes something which we might call a structure, although it is not something rigid. Finally, XX-AIR establishes regular flights between given points. Here organization comes into play. The aim of the company is to provide as good and reliable a service as possible. This is not automatically ensured by timetables and personnel. The staff have to face all sorts of unexpected difficulties. One day, at 08:10, the telephone operator at the airline offices of airport YY receives a call informing that the pilot of flight 71, due to leave from airport YY at 09:30, has a sudden attack of migraine; she immediately calls the traffic director. The latter makes a call to an incoming plane to know if its pilot can replace the sick one. The answer is negative, but further inquiries yield the good news that the pilot of flight 55, landing in five minutes at the airport of ZZ, twenty minutes' flight away, could do the job. Then, the traffic director calls headquarters to know if a service plane can be sent to ZZ to get the pilot to YY. Headquarters call a third airport where a plane is available for hire and arrange the trip. In the meantime, the passengers are informed that flight 71 will be delayed for ten minutes. Headquarters send their OK. The pilot arrives at 09:10. At 09:40 flight 71 is off to its destination.

Real situations, of course, are often more difficult than the one I have described, if nothing else because costs are a serious limitation. But our idealized example is probably sufficient to illustrate the essential point that organization is a dynamical cooperation of parts aimed at performing a given task. If the men involved had not known what to do in an exceptional situation, if there had not been the right competences and powers of decision at the right places, the existence of a structure consisting of different elements differently distributed according to a precise scheme would have been useless. But what was the essential reason why organization was required? Simply that, like a living organism, an airline is a unit composed of many elements, and is expected to perform as a unit a specific task in a variable environment. What task? You may call it ensuring self-survival, or more generally protecting its own identity in the face of a changing environment, by adjusting all the time to external (and internal) fluctuations and disturbances. That is to say, the airline tries to assure its service as scheduled on its entire network, because that is precisely what makes it an airline.

The application to living beings and sophisticated machines is evident. A more detailed reflexion will show that the general notion of coherence is indeed applicable to our example, but that special sort of coherence that is organization is so far from the sort of coherence which you find, for example, in laser light, that no specific application to organisms of that concept alone seems possible.

As is suggested by our example, organization appears to be a necessary condition for the result-oriented behaviour of a system acting as a whole in a variable context. It is precisely that kind of interdependence of the parts of a whole that makes it possible for the given system to adjust its behaviour and its internal activity to changes in the environment, perceived as external stimuli or input signals, as well as to internal changes, so as to ensure preservation of its identity or execution of a pre-established programme. Typical examples are selfdefence and immunity responses of a living being, automatic route corrections of space probes.

Organization may be seen as a high-quality sort of "information". This is why the latter is the other fundamental notion in the analysis of biological complexity we wish to review in this section. The history of information theory is well known. Suffice it here to remind the reader that, though originally strictly associated with communication under the impact of the new fields of inquiry opened by the discovery of the chemical basis of genetics, information was first given back its etymological meaning of reception or acquisition of a "form" in the Aristotelian sense, and then became the modern, scientific name of that notion. Consider the clay vase given by Aristotle as an example. The potter gives a piece of clay a particular shape or form: that action is "information", and is equivalent to writing a message, to transmitting to matter an idea in the potter's mind. But the information given to the original lump of clay remains in it, possibly for thousands of years; thus it may be seen as that resident property of the vase which makes it a vase and not something else[1].

The modern tendency to rediscover old notions from scratch may, however, play tricks. A case in point is when the "quality" of information, viz. its semantic value, is confused with the "relative information content" C_i, introduced[2] at the dawn of information theory (around 1945) as a measure of the fidelity to the source message of a text received through a communication line. The quantity C_i was given by the well known expression $C_i = - \Sigma_n\, p_n \log p_n$, where the summation runs over the probabilities (defined as relative frequencies) $p_1, p_2 \ldots p_N$ of the various symbol sequences compatible with the form in which the message under consideration has been received. Since the probabilities in question may be misinterpreted, let me illustrate by an example what they mean.

Consider the incomplete message "pr.y f.r me". It may correspond to 27 different complete messages, if the dots are assumed to be low-case letters of the English alphabet. If it cannot contain but English words, then the possible complete messages compatible with the sequence received are just eight: "pray for me". "prey for me", "pray fur me", "prey fur me", "pray far me", "prey far me", "pray fir me", "prey fir me". Now, the probabilities appearing in the Shannon expression are data derived from available evidence, e.g. from the examination of repeated transmission of the same message. We say that the letter "a" has a probability 75% if it appeared at the location of the first dot in three sendings out of four. Let us now suppose that in a certain case the probability (thus calculated) that the first unknown letter is an "a" is precisely 75%, and the probabilities that the second letter is an "o", an "i", an "a", a "u" are 25%, 35%, 22%, 18%, respectively; then the probabilities of the eight messages are 18.75%, 6.25%, 13.50%, 4.50%, 16.50%, 5.50%, 26.25%, 8.75%. Applying the above formula, we obtain for its relative information content $C_i = -1.9177$. This quantity would be zero if all the letters were known; if only the first letter is unknown, with the probabilities given above, it would be -0.5623.

What do these results mean? Without its minus sign, the quantity just

[1] This consideration explains why the Nobel laureate MANFRED EIGEN entitled a paper on the origin of life: "How does information arise?" - *Wie entsteht Information? Prinzipien der Selbst-Organisation in der Biologie*. Berichte der Bunsengesellschaft, 80, 1059 (1981).

[2] C.E. SHANNON and W. WEAVER, *The Mathematical Theory of Communication*, Chicago, The Univ. of Illinois Press 1949.

computed actually represents the uncertainty the message received leaves regarding the message transmitted. With the minus sign, it does represent the information retained by the output message with respect to "full" information, provided one gives conventionally the value zero to the information of a completely faithful message and refers to a message of a given length. This is why we have called it a relative information content. If so desired, it is also possible to assign an absolute information content to a received message. In order to do that, one must include the length of the message in the computation. Consider, in the case of our example, all the triplets one can make out of a dictionary of 45,000 words. Their number is $N \approx 15186.4$ billions. The absolute information content of such a message, if entirely unambiguous, may then be taken as the natural logarithm of that number, i.e. $C_0 = 30.3514$. Therefore, the absolute information content of the message received in the above example is $C_0 + C_i$, namely 28.4337 in the case of two letters missing, and 29.7891 in the case of just one letter missing.

I have dwelt on the famous expression of information content in information theory in order to make it clear that the semantics of the message, that is to say what the information means to a man, have little to do with the number obtained, unless a re-interpretation of the whole theory is carried out[3]. Even less is it related to the quality of the information; Shannon's quantity is by no means intended to distinguish between a line of Dante's Comedy and a nonsense line made with the same number of words or letters. In other words the standard information content of a message is a notion constructed to provide an objective measure of the degree of certainty about the correspondence between the signals transmitted and received, and is invaluable precisely in that role.

The same is true of information seen as that resident property which characterizes an object. We can associate to it the standard expression by referring to the message needed to describe it. In this case the message certainly has a meaning and may well have no uncertainty going with it, so that its information content is just C_0. But it should be evident that a Greek vase and a lump of clay may require descriptions of the same length, and thus have the same information content, despite the difference in quality[4]

---

[3] R. CARNAP and Y. BAR-HILLEL, *An Outline of a Theory of Semantic Information* (Technical report 247, Research Laboratory of Electronics), Boston, MIT 1952.

[4] This point was stressed in this Academy a few years ago by J. LEJEUNE, *Existe-t-il une morale naturelle?* in: *The responsibility of science* (C. CHAGAS and F. ROVERSI MONACO eds.), Vatican City: Pontifical Academy of Sciences 1990, pp. 97-103, note 2.

and the fact that the former embodies an idea of the artist who made it, the latter is just a product of the random action of all sorts of forces.

It might seem that this consideration disposes of information content as a tool for the study of complexity. This impression is mistaken. In fact, that notion makes it possible to solve in a modern way an interesting puzzle encountered in the application of the Aristotelian ("holistic") analysis of reality and change to wholes such as a molecule, a cell, a human body. Our science describes those wholes as resulting from the coherent cooperation of parts (i.e. organization), each part contributing to the properties of the whole by its specific properties, which are the same as if the part were alone. In other words, an integral part of a whole is seen as *actually existing* in the whole. In traditional Aristotelianism this prompts the following question: how can a whole have a unitary nature which is distinct from and richer than the sum of the properties of the parts, if it consists of nothing but its parts, and each of them participates in it with its own characteristic nature?[5]. The preoccupation underlying this question is by no means specific to Aristotelian philosophy; because, in the absence of a satisfactory answer, the overwhelming evidence we have that parts exist as such in the whole would support either mechanistic reductionism, which is philosophically untenable, or vitalism, which is scientifically untenable.

The new notions of emergence of information and of level of complexity solve Aristotle's problem in a way consistent with the scientific knowledge of our time.

The concept of emergence is contained in a remark that still surprises many scientists who tacitly assume that prediction and explanation are the same thing: already at the molecular level, knowledge of the parts (atom cores and electrons, in a molecule) only allows prediction of the properties of the whole *if the existence of those properties is already known;* and this is not possible for all properties, because formation of the whole leads to the *emergence* of properties not present in the parts.

Now, when you ask science to say what an object is, you are asking for the shortest possible list of the properties which identify it unambiguously.

---

[5] Aristotle wrote: "an animal is a sensible object and cannot be defined without reference to movement, i.e. without reference to its parts and to their being in a certain state." And later: "...in the question 'what is man?' The interrogation is a simple one, not analysed into subject and attributes; we do not ask expressly 'why do these parts form this whole?' We must first make our question articulate..." (ARISTOTLE, *Metaphysics*, Book Z. Edited and translated by JOHN WARRINGTON. London: Everyman's Library 1961, pp. 206, 194. Cf. also *Physics*, I, 2, VIII, 6 etc.).

The what-is-it of that object can therefore be treated as the information contained in it. The information resulting from the combination of the information about the individual parts will therefore be insufficient, for the nature of the emergent properties of the whole will not be contained in it. That is to say, the whole has a greater information content than the sum of its parts; it is a collection of the parts, *but not just that;* and the properties of the parts describe the reality of the object under consideration only to a limited extent.

Until recently, the physicists had views which did not admit the notion of emergence. They had in mind systems such as an ideal gas, whose pressure (as other macroscopic properties) is just the sum of the forces applied by colliding molecules to the surface of the container. Attention has only recently been focused on the fact that nonlinearities in the equations of motions would make such simple additivity invalid. As I have already pointed out, knowledge of those equations does not imply the ability to tell what the possible new properties will be, but the recognition of the significance of nonlinearities imposes the admission that (as the distinguished metallurgist A. H. Cottrell[16] pointed out when physics was discovering complexity) "the whole is more than the sum of the parts". The prediction of the nature of the emergent properties is one of the aims of the general theory of systems, but is still a little studied and practically unsolved problem. As I have mentioned, that problem already arises in the case of molecules, considered as edifices of atoms. It is possible to show that quantum physics accounts for the existence of binding in diatomic molecules and correctly describes the structure of individual polyatomic molecules, but no procedure has been found for deriving from its equations those general characteristics of molecular structure that have led to the notion of chemical bond.

If we consider next such living beings as the vertebrates, we are confronted with systems by many orders of magnitude more remote from atoms than molecules. Therefore, the task of bridging the gap between knowledge of the non-linear behaviour of systems of particles and the rules which the parts of a living organism obey in forming a unit seems to be next to impossible.

Another fundamental question arises from the notion of emergence: if new properties emerge when interacting systems are brought together, is it legitimate to define a complex, unitary system as a collection of those

---

⁶ *The Natural Philosophy of Engines.* Contemporary physics, vol. 20, p. 1, 1979.

constituent systems? The traditional answer is negative. But then it should also be true that you cannot derive a complete formal description of the behaviour of your complex system only using information about the parts, whereas today's science claims that you can, e.g. by solving the appropriate Schrodinger equation. The notion which reconciles the two views is that of "level of complexity", which can also be seen as a "level of reality"[7]. Let me illustrate it first of all by an example. We often say that a given cell *is* a collection of atoms, is a collection of molecules and macromolecules, and is a whole cell. The three statements are all equally valid, but at the first level the information about the cell is largely potential, since the reality of a molecule, as I have pointed out above, can be described explicitly only if new concepts are introduced; at the second level the situation is but slightly better, since knowledge of the properties of the constituent molecules and macromolecules permits in principle the prediction of the properties of the cell only if one already knows what to look for, that is to say, if those emergent properties are known from some other source.

This example suggests that, in general, the descriptions of a system in terms of parts of different types and structures given by the various disciplines - microphysics, chemistry, molecular biology, etc. - form a hierarchy of descriptions of the reality of that system associated to different measures of actual information and different degrees of complexity. These descriptions correspond to what we have called different levels of complexity and of reality.

To see the full purport of the new approach to the problem of the relations between the whole and the parts implied by the concept of complexity level, take the most complex system we know, man. You can say that he is a collection of molecules, but then you are not describing his whole reality, for a man is more than that by far; you have been looking at a low level of complexity (or of reality), but there are higher ones. A less incomplete description of what a human being is emerges when you consider fewer parts, each much more complex than a molecule, as when you say that a man is the organized ensemble of his organs. Then you are

---

[7] This concept has been around for a long time, but only recently has it become a subject of rigorous analysis. One of its major advocates is B. NICOLESCU. For a brief summary and references see his paper *Complexité et Niveaux de Réalité*, in: *Simplicité et Complexité* (M. CERUTI and E. MORIN, eds.), Suppl. to "50, Rue de Varennes", March 1988, p. 38, and his contribution to this volume. Although not using the term "level", DAVID LAYZER beautifully explains the same idea in *Cosmogenesis: the Growth of Order in the Universe*, New York: Oxford U. Press, 1990, pp. 32 ff.

dealing with a much higher level of complexity. On climbing the ladder, you come to the level that has been the object of studies by philosophers and thinkers of all ages: that where you consider as 'parts' the mind and the emotional psyche, νοῦς and θύμος. But that too falls short of completeness, because from the interaction of emotions and detached contemplation a whole new set of properties and activities of human beings emerges: think of poetry, think of what is revealed about man's reality by the very existence of works like Dante's Comedy and Eliot's Four Quartets. You thus come to the uppermost level of complexity, where you can only speak of a man's characteristics or properties and not of his parts, where you look at a human being as a fully integrated whole, as a unit. This is the level at which you must place yourself if you wish to deal with what a man really is. This point was emphasized by Aristotle and by Aquinas long before the birth of our science, and evil has ensued whenever it has been forgotten, as recent history teaches. But much has been gained by the modern scientific approach also in this connection after the collapse of reductionism, for we now know that complete knowledge of man (or, for that matter, of any object in spacetime) requires consideration of all the pertinent levels of complexity.

2. Autopoiesis: spontaneous appearance of biological information

According to Simon Hadlington[8], "the term 'autopoiesis' was coined in the mid 1970s to characterize a 'living' system as a structure defined by a boundary within which occurs a series of interdependent reactions that regenerate the boundary and its components which then assemble in the structure itself. By this definition autopoiesis is broader than simple self-replication, and none of the earlier systems could be classified as autopoietic. Pascale Bachmann and colleagues[9] have apparently successfully devised a series of elegant systems that would appear to fulfil the criteria of simple chemical autopoiesis."

That is to say, autopoiesis just means spontaneous formation of a complex object having some kind of autonomous behaviour (otherwise the

---

[8] S. HADLINGTON Chemistry in Britain, Jan. 1990, p. 10.

[9] References to this work and information about its sequel is given P. A. BACHMANN. P. L. LUISI, J. LANG, Autocatalytic self-replicating micelles as models for prebiotic structures, Nature 357, 57-58 (1992).

prefix "auto-" would not apply); its meaning is close to that of "self-organization", and in our case can be identified with it. It is a scientific notion, because science nowadays accepts the view that ordered structures and organized systems can emerge from chaos and "grow", at least under certain conditions, without the aid of external factors. Its mechanism, in our case, can be thought of as spontaneous assembling consisting (at least in the first stages) in the repetition of three steps:

1- the casual encounter of two parts (say, molecules) A, B having certain properties;

2 - the establishment of an interaction between the two parts resulting in a unit AB capable of forming a more elaborate assemblage upon casual encounter with (or by exerting an attraction on) a part C of a type in general different from A and B;

3 - encounter with the part C.

These steps also apply to the formation of a crystal from a solution; however, if the product shares at least some characteristics of life, at each stage the structure formed will actively participate[10] in orienting and promoting development towards a system which will not be at equilibrium, but in a steady state dependent on continuous exchange of energy and matter with the environment.

With an appropriate chemical composition and under appropriate conditions, repetition of the above steps could give rise to extremely complicated units, provided the same conditions persisted for a sufficiently long time and there were no tendency to disruption of the units formed. In fact, the inevitable existence of some tendency to disorder will put a limit to the increase of order by the above steps.

At a certain stage, further development of a unit formed by successive random encounters with suitable parts must take place by some more elaborate mechanism not relying on chance at all, such as the growth of a living organism. This later stage may be seen as an interaction with the environment that will favour growth until adjustment to the environment has reached an optimum. Reflection on this extremely important stage of self-organization requires a discussion of evolution and environmental equilibrium, in particular the competition between individuals of different species. There are, of course, causes acting against the construction of order,

---

[10] The 'activity' is alluded to in the term "poiesis", the fact that it belongs to the very system which is being formed is indicated by the prefix "auto".

particularly spontaneous tendency to disruption of coherence and competing processes in the environment. They can explain why natural organized systems reach a certain degree of complexity and stop there, or indeed begin an inverse process of decay, such as aging. In this respect, the ideas of the pontifical Academician and Nobel laureate (for work on chemical reactions) Manfred Eigen on the mechanism by which, after the steps listed above, nonliving matter gave birth to the first most elementary living structures are especially interesting for the general theory of complexity. Before attempting to summarize what I deem most important in Eigen's ideas from the conceptual point of view, let me set up in outline the scenario to which those ideas apply.

In the reducing atmosphere of the earlier Earth, electric discharges catalyzed formation of amino acids and purines, the fundamental building blocks of living matter. Water containing a solution of those primordial chemicals accumulated in small recesses in the rocks and was protected by them from the violent temperature changes taking place in the open air. Temperature changes, particularly those between day and night would not reach those natural reaction vessels with their full strength, and would cause them to heat up or cool down by just a few degrees centigrade, enough to speed up certain reactions and to slow down other reactions, but not enough to bring about dramatic changes in chemical composition. Thus, the conditions were realized for the spontaneous formation of chains of small molecules which in the long run would yield self-catalysing or self-replicating systems, possibly contained in droplets capable in their turn of spontaneous multiplication. The self-replicating systems would be large molecules or groups of molecules of a very special type, carrying a highly specific and comparatively large amount of information, such as the length of the chain, the nature of the atoms forming it, the number and arrangement of atoms and bonds. Their ability to catalyse the assembling of copies of themselves meant that they would hand on to other groups of atoms the information they embodied, and that information would 'survive' far beyond the limits of their natural lifetimes. If they could do that despite disturbances capable of disrupting the ideal incubation conditions in which they had been formed and despite the inevitable replication errors, then they would constitute the first timid premonition of life, the *prebiotic* molecules. Eventual formation of more and more complicated molecular and supra-molecular systems having the same characteristics would then lay the way to the appearance of life proper.

The scenario just outlined is of course largely hypothetical, but to most experts in the field it seems consistent with all that is known about the history of the Earth, and deserves systematic assessment. Its substantiation according to the usual procedures of science is in progress, and, if it keeps its promises, it will constitute a great advance in our grasp of the created spatio-temporal order of things.

One of the most important points to be clarified is the mechanism by which a self-replicating 'prebiotic' system would be formed and the persistence and enrichment of the corresponding information ensured[11]. This is where Eigen's ideas come into play.

To simplify matters, think of those molecules, such as purines, phosphates, sugars, which could be building blocks of the simplest self-replicating molecular systems as letters, and of the resulting systems as words. Any chemist would expect that many different sorts of 'words' could be formed with the same 'alphabet'. Why is biological information only contained in words made with a limited number of the available letters, arranged according to very strict rules, namely self-replicating molecules of the DNA type? To answer this question, Eigen considered that in fact a variety of self-replicating systems (the 'words' A, B, C, ...) was formed in the primordial soup, but a sort of Darwinian selection operated. Consider for example the population of words A. It would increase as a result of spontaneous formation[12], faithful replication of A and errors in the replication of other words, and would decrease as a result of spontaneous dissociation and errors of replication of A. Therefore, under steady external conditions, the various populations would change until equilibrium concentrations were reached. With appropriate parameters, the system of differential equations describing the approach to equilibrium admits a solution where only one 'word' has a significant concentration.

Thus, although a number of important questions remain anyway open, it would seem that a mechanism has been found to explain why — despite its appearance out of a chaotic situation — the molecular basis of life is the same for all living matter, at least on Earth. However, there is a crucial objection, which Eigen and Schuster have pointed out and attempted to

---

[11] The comparatively novel notion of molecular self-replication has been briefly reviewed by L. E. ORGEL, *Molecular replication*, Nature 358, 203-209 (1992).

[12] Here 'spontaneous' refers not only to successive random collisions between smaller systems, but to the action of external factors, such as γ rays.

overcome by the "theory of the hypercycle"[13]. The objection is that the number of errors of replication would increase with the size of the replicating system, so that Darwinian selection could yield significant concentrations of one or a few comparatively short 'words', but would not explain the formation of self-replicating systems much richer in information. Eigen and Schuster's hypercycle can be illustrated by considering five words, A, B, C, D, E. Suppose that the system of these words is self-replicating not (or maybe not only) because each word is self-replicating, but because A catalyzes the replication of B, B that of C, C that of D, D that of E, and E that of A. Then a steady state can be reached where a particular 'sentence' ABCDE is dominant and capable of self-replication.

Thus, by a very simple model, Eigen and Schuster have shown that the gradual complexification of living matter starting from a chaotic mixture of its non-living component is perfectly compatible with the laws of nature, indeed may correspond to a potentiality present in non-living matter such as it is at a certain stage of the history of a planet. Of course, there is a large gap between the presentation of the outline of a possible mechanism and the proof that things actually went that way; but work is being done in that direction[14], and at any rate the hypercycle hypothesis is sufficient to show that some of the most popular objections of principle to the thesis that life appeared spontaneously from nonliving matter are untenable.

3. The role of chance

The role of chance in autopoiesis deserves special attention. The claim that a spontaneous process could never yield such a highly organized system as a living being or that at least it requires some external "ordering influence" (such as the regular heat flow leading to the appearance of Bénard's structures) cannot apply to chemical and biological processes.

On a smaller scale, such random processes are the everyday experience of chemists. Every chemical synthesis relies on chance encounters of molecules. For example, in Grignard reactions certain molecules X (alkyl halides) form molecules of a relatively stable complex Y (the Grignard

---

[13] M. EIGEN and E. SCHUSTER, *The Hypercycle: A Principle of Natural Self Organization.* New York, Springer 1979; M. Eigen, op. cit.

[14] For a commentary on the state of the problem, cf. R. M. MAY. *Hypercycles spring to life, Nature*, 353, 607-608 (1991).

reagent) as a result of random collisions with grains of magnesium, and the X molecules sooner or later collide with molecules of type Z, also present in the reaction vessel), forming new complex molecules, which then dissociate to yield the sought-for XZ molecule and a magnesium salt. All this takes place in a solvent under appropriate conditions. The objection might be raised that this example is limited to the production of molecules far smaller than the building blocks of living matter. But it points to the right direction inasmuch as it shows that molecules whose formation is extremely improbable by direct collision can be formed by indirect mechanisms.

As to questions of size and structure, convincing evidence is accumulating in that connection too. For one thing, rather complicated self-assembled molecules have been produced in recent times[15].

Thus, the generation of ordered structures from chaos is by no means a novel discovery. The novelty has been introduced by Ilya Prigogine, the Nobel laureate for chemistry who has introduced the notion of dissipative structure into science. He has pointed out something more, namely that the ordered structures formed as a result of random collisions may be out of equilibrium, and yet kept in a steady state by a continuous inward and outward flow of matter and energy[16]. Living beings are extremely sophisticated examples of such systems, more specifically of a particular class of them: those which maintain their identity and indeed produce new order by growth and reproduction in virtue of an internal organization which "metabolizes" energy and matter coming from outside in an extremely efficient way according to specific built-in rules. But there is no logical gap between chemical reactions followed by stages of molecular selection and organization and subsequent complexification leading to higher living beings.

The arguments for the emergence of life by a sequence of chemical reactions and associations from a completely disordered "primaeval soup" are thus seen to be scientifically plausible. As I have mentioned, supporting evidence obtained by ad hoc experimental studies is gradually accumulating. Nevertheless, in view of the high selectivity of the processes leading to the emergence and growth of order and organization, it is perfectly legitimate to

---

[15] E. C. CONSTABLE, *Molecule, assemble thyself*, Nature 362, 412-413 (1993).

[16] Apart from his merits in the special field of the thermodynamics of irreversible processes, the contribution of I. PRIGOGINE to present scientific thought is extremely important. Perhaps the best compendium of his contributions to science and of their relation to culture is to be found in his book with I. STENGERS, *La nouvelle Alliance: Metamorphose de la Science*. Paris: Gallimard 1979.

be perplexed by the notion of chance as used in hypotheses about the origin and evolution of life. Confusion between metaphysical and scientific issues has even brought about unjustified emotional overtones. Therefore, I shall briefly review the question, asking the reader to refer to my own previous studies for additional references and details[17].

From the scientific point of view, it seems clear that what is claimed by the advocates of the spontaneous emergence of life actually amounts to saying that, given the right mixture of chemicals and the right environmental conditions, sooner or later, in virtue of the random motion of molecules, self-replicating molecules would be formed, just as surely as a new molecule is formed by random collisions in the reaction vessel of a chemist. Now, astrophysical studies strongly suggest that the development of a planet such as Earth was bound to produce on cooling the reducing atmosphere required to form the building blocks of biomolecules and local conditions whereby precisely such self-replicating molecules as we know would be formed. Furthermore, according to a view now considered practically certain, the changes in the geological and atmospheric conditions of Earth began precisely as a result of the action of living organisms, most dramatic of all being the transition from a reducing atmosphere to an oxidizing one, such as we have today, between three and two billion years ago. This means that, when life was not present, Earth would most probably stay as long as necessary in a condition favourable for life formation until the latter took place.

Thus, only the time and place of the formation of life would be left to chance. Even the question: "why precisely on Earth?" loses significance. There are hints suggesting that the formation of a planet with the characteristics required for the spontaneous appearance of life was in the nature of things at least somewhere in the Universe, and our Earth happens to be precisely that planet. In short, the significance of chance in the history of the Universe would only be great if, for the first random steps of self-organization to take place, there was only a limited time, in a place which might or might not be formed during the life of the Universe; and these limitations do not seem to be supported by any significant evidence.

---

[17] G. DEL RE, *L'organisation, l'auto-organisation et l'image de l'horloge*. Epistemologia, vol. IV, special issue pp. 53-72. 1981; *Frequency and Probability in the Natural Science*. Epistemologia, 7, spec. issue p.75 (1984); *Cause, Chance, and the Slate-Space Approach*. In: *Probability in the Sciences* (E. Agazzi ed.). Dordrecht: Kluwer Acad. Publ. 1988. pp. 89-101; *The case for finalism in science*. In: *Poznan Studies in the Philosophy of Science and the Humanities: Intelligibility in Science* (ed. C. Dilworth). Amsterdam: Rodopi 1992. pp.161-171.

It is perhaps less easy to apply the same argument to speciation. There, it may be said that evolution is a random exploration of all the creative potentialities of life as time goes by and the conditions of the biosphere change. Yet, it seems reasonable to claim that, given the rules of the game, even such a complicated living being as man was bound to appear, sooner or later, on the stage of the Universe. Such a claim is in line with the general views underlying the widely debated "anthropic principle"[18].

In sum, it is one thing to claim that spontaneous emergence of complex organized systems from a disordered situation is perfectly compatible with the laws of science and is indeed exemplified by the appearance of life, it is another thing to declare that certain events had a significant probability of never taking place. A careful examination of the significance and actual role of chance in the scientific description of those events is an indispensable preliminary to such statements. In particular, extreme care should be taken in using such expressions as "life appeared by chance" for they place emphasis on what might be a minor feature of the extremely complex phenomenon under study.

A digression is necessary here, even though it touches on a point lying outside science. The above remarks are intended as a contribution to the philosophy of science, not as hints for coping with the false problem which has given rise to the debate between creationists and anticreationists. In fact, a serious metaphysical error is contained in the idea that if life actually appeared by chance (in the strong sense of the word) the existence of an intentional design of the Universe is ruled out. As Augustine of Hippo pointed out fifteen centuries ago[19], God is above time, and has created as it were *tota simul*, everything at the same time, what was, what is, and what will be. In other words, God is not an engineer who has set up a mechanism, and possibly modifies and adds pieces to it as time goes by; rather, he has conceived and realized the whole creation by a single act[20]. The history of the physical Universe such as is being laboriously reconstructed by science is like the description of a line, which we see in its entirety at one glance, by a one-dimensional being who advances along the line, and therefore meets its points one after the other.

---

[18] J.D. BARROW and F. J. TIPLER, *The anthropic cosmological principle*, Oxford: Clarendon 1986.
[19] Conf. 11, 4,6-6,8)
[20] This does not exclude 'miracles', i. e. exceptional interventions of God in space-time: to be above time is not the same as being outside time. Nor does it exclude free will: to have taken our choices into account in the eternal instant of creation is not the same as having determined them.

Thus, for a metaphysician referring to the God of Christianity, the very existence of life is a sufficient proof that God has chosen it to be; the process by which it has appeared within time is only relevant to man - and to God when he deliberately places himself at our level.

Therefore, neither creationism nor anticreationism can gain anything by proving or disproving that life appeared by chance. If anything, one can say that, for those who believe in God the Creator, the artistic side added by the failure of Laplacian determinism to account for the history of the Universe makes the new Weltanschauung suggested by science much more open to the glory of God revealing itself in the physical universe[21].

---

[21] An extremely illuminating passage on this point can be found in: CHARLES JOURNET, *Le Mal. Essai Théologique*, Paris: Desclée De Brouwer 1961, ch. 5, sec. 2.

DISCUSSION
(B. Pullman, chairman)

ARECCHI: I think that your definition of complexity is the number of unforeseen situations a system can face. You have said that it is more or less equivalent to Aristotle's form. It seems to me that this is the point of view of somebody who looks at a complex object as an "agent". I tried to stick to the point of view of a scientific observer who makes taxonomy, who cannot include forms within his programme. In fact this is a limitation of science, the impossibility to grasp Aristotelian forms.

That was my first point. My second point is about autopoiesis. You say that at variance with heat flow, in chemical processes organization can arise spontaneously without the action of an external ordering influence. This may be a terminological ambiguity, because, being active in this field, I have scanned all the existing literature in order to give credit to previous authors, and I am not aware of any single piece of experiment of spontaneous formation without boundary influence, not only in heat flow or in fluid flow, but not even in chemistry. So presumably what you attribute to Eigen is what we call in an informal way "wishful thinking", in the sense that it is expected to be realized in the future, but so far has no experimental counterpart. Moreover, you correct your autopoietic position at the end, when you dedramatize the role of chance, and then you resort to some external factor such as the role of the atmosphere, etc. So, presumably, either there is a terminological ambiguity, and you believe, like me, that there is no spontaneous formation or self-organization, or there is an inconsistency between your eulogy of autopoiesis and your dedramatization of chance.

DEL RE: Your suggestion that the complexity of a system acting as a unit, such as a living being, can be measured by the number of different situations it can face is extremely interesting. The question, in my opinion, is whether the notion of situation can be made to correspond to a discrete set, such as the states of a bound quantum system.

Concerning Aristotelian form, what I have briefly mentioned is the result of studies and discussion with philosophers I have carried out for about ten years. Its identification with the information specific to a system appears to be unquestionable, inasmuch as 'form' and *quidditas* are isomorphic, as explained by Aquinas in *De ente et essentia*.

Organization is that part of the information which is related to its behaviour as a unit, more or less what Aristotle called the 'soul'. In this connection I must emphasize that chemists and biologists understand organization in a way different from physicists. They do not take as parts of a system elementary particles, which are components having invariant properties, but subsystems having a variety of internal degrees of freedom, to the point of being capable of varying their properties according to their environment. Therefore, organization for a chemist or a biologist is the interdependence of parts such that the state of each is determined by the states of the other, and all co-operate to determine the behaviour of the whole, whose main characteristic is to be a steady-state protecting its own identity against the aggressions of the environment.

Concerning the question of autopoiesis, the fact that you have not found a specific mention of the point I have made is that it is taken for granted by chemists and biologists. In an *in vitro* chemical reaction, molecules are formed by random collisions in the absence of an orienting external force. New order appears because you obtain rather complicated structures from very simple atoms — e.g., aminoacids from ammonia, methane and water. What determines the result is not fields or forces, but what a physicist would call 'selection rules'. Now, these selection rules also act in the growing of an organism, which implies emergence of new information; and, as you know, Aristotle considers growing as a process only determined by the *nature* of the organism in question. Since the prefix 'self' applies to growth, which is essentially controlled by selection rules, I do not see why I should avoid the same prefix for the prebiotic case. Moreover, esperimental cases of autopoiesis have been presented. References are to be found in the final version of my paper.

Finally, I have dedramatized chance because the scientific role of selection rules and the like has been largely ignored especially in connection with quantum mechanics. I believe that chance does play a significant role in speciation, because many species possible in principle were probably never formed. But there again, I emphasize that no species violating the appropriate selection rules could be formed.

At any rate, suppose Monod and others were right in assigning chance a great role in the origin of life. Then indeed, to think that this would put in question the existence of a Supreme Intelligence having willed not only matter, but each living being is a pathetic sign of attachment to an anthropomorphic picture of God: God created past and future at the same time, and a significant role of chance only proves that our deterministic science is incapable of 'reading' the whole of reality.

ARECCHI: So, this confirms my suspicion that in order to have a spontaneous process, — let us say, formation of micelles — as you said, in order to have life it was necessary to have a given environment. So it is just a semantic difference of a sort, but we are saying practically the same thing.

DEL RE: We are indeed, if you admit that those processes which we call self-organization, though spontaneous, are subject to conditions which only allow one or a few final outcomes; we are not, if you mean that the evolution of a self-assembling system is a necessary process caused by external forces.

MALU: J'ai apprécié le fait que le Dr. Del Re ait insisté sur l'importance d'une bonne appréhension de la notion de complexité. "Complexe" ne signifie pas "compliqué". Un système complexe est un système qui n'est pas justifiable d'une stratégie de simplification déterministe ou probabiliste. La raison est qu'il contient un nombre important de variables ou de paramètres significatifs, qu'il n'est pas possible d'ignorer sans dénaturer complètement le système.

Cette remarque m'amène à dire que je suis quelque peu déçu du déroulement des présentes assises. Je m'attendais à voir les intervenants à la présente session identifier, dans chacune des disciplines scientifiques retenues, des exemples de systèmes complexes. Tel n'a malheureusement pas été le cas pour la plupart.

Ceci étant dit je souhaiterais voir le Dr. Del Re commenter sur l'affirmation suivante avancée par de nombreus biologistes: si la vie était apparue sur la terre de façon aléatoire, le processus aurait pris un temps supérieur à l'âge de la terre.

DEL RE: This point is briefly mentioned in my complete paper. At any rate, the mistake of those who make such probability calculations is that they assume that the various particles must meet simultaneously. In that case, their estimate is correct. But in fact, as happens in many complicated chemical reactions, the process we are considering is stepwise, and at each step only a pair of parts or partial systems is expected to form.

Probability theory proves that the overall probability in the latter case is enormously larger than in the former. Without making a calculation, we can use an analogy: those who speak of very small probabilities are thinking like people who determine the probability that a city like Rome would arise by the simultaneous moving in to it of three million inhabitants. This is, of course, a probability so small as to be close to impossibility. But if you think of the real process, you will consider that first a few men (some say, brigands) moved in and founded the city, then other people were attracted to the place, babies were born, and so on. Then, considering the geographical and agricultural conditions, you might even say that the appearance of a great city in that location in those centuries was extremely probable. I believe this analogy shows the difference between the estimates of probabilities by non-chemists and biochemists, whom I represent here, so to speak. I have given a few references in my final paper.

FONDI: Among the many problems existing in the field of biological sciences, that of the origin of life is undoubtedly the most tremendous. Personally, as a biologist, and not as a physicist or chemist, I feel that the opinions of authors like Ilya Prigogine or Manfred Eigen concerning the spontaneous origin of living systems and biological information are only good speculations, I find it certainly interesting, but speculations and nothing more. It is very significant that such an eminent author as Sir Francis Crick, the discoverer of the double ellipse structure of DNA, declared some years ago in his book *Life Itself*, rather than believe in spontaneous formations of life from a primaeval soup, it is preferable to believe that life arrived on our planet by means of an extraterrestrial spaceship! My question is the following: considering that the most ancient rocks of the earth were formed more or less millions of years ago and that the first fossil bacteria were present in rocks dating 3,8 billions of years ago, which is only two hundred million years after the depositions of the most ancient rocks, considering these effects and the short time at our disposal, do you think that the autopoiesis hypothesis can be equally considered useful and sufficient?

DEL RE: I have already summarized the answer to objections based on probabilities in my answer to Professor Malu. I cannot give estimates, but my experience with chemical reactions suggests that each step could have taken one day or a year, thus giving ample time for the first appearance of "protobions". Afterwards, as I have tried to explain in my full paper, the same general picture of the origins of life admits that self - organization became a programme-controlled process as is the growth of any living being, since the first DNA or (RNA) was already there. At that stage life started to influence the conditions in which it developed, and chance only continued to play a role in speciation. Concerning the various possible hypotheses, of course Crick's hypothesis is an alternative that must be (and has been) taken seriously. As to Eigen's ideas, I do not know what he personally believes, but I consider his theory as an argument showing the *plausibility* of the hypothesis that life formed spontaneously. It is useful, because it inspires new studies. Whether or not it describes what really happened is a question to be answered in the future. So far, we have only a number of hints. The main question, in my view, is the arrival at self-replicating RNA-like molecules. That point has been recently reviewed by L. E. Orgel.

DALLAPORTA: The example you have shown us so clearly is rather the formation of structures, or even perhaps organization; I would be rather cautious to go beyond. The mechanism is a cyclic catalysis of A on B, B on C, etc. That sort of interdependence is a very peculiar property, it is a potentiality that belongs to these five molecules. The resulting structure is a kind of idea, a kind of form, as you said. I quite agree that in this respect this is the real reason why life can originate: the fact that, when those molecules get together, the random aspects of the process are completely irrelevant. Thus, my point of view is that such phenomena as structure, and life itself exist because there exists in some sense a potential, *a priori* structure, and they form because of that, not just because they meet. My question is: you mentioned that there is a single way in which A, B, C ... can be bound. I would like to know if you mean that A catalyses only B, B only C, and so on, so that only this single cyclic system is formed, not a random ring.

DEL RE: What you have said summarizes the main point of Eigen's view of chemical self-organization very well. As to your question, there are two different, but similar aspects of the story. One is the formation of chemical bonds proper, and that is highly selective: I shall not say that in general two atoms or molecules can join to form only one new combined molecule, but certainly only two or three different ones will be formed, and most often only one, depending on the chemical nature of the partners. What you are pointing out concerns chain catalysis. I do not think a complete answer can be given from a general suggestion, and Eigen has not specified the molecules or group of molecules involved in the real process. That is a problem for the future. However, from my experience as a specialist of chemical reaction mechanisms, I should say that such a complicated system of molecules catalysing one another is possible only with a very special choice of the individual members. Thus my answer is: yes, with the chemicals present in the primaeval soup, there would be only one choice for the stable hypercycle to be formed.

PULLMAN: I would like to comment on two subjects raised in your lecture and referred to also in previous comments.

The first refers to the question: how is it that we find practically the same fundamental building blocks, performing the same mechanisms in all living creatures? This is a most interesting question in connection with which I would very briefly like to relate an intellectual adventure which happened to me many years ago. When I was a student of bio-chemistry learning by heart the formulas of biologically important molecules, I was very much impressed to observe that porphyrins for example were present in so many different types of biological activities: porphyrins take part in photosynthesis, a very fundamental activity — they are involved in electron transfer by oxydation-reduction enzymes and they form a central block in hemoglobins for the transfer of oxygen. I was very puzzled about how it happens that the same molecule is so extensively used in so many different ways. I made a similar observation again when I was a student, about the omnipresence and the importance of adenine, a molecule which exists in the purines of the nucleic acids, which is a fundamental component of ATP, which is the basic currency in bioenergetics, which is present in a number of coenzymes etc. Now, five, ten years later, when I started to do research, I came across a paper which indicated that porphyrins — although very complex structures — existed already three billion years ago: you can find them in paleobiological fossils of that period, which means they originated very early and have persisted since. And then one day I came across a possible solution to this query in my own studies on the electronic structure of molecules. Porphyrins can be shown to be exceedingly stable molecules.

As they are built of carbon, nitrogen and hydrogen they can be processed from the primitive atmosphere which was there three billion years ago, but their enormous advantage of persisting is to be exceptionally stable. The explanation of the origin of this stability is of quantum-mechanical nature, namely that there is a very strong *resonance* between the different structures which you may write for them, or in the language of the molecular orbital theory, you may say that there is a very strong delocalization of the π electrons, the electrons of the double bonds. This is the source of their great stability. There is more to it, because these and other primitive constituents of nature, which were formed in the primitive soup by the action of radiation or light or heat, must resist the destructive effect of the same factors which produced them. Light which can make molecules can just as well destroy them. Heat can do just the same.

Now, the stability or porphyrins extends to their excited states which are resistant to radiations, ultraviolet light etc. This enormously increases their chances to persist. And then I think nature just made use of what it had, what was at its disposal, and adapted it to many different uses, in photosynthesis, hemoglobin etc. This is a simplified answer to the first point. Now, of course, the whole story is much more complicated, because you do not just construct life out of these sole elements. They have to get involved in more complex structures. This is the place where the second problem which you have raised intervenes, namely possible autocatalysing effects of biological macromolecules. This can also be envisaged, in principle at least, in a simple way. I don't have time to dwell on it here, but one may show, in fact we have shown, that a fusion of two binding blocks of, say, proteins (amino acids) or nucleic acids (nucleotides) may very well facilitate, thus autocatalyse, the addition of a third block by the effect of increasing the overall molecular electrostatic potential in the two-membered block with respect to that of its constituents. And this effect continues in higher oligomers and polymers. Finally, I would like to make one more small remark. You are speaking all the time about a phenomenon to which you don't give the appropriate name. It concerns the interaction of molecules. You go to great trouble to describe a phenomenon which has a name: it is specificity in biological interactions. It is one of the most fundamental problems in all biological reactions. A few years ago in this Academy I organized a symposium which is published as "Specificity in Biological Interactions".

THOM: Just a question on terminology. If I have understood you, you would reserve the term "organized" for structures for which you could recognize something like a biological finality.

DEL RE: Of course, the choice between the terms structure and organization is free, but I believe that a difference exists and is represented in our languages by the two words. I have reserved the word organization precisely for the meaning you have mentioned. What you have called finality, however, may be nothing more than an active cooperation between parts aimed at performing some tasks, or a certain kind of activity which is related to some characteristics of the system. That is to say, the system has a certain identity, and organization makes it capable of acting in certain specified ways even in a changing environment.

THOM: For you a corpse is not organized?

DEL RE: No, it is not.

THIRRING: I wanted just to follow up the question which was asked by Professor Malu: can you give a reasonable estimate of the order of the magnitude of the time one needs to make life by chance? Because apparently it took about three hundred million years, which is 10^{16} seconds. Now, if you make it step by step, and such an encounter between molecules takes 10^{-10} seconds, then you make 10^{26} encounters and the question is: would this be enough to reach the goal, if you just make a rough order of magnitude estimate?

DEL RE: By all means, if you consider that the actual probabilities of the various steps are the single encounter probabilities multiplied by the concentrations of the particles under consideration in a droplet of the primaeval soup. Now, processes are known whereby the concentrations of chemical species coming from the atmosphere increase to large values just in a few months, particularly as a result of the interplay of solubility, diffusion, temperature, etc. The importance of the stepwise scheme lies in the fact that sufficiently high concentrations of, say, the first pair, can be reached within comparatively short times. Thus, the whole story hinges on the concentrations of the building blocks of the final self-replicating molecules as well as on the probabilities of single encounters, and the experimental evidence of millions of chemical reactions supports the claim that the times for each step may be of the order of days or months. One must allow for much longer times and assume that a large number of droplets were formed approximately at the same time because the actual conditions were not as strictly controlled as those in the laboratory, and because, at any rate, the concentrations of the intermediate chemical systems would be smaller and smaller as the process approached its final stage. A rigorous analysis of the pertinent probability considerations, with all sorts of examples, is given in treatises on chemical kinetics. One of Eigen's assets in connection with his contribution to the theory of the origin of life has been precisely his great experience with research in this field, which has won him a Nobel prize.

ODA: I may have misunderstood you, but, if not, do you not imply that whenever there are conditions very close to those of the appearance of life on Earth, life should be created and would develop like ours? If so, the chance of the existence of extraterrestrial intelligence, EI, may not be zero. Then, the bearing of this on the future of mankind might be very deep and important. I personally have never been eager to reflect about the probability of EI. This has been taken seriously only to a very limited extent by scientists. But, should the issue be taken as a really serious one, physicists could think of improved ways to search for EI.

DEL RE: I believe you have correctly seen one of the implications of the picture I have described. Of course, I am not an expert on everything, and I can only make an educated guess about the chance of EI based on the theory I have expounded. I could say that, given the Sun at the beginning of the history of the Solar System, there was a sequence of events which was somehow bound to produce Earth, certain conditions on Earth, and eventually life as we know it. If there has been, there is, or there will be in the Universe another solar system closely similar to ours, then my opinion is that life similar to ours should emerge. But the problem is then shifted to the origin of our Solar System.

ODA: Then your answer is yes?

DEL RE: It depends. If we are referring to life in a broad definition, I do not think we can say anything, because we have no idea of what material basis a different sort of life could have. But as to the same life, I am inclined to answer no. This is because extremely small changes in the conditions at the very beginning could lead to a different kind of life, or to a situation where the emergence of any form of life is not possible. Therefore, I do not really expect to meet other human beings in the Universe, although I may expect that life also exists in other parts of the Universe.

ODA: That is deterministic chaos, I think. Thank you.

PAVAN: We are talking a great deal about the origin of life. Could you clarify a point on the origin of death? You said that during the self-organization process a large molecule originated, lives and dies. Now, how can a self-duplicating molecule die? Would it not be like any DNA molecule or like an ordinary virus? What do you mean when you say a molecule is born, lives, and dies?

DEL RE: I use that language metaphorically, of course, but anyway in analogy with what is also used for particles. At a certain moment an event takes place by which a new entity appears. In the case of molecules this consists in general in the collision of smaller molecules, say just two, for example ammonia and methylchloride, which meet to form methylamine. That is the birth of the new molecule, methylamine. Its life is of course the time between its birth and its death. The latter would be a collision, say with another molecule, which would break it down, or even spontaneous dissociation, by which it would give two different molecular species. Of course, methylamine has such a long life span that it is almost immortal, but bigger molecules have a shorter lifetime, in the sense that sooner or later they can either spontaneously dissociate, which is a sort of natural death, or be killed by a collision. The terminology, is of course, metaphorical.

PAVAN: I would agree with "can die if you kill it", but not die naturally, I cannot see any mechanism that makes a virus die just for its own biological or chemical structure.

DEL RE: But why would you not accept that, for instance, unstable elementary particles die? I think the physicists would accept that...

PAVAN: If you are talking about chemistry I agree with you, but I thought that you were talking about living organizations, or living molecules, in self duplicating systems.

DEL RE: I honestly doubt that single molecules of DNA or single individual virus are immortal. There has been much talk about DNA strands aging and beginning to make more and more errors of replication. Even cells are immortal only because they renew their material all the time, but the macromolecules present at a given instant may get out of the game for a variety of reasons. I can think of that possibility, and I believe it is a real fact, but of course I do not know whether tests, say with radioactive labelling, have been made to check this aspect of the story.

PULLMAN: There is an extremely good paper by George Wald, a Nobel Prize winner, in the proceedings of the U.S.A. National Academy of Sciences, published a number of years ago, which has the title: "The origin of death". George Wald being a very clever and witty person, you will enjoy reading it, it's witty, very amusing . Now, just to answer your question, in fact not to answer it, but in connection with it, I wish to recall a remark by Diderot, a famous French philosopher in the 18th century. He said: "Vivant, je suis une masse. Mort, je suis un amas de molécules".

LEJEUNE: J'ai été assez intéressé par le développement des discussions, où nous avons péniblement démontré que les lois physiques ou chimiques connues aujourd'hui n'interdisaient pas l'apparition de la vie. Je crois que c'est une très bonne découverte, car si les lois physiques ou chimiques que nous connaissons aujourd'hui interdisaient l'apparition de la vie ce sont les lois physiques ou chimiques qu'il faudrait changer. Mais je ne crois pas qu'une démonstration de non-incompatibilité soit une démonstration de causalité.

DEL RE: Je suis tout à fait d'accord avec vous. Ce n'est pas une démonstration de causalité. La vision du monde dans laquelle s'encadrent les questions que j'ai présentées n'est pas déterministe au sens Laplacien du terme. C'est là la nouveauté introduite par ce recours au hasard qui, en soi, comme j'ai essayé de montrer, veut dire très peu, mais qui met l'accent sur l'absence de causes efficaces. Il y a un flux naturel des choses, et ce qui arrive est le fait de règles de sélection plutôt que de "forces" ou de décisions préala-bles. Si ce à quoi vous faites référence est le côté métaphysique de la question, je ne peux que répéter ce que j'ai déjà dit dans ma réponse à Tito Arecchi: la création n'est pas mise en question par la nouvelle manière de voir l'origine de la vie. Au contraire, elle montre que nous n'avons pas le droit d'attribuer au Créateur une création dans le temps selon les modalités de notre agir, car ce serait de l'anthropomorphisme.

CHAUVIN: Je ne suis pas spécialiste de ces matières, mais je suis tout de même biologiste, et quant à l'origine de la vie je vous rappelle qu'il peut y avoir plusieurs arguments en tous cas pour l'existence de la vie en dehors de notre planète. Je pense que beaucoup d'académiciens connaissent l'existence des chondrites charbonneuses comme le météorite tombé à Orgueil, un village de Tarn et Garonne il y a cent ans à-peu-près, le météorite d'Ivuna, et je crois qu'il y en a encore un ou deux autres, pas plus. Ces météorites ressemblent à des sacs de charbon et on a trouvè à l'intérieur des cires, des lipides supérieurs, des acides aminés, et (là vous allez peut-être sauter, mais je tiens la chose de Hoyle qui est un astronome quand-même assez connu et qui dans son bouquin "L'Univers intelligent" en montre des photos) des organismes qui ressemblent à des filaments, à des champignons, qui auraient été isolés par Pflug dans le météorite d'Ivuna, ou d'Orgueil, je ne me souviens plus exactement. Vous savez que actuellement des sondes sont en train de partir vers Mars, elles ne sont pas encore capables, je crois, de ramener des échantillons à la Terre, mais lorsque dans très peu de temps nous pourrons envoyer des robots ou des explorateurs sur Mars il se pourrait bien que nous n'y trouvions non pas de petits hommes verts, ça m'étonnerait, mais des fossiles, ça ne m'étonnerait pas. Alors, il y a deux possibilités, les fossiles sont pareils à ceux de la Terre ou ils sont très différents. Dans les deux cas c'est un énorme problème. Donc le problème qui nous amuse tellement sur l'origine de la vie, et qui est tellement important, pourrait rebondir prochainement, je crois.

PULLMAN: Je pense que, prochainement, c'est peut-être optimiste. Je voudrais signaler qu'il y a un membre de notre Académie, qui est extrêmement partisan d'une poli-existence de la vie. Malheureusement, il n'est pas là aujourd'hui. C'est le Professeur De Duve. Il a même écrit un livre à ce sujet. Je ne peux pas dire que je suis très en accord avec lui.

CHEMICAL ASPECTS OF EVOLUTION

G. GERACI

Department of Genetics, General and Molecular Biology
University of Naples "Federico II"

Introduction

The fundamental chemistry of the genetic information and the mechanisms through which it is replicated are both well established. It is also common knowledge that mutation, the changes in chemical composition occurring in the genomic DNA of an organism, is an expected activity. Mutations are caused both by the external action of physical and chemical agents and as consequences of internal activities such as spontaneous loss of purines from DNA, errors during DNA replication, exchange of DNA filaments and others. It is possible now to chemically modify genetic information or to chemically synthesize a gene and introduce it in a stable condition in the genomes of prokaryotes and eukaryotes. Nonetheless the mechanisms that control evolution are still mysterious. Darwin's mechanism of mutation and selection for adaptation and formation of species requires some further elements. The chemical alterations of DNA leading to the formation of mutant cells can be shown to occur spontaneously, generation after generation, in a population of bacteria or by exposing cells to mutagens in suitable conditions. I think, however, that before attempting to try any possible interpretation of the chemical bases that have produced the variety of forms of life existing today, it is necessary to distinguish between adaptation and evolution. On closer inspection these appear to be different phenomena.

Adaptation, and the consequent possible speciation, concern refinement of the already existing and are continuous activities. Evolution is production of new forms, a historical phenomenon, something that has happened at particular times, with very low probability of repetition and that generates an increase of complexity in the structure and organization of the living organism.

This is the aspect of evolution that I plan to discuss here. The differences between adaptation and evolution can be illustrated with a simple example[1]. The vastly larger number of individuals in a bacterial population and their very short generation time, as compared with hominids, make the expected number of experienced mutational events incomparably higher in bacteria. If mutation of a gene sequence and inheritance of the mutated genotype by the progeny is the mechanism for evolution, then it is not clear why bacteria of fossils show little or no difference with respect to bacteria of our times. There is no increase in complexity, in terms of formation of more complicated structures. However, it is difficult to establish how many types and species of bacteria exist, considering their enormous numbers and varieties colonizing all possible habitats, from deserts to ice to hot waters. Therefore, only mutation of the existing is not sufficient to justify the production of eukaryotes and their type of complexity. Other parameters should be considered. The comparison of structure and function of proteins provides data on the "local" implementation of the genetic information necessary for life, but provides no information on how the implementation occurred. Investigations at the level of the genomes may instead provide important clues. There are two major genetic differences between prokaryotes and eukaryotes: the number of expressed gene functions and the ratio between the amount of DNA coding for the actively expressed functions and the total genomic DNA. Genomes of bacteria code for several thousand gene units that are efficiently and tightly packed with no redundancy of DNA with respect to information. No repetitive sequences are found in the bacterial genome. The genomes of the eukaryotes code for several tens of thousand-gene functions. The amount of DNA is far more than necessary to code for the genetic functions. There is a correlation between the number of base pairs of genomic DNA and the complexity of the organisms. In the lowest forms of eukaryotes, the amount of genomic DNA is at least 10 times more abundant than in bacteria. In higher organisms the number of base pairs is absurdly huge with respect to the apparent informational content. It reaches values of a 1000 fold or more with respect to bacteria and there is no

---

[1] GRASSÉ, PIERRE-P., *L'évolution du vivant. Matériaux pour une nouvelle théorie transformiste.* Paris: Edition Albin Michel, (1973). Pag. 95.

correlation between amount of DNA and the evolutionary level of the particular organism. The genetic information is dispersed in an enormous mass of DNA. It appears as if evolution required redundancy of genetic material. The presence in the genome of eukaryotes of apparently non-functional DNA sequences is a point of major difference with respect to prokaryotes. This redundant DNA permits, in principle, to experiment new genetic assemblies: from modifications of the previous information to production of novel sequences without necessarily interfering with the viability of the organism. This is true if these activities are restricted to the redundant parts used as a scratch pad, with no necessary implication for, or modification of the structural and functional organization of the active units. There is a wealth of analyses made on eukaryotic genomes that have shown the presence of different types of DNA sequences. Short and long repetitive sequences are present, repeated in tandem millions of times or dispersed in the chromosomes in short clusters containing a small number of repeats. Pseudogenes have been found, which are sequences that resemble a gene but are non functional because of some difference in nucleotide composition with respect to the functioning gene. In the eukaryotic genome the information to generate a protein is almost invariably dispersed on the DNA in a number of fragments to be assembled after the gene has been transcribed into RNA. In these cases the primary transcription process generates RNA molecules containing sequences that must be eliminated to obtain the final operative molecules. An extreme case of dispersion of coding sequences on the DNA is that of the Dystrophin gene. This is coded in over 2 million base pairs[2] of which only about 4% are used as RNA and about 96% discarded. There are cases in which, from the same primary RNA transcript, different functional RNA molecules are obtained, coding for different proteins. The variety of gene functions in the eukaryotes is then increased also at the level of the "choice" of the RNA sequences of the primary transcript to be put together in the final molecule. Most interestingly, there are sequences that are control sites for the expression of others. They turn out to be very crucial components of the mechanisms that generate cell determination and differentiation in eukaryotes.

---

[2] WORTON, R.G., RAY, P.N., BODRUG, S., BURGES, A.H. HU, X. AND THOMPSON, M.W., *The problem of Duchenne muscular dystrophy*, Philos. Trans. R. Soc. London Biol., 319, 275-284, (1988).

The contributions of molecular biology

The possibility of understanding the chemical mechanisms that might have led to evolution derives from the increased knowledge of the genome structure and organization of present day organisms, as provided by the modern approaches of molecular biology. The initial studies concerning the sequence composition of proteins and of nucleic acids allowed the analysis on chemical bases of relations among different organisms. These results allowed the verification and the conclusions derived from studies of comparative anatomy and zoology and the discrimination of situations that were not possible without molecular information. Analyses of DNA compositions in the different organisms made it clear that genomic DNA is continually changing. A gene coding for a protein, or another function, usually has different nucleotide compositions in different species. A gene may show nucleotide differences also in individuals of the same species. In man, the DNA variability in certain regions of the genome is so high that it is used to distinguish individuals with an accuracy better than that provided by fingerprint analysis. These findings, however, do not contribute much to the unravelling of the mechanisms that might have caused evolution. Comparative analyses are more accurate, because they are performed at the molecular level, but provide no new insight. In recent literature evolutionary trees, based on sequence data, are under reconsideration and subject to interesting criticism and alternative conclusions. A major problem for the presentation of evolutionary correlation is that it is necessary to assess the role of parameters that may have several aspects. For example, the relevance of genes for the survival of organisms is difficult to assess. An ongoing debate is still attempting to define all the activity functions, relevant to evolution, deriving from the gene, beyond the mere statement that a protein or another product, is the result of the gene expression[3]. As additional problems, there are the identification of the beneficiary of the gene's activity and the definition of parameters for "fitness", as a measure of the ability of an organism to propagate its genes. Both have several aspects and conse-quences. It appears to me that clues concerning the successful chemical mechanisms through which life evolved can be derived from the analysis of the recent results of structural and functional studies on the genomes of the organisms populating the earth, at present. These may be taken as positive

---

[3] DAWKINS, R., *The extended phenotype*. New York: Oxford University Press, (1982).

representatives of successful evolution. In these past 10 years studies concerning the structural properties of DNA and the mechanisms of gene expression in "in vitro" cell cultures, in developing embryos and in transgenic animals have taken great steps forward. These studies have provided the initial chemical evidence of the epigenetic aspects, peculiar to eukaryotes, so important for evolution. In particular, the results of these studies have shed a new light on the molecular mechanisms that control cell differentiation and the formation of orderly body plans in very distant organisms. They have shown that the possibility of cell differentiation is the result of the implementation of a combinatorial mechanism for gene expression. This mechanism permits the eukaryotic genome to produce different stable phenotypes by the proper utilization of different groups of genes. It is as if the complete genome were a master gene sequence from which a number of sub-genetic assemblies could be selected. From the complete set of n genes contained in the genome, different groups of k_i genes would be selected by each of the different cell lines composing the organism. The implementation of this combinatorial mechanism appears to be the chemical basis through which eukaryotes increase the number of expressible phenotypes, and hence body complexity, using a number of structural genes only ten times higher than that found in prokaryotes.

Evidence for the chemical basis of cell differentiation.

The mystery of how cells with different phenotypes are produced starting from a fertilized egg is on the way to being explained chemically. Multiple specific sequences have been found associated with eukaryotic genes. These sequences are the targets for the binding of specific factors that are important for gene expression. The genetic unit of eukaryotes is consequently formed by sequences that contain the information to generate a product and by signal sequences that contain the information to control the expression of the entire unit, in correlation with others. The consequences of this type of mechanism are a fine and detailed control of gene expression and the possibility of coordinating the functional potentialities of genes grouped in different genetic assemblies. A factor may be necessary for the expression of some genes, stimulatory for the expression of some others and not relevant for others. The first molecule of this type to be known, was the Nerve Growth Factor, identified in pioneer work by Rita Levi-Montalcini.

That molecule is a good example of a factor with multiple roles[4]. Many more factors have been isolated since then. The importance of factors has now been established also by experiments in chimeric mice. It has been demonstrated that it is sufficient to modify the DNA binding capacity of a protein factor to specifically cancel the ability of an organism to produce a particular cell line. The results of an experiment of this kind have shown that the binding of the factor called eryf-1 to its regulatory sequences is necessary to express the cellular erythroid phenotype in the mouse. In fact, cells containing the gene for eryf-1, mutated to impair its ability to bind to its regulatory sites on DNA, participate in the formation of all tissues of the organism except the erythroid lines. Blood cells are not produced though the genes necessary to produce them are there and are unmodified[5]. This means that the possibility of production of a particular differentiated cell type, in a particular organism, depends on the presence of a specific protein factor. These are gene functions that do not contribute physically to the architecture of a particular cell line but are necessary for its formation. This type of protein adds to the complexity of the expression of a genome. The production of such a type of protein is an "invention" that has opened new possibilities. Many protein factors have been found that, through different mechanisms, exert differential controls on the expressions of gene functions contributing to produce different cell lines having different determinations and differentiation. Analyses of the sequences show that most of those factors have similar organization in functional domains, several of which show homologies with each other. These homologies are indicative that once an active sequence has been produced, it is not only propagated to the progeny but, very likely, it is mutated producing variants that contribute to increase the number of different cell types and, therefore, to increase the body complexity.

A family of genes, the homeobox genes, appears to control functions at even more elevated hierarchical levels. Their activity is necessary for the correct formation of the body plan in organisms as diverse as cnidaria, insects and mammals[6]. The organization of these genes is in clusters. The

[4] LEVI-MONTALCINI, R., ALOE, L. AND ALLEVA, E., *A role for Nerve Growth Factor in Nervous, Endocrine and Immune system*, PNEI, 3, 1-10, (1990).

[5] PEVNY, L., SIMON, M.C., ROBERTSON, E., KLEIN, W.H., TSAI, S.F., D'AGATI, V., ORKIN, S.H. AND COSTANTINI, F., ERYTHROID, *Differentiation in chimaeric mice blocked by a targeted mutation in the gene for trascription factor GATA-1*, Nature, 349, 257-260, (1991).

[6] MURTHA, M.T., LECKMAN, J.F. AND RUDDLE, F.H., *Detection of homeobox genes in development and evolution*, Proc. Natl. Acad. Sci. USA, 88, 10711-10715, (1991).

number of clusters and the number of genes in each cluster appear related to the complexity of the organism. This is as if the increase in complexity of the homeobox genes in terms of increase of number of genes in a cluster and increase of the number of clusters on more than one chromosome, are prerequisites for more complex body plan organizations. The relevance of homeobox genes has been known for a long time both in insects and plants but only with the advent of molecular biology and production of transgenic organisms has it been possible to obtain direct chemical evidence of the correlation between a particular homeobox gene in a cluster and the particular region of the body influenced by its activity. The effects of alterations of homeobox genes were initially shown in the fruit fly *Drosophila melanogaster*, as reported in textbooks of genetics and molecular biology. Recently these genes have also been found in man. The demonstration of their relevance for the formation of correct body structures has been obtained in mammals with transgenic mice carrying mutated homeobox genes. Experiments performed on transgenic mice with gain-of-the-function and loss-of-the-function mutants show "in vivo" that homeobox genes have similar responsibility for body structures, with similar correlation as in *Drosophila* between influenced body structure and gene positions in the genome[7]. More important, it has been demonstrated that mouse homeobox gene Hox 2.2, (homeobox genes are called Hox genes in the mouse,) which is homologous to the *Drosophila* as the original Antennapedia gene, is active in *Drosophila* as the original Antennapedia gene showing that the gene has maintained its specificity and function between insects and mammals. Very recently homeobox genes have been identified also in plants. It seems reasonable to imagine that homeobox genes are among those genes that, when produced, contributed to the unexpected and sudden appearance of new life forms. Once a molecule of this type has become initially operative, it is not difficult to imagine how its diversification and dispersion in the genome, by means of the normal genetic mechanisms, may have contributed to the production of different new life forms.

---

[7] LE LOUELLIC, H., LALLEMAND, Y. AND BRULET, P., *Homeosis in mouse induced by a null mutation in the Hox-3.1 gene*, Cell., 69, 251-264, (1992).

Repetitive DNA of eukaryotic genomes

The mechanism for the control of gene expression based on multiple specific factors binding to specific regulatory sites requires, obviously, the "invention" of factors. It also requires the presence, at appropriate positions of the genetic units, of the suitable DNA control sequences to which control factors should specifically combine. Inspection of the base composition of the control sites in proximity of genes shows composition motifs. Interestingly, motifs such as the AT-rich sites, that frequently label the beginning and the end of the genetic units on DNA and that are recognized by components of the apparatuses that have the responsibility to transcribe the genetic information into RNA molecules, are common to eukaryotes and procaryotes. Recently a catalog has been published reporting the compositions of over 140 different DNA sequences to which vertebrate transcription factors combine[8]. These sequences show base compositions that are reminiscent of repeats present as clusters in different parts of the chromosomes. These similarities suggest that it is not impossible that the regulatory sequences might derive from parts of the chromosomal DNA that have been so far overlooked because considered superfluous and non informative. Amplification of short master sequences might have produced the repetitive clusters. Parts of these units might have migrated to other positions in the chromosomes, transported there by the usual genetic mechanisms forming the interspersed repetitive sequences. Some others, at appropriate positions near genes, became the targets for the binding of regulatory factors.

Considerations about evolution

Evolution is a term that requires specification. It can be considered from many points of view. At the beginning I pointed out that I would consider the aspects of increased complexity of informational content both in terms of quantity of DNA, expressed in base pairs, and formation of complex body structures. This was not to ignore evolution in prokaryotes but to stress that, in those organisms, evolution appears to have followed a different path, probably depending on a different approach to the problem of improving

---

[8] FAISST, S. AND MEYER, S., *Compilation of vertebrate encoded transcription factors*, Nucleic Acids Res., 20, 3-26, (1992).

genetic information. The increase of efficiency of genetic coding, the economy of the genetic information, the immediate flexibility and adaptability of the gene activity to the environmental situations, have made bacteria capable of colonizing the most diverse habitats. Their apparent approach to genetic improvement, not tolerating additional non-operative DNA sequences, is on the one side efficient, but on the other, limiting for innovation. It is as if bacteria had reached perfection in adapting themselves to the environment, to the cost of innovative solutions. In contrast, eukaryotes appear to have "chosen" to improve genetic information by increasing the amount of genome DNA, with no particular limitations to active units. This is a costly approach because cells have to take the burden of replicating and conserving large amounts of DNA apparently not involved in the maintenance of life. This DNA, permits the disposal of chromosome regions where genetic experimentation is possible with no substantial constraint due to the necessity of survival of the organism. When new sequences are produced, endowed of functional activities, capable of productive interaction with the pre-existing genetic information, they might add complexity to the genome by becoming components of the operative regions. This approach is the one leading to "inventions", to the production of gene functions not depending on, and not predictable from, the pre-existing gene assembly. In this picture eukaryotes would take advantage of two different mechanisms for evolution. The continual mutational activity of the functional parts of the genome, typical also of the prokaryotes, and the stochastic additions of newly "invented" functions assembled on the redundant DNA regions not directly responsible for the maintenance of life. The "inventions" of new functions, are obviously very unlikely and unique occurrences. In coincidence with these "inventions", a sudden wealth of appearance of new organisms is expected. Fossil evidence indeed shows that the appearance of new forms is episodic and explosive[9].

As a last point, it is illusory to believe that it is possible to control the quests for all possible applications of the new methodologies using modification of genes in a cell line and transgenic organisms. The innumerable desirable effects that can be easily visualized as outcomes of this type of study will urge researchers to explore all fields. I think that, rather than relying on boundaries possibly set to these research lines, which are difficult to implement, governments should participate in the research

---

[9] *Op. cit.* GRASSÉ, PIERRE-P.

activities as effective monitors and as a possible control on the consequences for humanity and future evolution.

Post Plenary Session Note

It is evident that DNA redundancy can only be a part of the mechanism through which evolution has occurred. As pointed out by Fondi, DNA redundancy cannot explain why evolution appears vectorial, proceeding only in one direction and, more importantly, at a rate much higher than expected on the bases of the occurrence of simple random events. I think that the results of a research line that we have been following for several years may be of interest in this respect. We have found that the random effects due to the action of physical and chemical mutagens on mouse erythroleukemia cells in cultures are strongly dependent on cell mechanisms. The influence of the cell activity on the outcome of a primary chemical DNA lesion, due to a chemical or physical agent, is such that the probability of producing clones transformed in a particular gene function can be modulated over a range of 1 million times[10]. If similar effects were operative also in whole organisms, the evolutionary outcomes of random mutational events might not correspond to the expectations. Large deviations towards high frequency values might find an explanation.

---

[10] FORESTI, M., GAUDIO, L., AND GERACI, G., *Selective gene mutation in MEL cells,* Mutation Res., 265, 195-202, (1992).

DISCUSSION
(B. Pullman, chairman)

Fondi: I found this paper very exciting. I was very glad to hear that it is necessary to distinguish between adaptation — with consequent possible speciation — and evolution. I was also very glad to hear that evolution essentially consents the appearance of novelties in eukaryots. In your opinion, however, the appearance of novelties can depend on the production by spontaneous mutation of specific protein factors and by molecular drive, in the sense of Gabriel Dover, from genetic material in homeobox genes. I think that these mechanisms are possible in theory, but extremely improbable in practice. Considering the extraordinarily sophisticated correlations existing among all the parts of the organisms and among the organisms themselves, I think that the mechanisms that you advocate can explain the evolutionary process only if coordinated by some organizing law and not solely driven by chance. Thank you. This is not a question, it is only a consideration, but if you care to answer I should be very glad.

Geraci: I agree that it is difficult to invoke chance for the evolution of mammals. They evolved from similar earlier forms in different ecological conditions in physically separated regions of the earth. If the driving force were only chance, one would have expected them to appear in one place and not in others. It seems that the emergence of mammals was a sort of vectorial phenomenon, proceeding in all cases and conditions in a particular direction. Speakers have presented evidence that random processes are influenced by external factors in the choice at bifurcations. External factors can be any type of physical parameter. I think that a factor that might have influenced random processes of evolution was the pre-existing gene assembly. It is not possible to add a function to a gene assembly unless it cooperates and adds to it with no conflict, otherwise the organism cannot survive. The pre-existing gene assembly imposes choices. This philosophy is kept in the development of the adult organism from the egg. There are exact changes and specifications that go step by step. The egg can produce only two blastomeres and these four and so on. For example an egg cannot make a blood cell or a neuron directly. It must undergo a series of events that are strictly specified. The processes driving development are random but their outcome is a very predictable and reproducible series of events.

MOSSBAUER: Why is genetic information distributed over chromosomes and what determines their sizes and number? Number and shape?

GERACI: That is a very good question. I can give an opinion, not an answer. My opinion is that cells are faced essentially with two alternatives. Either the genome becomes more specific and efficient and tries to get as much as possible from the flexibility of the genetic information by adding new sequences. These are, I would say, different methodological approaches, different philosophies. Bacteria "chose" the first one. They have undergone a very large number of cell generations, so that they have had the chance to refine their genetic ability to the utmost. In fact, they are near perfection. Eukaryotes, instead, "chose" the more heuristic line changing and adding sequences to the genome. Once the genome of an organism is structured in such a way that individual pieces can be put together and acquire a final meaning after some elaboration, then an avenue is opened that permits the addition "ad libitum" of whatever is desired. I showed a picture in which, when the amount of genomic DNA was 10 to the 9 base pairs, it did not make any more evolutionary difference if the value was higher. The amount of information is so great at such levels that it is possible to code for whatever wanted. If there is time I will comment further on the evolutionary meaning of the repetitive sequences, as I see them. I believe that they are one of the most important "inventions" of the eukaryotes.

LEJEUNE: It's a question for the future that homeoboxes have shown that Cuvier was right in thinking that there were a few "plans généraux" of the construction of organisms, and he would be delighted to have understood the homeobox system. There is a very simple phenomenon which would lead biology to numerology which is a fact that nature does count on its fingers one, two, three, four, five and stop there. In a sea star there are five arms and we have five fingers and most mammals have pentadactile systems. Then, I would guess that there must be a homeobox that is still acting in the sea urchin and in us, and this homeobox allows our system to count one, two, three, four, five. And to know what five is. At the purely epistemological level it means that calculus was not invented by men but is written in. We have just to decipher the language in which it is written.

GERACI: It will be interesting to compare how the activity of the homeobox genes in sea urchins compares with that of other species.

SEIFERT: You spoke about chemical aspects of evolution but the concept of evolution itself I think is not precisely considered in your papers, so I wonder what exact meaning you give to evolution and whether you regard it as if it were something demonstrable or something even in certain aspects mistaken or problematic; or do you introduce it as an obvious notion? To me it seems, at least philoso-phically, an extremely confused notion and has quite opposite, quite distinct meanings as well as a certain number of problems. But I would be interested in knowing how you see it from the point of view of chemistry.

GERACI: I agree that evolution can have many aspects. I am so much aware of the fact that evolution may have different meanings, that I specified it in my presentation. For me bacteria are more evoluted than man. But I specified that I intended to talk of evolution in terms of formation of more complex, complicated, elaborated structures and accumulation of large amounts of genetic information. I used "elaborated", avoiding "complex" because it would involve a different semantic discussion. It is out of question that the number of genes in prokaryotes and in eukaryotes are different by at least a factor of 10, may be even larger because we are just grasping the genetic organization of man. There is the human genome project that will be of help to this aspect. It is a formidable project and one of its major problems is how to cope with the outcome of those studies. The amount of information will be difficult to use and mathematicians have to provide the proper instruments to make use of it. We do not know the meaning of the different sequences present in the human genome. So much so that excess DNA, redundant DNA is considered junk or garbage. No organism would spend energy to maintain garbage. Organisms maintain repetitive sequences with such an accuracy that it is possible to distinguish apes from one another. They have, as we have, some common and some specific repetitive sequences that make it possible to distinguish species not on the basis of the active genes but on the sequences that are supposed to mean nothing.

PULLMAN: Well, it's amusing because you just made a comment saying that there was not much evolution in this paper. I would make a different comment saying that I don't see much chemistry in this evolution. I have spent my life being a chemist and chemistry means speaking about molecules and atoms. DNA is a molecule, of course, and so are proteins. A discussion of chemical evolution must take into consideration evolution in terms of explicit chemical constituents or compounds. So obviously for some people you didn't speak of evolution for others of chemistry, but it was a presentation on some aspects of the problems involved.

GERACI: The chemical aspects that I pointed out concerning the molecules required to have the type of increase in complexity typical of eukaryotes, such as the factors that coordinate the expression in terms of mechanisms of interaction between the factors controlling gene expression and DNA. I discussed it considering the mechanism of increase in complexity due to the addition to the genome of gene functions such as the homeobox genes. This is one aspect of the chemistry. Of course it requires the detailed studies of the mechanisms of action and these can be provided by the chemistry that I have not discussed.

PAVAN: I very much enjoyed hearing from you that bacteria are superior to man. That's interesting. Yes, perhaps they are, because I've a sore throat now and I feel that some bacteria are affecting me. The second thing: when you answered Dr. Fondi, did I understand that you were saying that evolution in prokaryotes is different from eukaryotes? And the third point: I think that you should emphasize that evolution is a reality. On details of evolution, I think, there may be some questions. But I have no doubt that evolution occurred. The way you put it looks as if you had some doubts. Am I right or not?

GERACI: First question: superiority. I said that bacteria are more evolved than man from certain points of view. As I illustrated in my presentation, if it is assumed that evolution derives from continuous mutation and inheritance of the mutated genotype, generation after generation, the number of generations that bacteria have undergone is so enormously larger than the number of generations of higher eukaryotes that they have certainly acquired the most evolved level according to that kind of mechanism. But this does not mean that they are superior to us. I have never said that. I have never had that thought. If I gave that impression I made a mistake. Again, evolution is a difficult term to define. I have tried in my previous answers to define the perspective under which I was looking at evolution, otherwise I would have felt lost.

PAVAN: Sorry, you didn't answer about prokaryotes and eukaryotes, do you think they have a different kind of evolution or the same?

GERACI: Evolution is evolution. It is the same in prokaryotes and eukaryotes but the directions in which the two groups of organisms are evolving are different depending on the mechanisms that are operative in the two situations. A similar primary action may result in different effects.

SELA: I just want to comment on Prof. Pullman's remark concerning chemical aspects. Dr. Geraci didn't talk about it in his written statement, but the chemical basis of evolution is clearly implied in terms of comparison of sequences of proteins, sequences of DNA. This doesn't add any additional proof of evolution but is very helpful in confirming or correcting genealogical trees. I think that there has been a lot of work concerning the chemical level. It may not help to produce basic philosophical notions of evolution but it helps to correct existing ones and adds a lot of additional knowledge.

TUPPY: One aspect of evolution is the emergence of new species. Could you devise experiments, so to say, for transision from one species to another, on the basis of present knowledge? Would you suggest what would be the mutations in structural genes? How many changes would be necessary to bring about a new species? Do you have any idea of what would be necessary for passing from one species to another?

GERACI: I do not know if I have an answer to this. I can give my opinion. Genes of the different organisms are essentially similar. The situation is similar to that in computers: by changing the programs different outcomes are obtained. I think that with the wealth of types of genes that are in animals — the latest estimates for man is of about one hundred thousand different genes — everything could be built, and also the opposite of that. The real problem is coordination between genes to produce a particular team of functions. The selection and assortment of genes produces different new phenotypes. We have found (in work that

has not yet been published) that some simple homopurine/ homopyri-
midine sequences are present in the cluster of genes coding of the early
embryonic histones in the sea urchin. There are three such sequences,
with different compositions. This type of sequence can undergo a
structural transition from the double helix B to a triple helix structure
(DNA H) under torsional stress. the interesting aspect is that these
sequences are present also in the cluster of histone genes that we are
studying in the marine worm *Chaetopterus variopedatus*. Quite
unexpectedly, these sequences, one to two hundred base pair in length,
are more conserved than the structural part of the H1 histone gene! It is
intriguing that these simple sequences are so conserved in an Annelid
and an Echinoderm, that are in different evolutionary lines, unless they
have a very special role. We started investigating in the gene banks for
the occurrence of this type of sequence close to gene functions. We were
submerged by sequences. I wish to clarify here some aspects of what I
have just mentioned in my answers to Prof. Mössbauer and to Prof.
Seifert. The highly repetitive sequences that are typical of the different
organisms may have constituted pools of master sequences that,
migrating through the chromosomes, might have acted as signal
sequences when positioned in the vicinity of genes permitting their
assortment in groups with coordinated expressions. This is, it seems to
me, the most important chemical mechanism through which eukaryotes
have really acquired their possibility to produce different cell lines.
Repetitive sequences are not present in prokaryotes. I believe that this
DNA, once considered useless, is a most important component because
of genes.

LAMBO: Obviously, as you mentioned earlier in your lecture, proteins and
aminoacids play a very important role in the activity of genes. If you now
think in terms of major nutritional deficiencies of clinical importance, I
wonder what effects these might have on genetic evolution or genetic
activities such as mutations in performing new functions and
rearranging structures, etc..

GERACI: I am not an expert on these aspects, but I would suppose that they
would exert quite big effects.

LEJEUNE: This is not a question, it's a reflection, I think the most important
question was asked by Professor Mössbauer, why are there

chromosomes? Because there is one piece of evidence in evolution, in the classification of living beings which is: each species has its own string of chromosomes, its own karyotype. When I wrote that around twenty years ago nobody noticed. Now we know that. It is a fact extraordinarily much bigger than all the rest. We know that the splitting of the living world into different species can be done just by looking at the shape and at the number of chromosomes without knowing that is written inside. That cannot be by chance. It must mean something very fundamental.

GERACI: I agree completely.

DAY FOUR
30 OCTOBER 1992

LES THÉORIES RÉDUCTIONNISTES EN BIOLOGIE

RÉMY CHAUVIN

Professeur honoraire à la Sorbonne, Paris

Liste des concepts

1. Le réductionnisme darwinien est insuffisant pour expliquer l'évolution.
2. Ses principaux concepts sont sujets à de graves critiques, par exemple la notion de sélection naturelle, d'avantage sélectif, etc...
3. Le terme d'adaptation n'a aucun sens défini; il est urgent de constater l'existence de certain traits de l'évolution: elle a une certaine direction, elle produit dans tous les phylums un gros cerveau; elle construit des appareils capables de voler dans presque tous les phylums, elle tend à la complication maximum, etc...
4. Une théorie holistique n'est pas prête toutefois à remplacer le darwinisme, peut être tout simplement parce qu'on ne l'a pas cherchée.

Réductionnisme et biologie

De tout temps les hommes ont été frappés par la diversité de la vie, par sa richesse et les prodigieux mécanismes qu'elle déploie sous nos yeux. Mais la notion de mécanisme est évidemment très récente, pour une bonne raison; c'est que les machines de l'Antiquité n'étaient point absentes, mais en nombre très réduit, et qu'on n'y attachait pas d'importance: elles avaient été créées par des ingénieurs, sortes d'ouvriers spécialisés dont le statut était bien inférieur à celui des philosophes; et pourquoi se soucier des machines puisqu'on avait des esclaves, dont il faudra toujours, disait Aristote, sans quoi les navettes devraient marcher toutes seules, ce qui est évidemment impossible. Pour ce qui est de l'apparition de la vie, on s'en remettait aux dieux, ou dejà, et c'est tout à fait remarquable, au jeu aveugle des atomes qui, selon Epicure, pouvait créer les bêtes, les hommes et même les dieux ... Ce n'est que plus tard, infiniment plus tard, que les homme s'intéressèrent aux machines; et pourtant sur le plan pratique, elles s'étaient déjà diversifiées, elles étaient devenues puissantes. Démétrius Poliorcète (le preneur de villes) avait beaucoup dévéloppé des appareils énormes et très précis qui servaient à démolir les remparts des villes. Plus tard un édit du Basileus interdit aux patrons d'utiliser les machines qui privaient les ouvriers de leur gagne-pain: les Grecs ingénieux avaient en effet imaginé tout de suite de multiples applications d'une roue que le courant de la rivière fait tourner: on cardait la laine, on la foulait, on écrasait le grain, on pouvait même forger les métaux. Mais cela n'intéressait pas les gens sérieux.

Je sais bien que le but de la science n'est pas de determiner si le Programmeur existe (mais ajoutons tout de suite qu'il *n'est pas plus de déterminer qu'il n'existe pas*) et restons sur le terrain mécaniste puisque c'est la mode; quant à savoir ce qui s'est formé le premier, du code ou de la chose à coder, des controverses infinies qui seront mieux exposées par d'autres que moi même se sont développées et restent loin d'avoir abouti à un consensus.

Je voudrais me tenir sur le terrain de l'évolution qui est le mien sans oublier qu'il est forcément lié, et de plus en plus, à la compréhension du code génétique; mais il est bien difficile d'être à la fois spécialiste des animaux dans leur milieu et généticien: on le voit quand les biologistes moléculaires s'inspirent de modèles darwiniens d'une naïveté touchante et se font de la sélection naturelle une idée pour le moins simpliste.

Voilà donc le grand mot lâché: on ne peut traiter du réductionnisme en biologie sans évoquer Darwin. Il vient de nous arriver depuis peu un événement qui nous a beaucoup étonnés, nous autres biologistes: nous nous sommes aperçus qu'il était mort, et depuis cent ans ... On ne le croirait pas a voir la rage et le fanatisme quasi religieux (au très mauvais sens du terme) avec lesquels ses sectateurs défendent des positions de plus en plus menacées. La durée moyenne des théories en biologie est d'environ vingt ans; il est frappant de constater que celles de Darwin ont duré quatre fois plus.

On peut trouver à cela une foule de raisons:

1) elles flattaient au maximum notre goût réductionniste et mécaniste qui était si fort à la mode et l'est encore;

2) c'étaient des théories d'une suprême élégance, comme celles de la gravitation universelle, à laquelle on les a un peu exagérément comparées:

remener tout à un principe très simple, celui de la sélection naturelle, qui trie des atomes biologiques s'agitant au hasard dans le vide, comme disait ou presque, Epicure, voilà qui est beau!

3) elles permettaient d'échapper aux spectres du finalisme et du vitalisme dont on disait qu'ils stérilisaient la recherche (ce qui était vrai) et qu'ils étaient naïfs (vrai encore, en ajoutant que le darwinisme ne leur cédait en rien la palme de la naïveté, mais il a fallu plus longtemps pour s'en apercevoir);

4) enfin — soyons justes — elles étaient au début très efficaces; Darwin a présidé à la naissance de la biologie moderne. Si les darwiniens m'agacent, je m'incline avec respect devant Darwin. Sa théorie de la sélection naturelle a donné une impulsion énorme à la biologie; elle était en grande partie fausse, mais quelle importance? Beaucoup de théories fausses ont été très efficaces. Les conceptions astronomiques de Ptolémée étaient archifausses, mais elles permettaient de prévoir les éclipses. Nos premières théories de l'atome, dont il ne reste presque rien, on fondé la physique moderne. La science ne découvre point la vérité tout d'un coup, elle procède par approximations et souvent par révolutions successives. Et quand une théorie est usée, on la jette; c'est ce que comprennent difficilement les philosophes: une théorie scientifique est moins un système de pensée qu'un outil plus ou moins efficace.

Mais rien n'est pire qu'une théorie usée que ses sectateurs veulent prolonger. C'est arrivé un grand nombre de fois dans les sciences, si ce n'est toujours ... Les novateurs sont l'objet de critiques où parfois la violence le dispute à la sottise; c'est qu'on ne fait pas de la science avec son cerveau mais avec tout son corps, tout son tempérament, tous ses préjugés et tout son orgueil ...

Il n'est pas question de faire ici l'historique très compliqué du darwinisme, et une foule d'auteurs l'ont écrit infiniment mieux que je ne saurais le faire; de plus, en cent ans, il s'est beaucoup modifié (songeons que Darwin ne connaissait par les chromosomes!); on lui a intégré la génétique et il est devenu le néodarwinisme. Je crois toutefois qu'on peut le ramener à un certain nombre d'assertions qui ressortent plus ou moins clairement dans les écrits des darwiniens:

a) l'évolution n'a ni sens ni direction; elle procède uniquement de mutations hasardeuses dont quelques-unes peuvent être utiles;

b) si par hasard le porteur d'une bonne mutation est en même temps plus fécond, il donnera plus de descendants: c'est la *sélection naturelle;*

c) ce mécanisme suffit à expliquer toutes les adaptations et tous les mécanismes comme l'oeil et l'aile;

d) l'organisme a donc été formé par le jeu hasardeux des mutations à partir d'organismes plus simples; le plus compliqué s'explique nécessairement par le plus simple, et le tout est égal à la somme des parties.

Quelques unes de ces idées sont à vrai dire fort anciennes et le concept d'évolution sous une forme embryonnaire pourrait bien remonter à St Augustin; quant au tout qui est égal à la somme des parties, on y voit facilement un avatar du cartésianisme ... Mais peu importe, le principal est de se demander si les assertions fondamentales ont quelque fondement, ou s'il ne s'agit que d'un réductionnisme forcené, négligeant les faits innombrables qui viennent à l'encontre de ses théories.

Est-il vrai que l'évolution n'a ni sens ni direction? Tout biologiste impartial doit convenir que *c'est le contraire qui est vrai:* car en gros les poissons sont venus d'abord, puis les batraciens, les reptiles et les oiseaux, enfin les mammifères et l'homme: mais *jamais on n'a vu l'inverse,* c'est à dire une grenouille redevenir poisson ou un mammifère redevenir reptile ... Les darwiniens répondent à cela que les mutations ayant donné naissance à chacun de ces grands embranchements sont si nombreuses et si compliquées qu'il est invraisemblable que la marche arrière puisse être enclenchée: on saisit ici un des paralogismes darwiniens favoris: pour échapper à une constatation inéluctable ils proposent un mécanisme hypothétique; même s'il etait conforme à la réalité cela n'infirme en rien la constatation que *l'évolution a un sens.* Quant à la sélection naturelle, c'est la notion la plus galvaudée qui soit: à l'origine, elle ne correspond à rien d'autre qu'à la fécondité différentielle, c'est à dire que les espèces ou les individus n'ont pas la même fecondité. Ensuite, pour sortir de cette constatation simpliste que les plus féconds ont plus de chance de survivre dans leur descendance que les moins féconds, on suppose que les plus féconds peuvent porter en eux un avantage quelconque, par exemple un oeil plus ou moins développé; puis pour expliquer le fait incontestable de l'evolution de l'oeil, qui commence par un petite tache pigmentaire et finit par l'organe très développé des mammifères et des oiseaux on est forcé de supposer que ce grand développement de l'oeil apporte un avantage à son heureux posssesseur. On expliquerait la genèse des organes les plus compliqués *en supposant qu'un avantage de fécondité au cours des millions d'années accompagne* toujours un progrès dans *le développement progressif de l'oeil;* et qu'un progrès même minime dans ce développement correspond à un avantage sélectif. Peu importe s'il tombe. Il suffit de postuler que c'est une nécessité logique, donc l'avantage doit être là, même, dit Fisher, s'il est trop petit pour nos méthodes de mesures: c'est une remarque inouïe pour un homme de science que d'admettre qu'un phénomène existe même si on ne peut pas le vérifier: les sophistes grecs ne faisaient pas mieux.

Mais venons-en à la notion d'avantage qui revient sans arrêt dans le discours darwinien. Je choisirai à dessein les exemples les plus massifs: nous allons voir s'effriter de fausses évidences.

Il n'est pas vrai que la vue soit obligatoirement un avantage ... Je ne citerai guère que les termites, pour la plupart aveugles, et dont l'abondance en Afrique par exemple, tient du prodigieux: et, je m'excuse de cette lapalissade, mais s'ils sont abondants, c'est qu'ils ont réussi. De même pour les fourmis dont on trouve, dans nos bois, côte à côte, des formes à yeux développés et des espèces aveugles sans qu'il soit possible de discerner laquelle des deux à le mieux "réussi". Il n'est pas vrai que le sabot de l'antilope favorise sa course, car si c'est effectivement un des mammifères les plus rapides, une glande située entre ses sabots laisse sur le sol une marque odorante, qui permet à n'importe quel carnivore de la suivre aisément à la trace: si bien qu'on ne saurait dire si l'avantage surpasse le désavantage; enfin, remarque plus inattendue, *il n'est pas vrai que le développement du cerveau chez les singes soit un avantage:* car les grand singes anthropoïdes chimpanzé et gorille, son beaucoup moins nombreux que les macaques au cerveau bien plus petit que le leur! On pourrait continuer ainsi et à chaque exemple d'"avantage" susciter un contre exemple. Ce qui nous amène à constater *qu'il très difficile sinon impossible de mesurer l'adaptation* à un milieu car si un animal vit et se reproduit dans un milieu c'est qu'il y est adapté, comme dirait Monsieur de la Palice; ensuite, dans la nature, si un animal se trouve mal dans son environnement, il a dans la plupart des cas la possibilité de le quitter, ce qu'il fait bel et bien. Cette notion d'adaptation, si couramment employée par les darwiniens, est la moins définie qui soit: elle ne résiste pas au moindre examen. On est stupéfait que tant de biologistes adhèrent encore à des théories aussi vagues: c'est sans exemple dans la science, on ne trouve ce laxisme des concepts et des raisonnements qu'a propos de l'évolution – vue sous l'angle darwinien, bien entendu.

Que des évolutionnistes ne s'en soient pas aperçus serait impensable; en fait des murmures se font entendre dans le temple, même parmi les membres de la stricte observance. Pour n'en citer que quelques uns, Kimura fait remarquer que parmi les nombreuses mutations, piliers centraux de l'argumentation darwinienne, un grand nombre, l'immense majorité sans doute, sont neutres et n'influencent pas sensiblement l'organisme. Un esprit brillant qui se dit darwinien (mais alors je le suis aussi!), Stephen Jay Gould, paléontologiste dont les travaux sont universellement connus, tient pour des changements brutaux dont on a même fixé le rythme: tous les vingt-six millions d'années à peu près, des catastrophes dévastent la faune et la flore, et on repart à zéro. Une de ces catastrophes a sans doute mis fin à l'explosion cambrienne, à laquelle Gould a consacré un ouvrage captivant: floraison inouïe d'animaux des plus étranges, dont il ne reste rien aujourdhui; et ne parlons pas de l'extinction des dinosaures. Si bien qu'il faudrait sans doute employer plutôt que le terme évolution celui de révolutions, et revenir au bon vieux catastrophisme de Cuvier (avec quelques aménagements tout de même); dans le même ordre d'idées, dans un ouvrage intitulé "Des mauvais gènes ou pas de chance?" (Bad genes or bad luck?), Raup se demande si vraiment l'extinction des espèces provient dans tous les cas de la sélection naturelle, ou simplement du fait qu'on se trouvait là au mauvais moment, etc.

Et cependant il reste des darwiniens en grand nombre, et (plusieurs me l'ont avoué) en désespoir de cause: parce qu'il n'y a pas d'autre théorie sur le marché. Le vieux réductionnisme a fait faillite, ils le sentent bien. Les exceptions à la règle, si règle il y a, sont trop nombreuses... Mais quoi d'autre?

Y a-t-il une alternative au réductionnisme darwinien?

Non; il n'en existe pas de nettement constituée à l'heure actuelle; tout simplement parce qu'on n'en a cherché aucune pendant des décennies. On était naïvement convaincu que la théorie darwinienne était définitive (croire une théorie définitive constitue dans les sciences la faute la plus radicale). D'autre part tout système qui suppose que le tout est "la somme" des parties exerce sur l'esprit du biologiste un attrait presque irrésistible; alors qu'il serait plus juste de dire que le tout est la "combinaison" des parties: la montre est-elle l'addition de toutes ses roues dentées? ne sont elles point assemblées dans un certain ordre, une certaine combinaison? de même que l'organisme n'est point la somme mais la combinaison de toutes ses cellules? Il se trouve pourtant des neurophysiologistes comme Changeux, pour admettre plus ou moins clairement que le cerveau est "la somme" de tous ses neurones. Cet état d'esprit règne dans les branches de la biologie les plus diverses: par exemple en biologie moléculaire, on n'enlèvera pas de l'idée de certains que les divisions de la cellule épuisent toute l'évolution de l'organisme; en oubliant allègrement une vieille science, bien délaissée et pourtant riche d'enseignement, l'embryologie (Chandebosi); qui montrait très évidemment qu'une cellule, bien que son génome soit commun avec toutes les autres, évolue pour former des organes tout à fait différents *suivant la place* qu'elle occupe dans l'embryon. La chose est si évidente qu'on cherche maintenant quels sont les "gènes du plan" qui président par exemple à la construction d'une drosophile.

Je ne puis donc evoquer une théorie, holistique ou autre, pour remplacer le darwinisme, puisqu'elle n'est pas encore née. Mais pourrait-on trouver quelques directions générales suivies par l'évolution? sans préjugés, sans vouloir les rattacher à une théorie d'ensemble qui n'est pas mûre, en les énumérant comme elles nous viennent à l'esprit? Nous retrouverions ainsi une série de faits bizarres et connus depuis toujours mais occultés par l'idéologie dominante.

1) Nous avons déjà vu que *l'évolution a une direction;*

2) l'evolution veut[1] faire la conquête de l'espace, et pour cela elle a créé l'aile. Mais c'est le but qu'elle "veut" atteindre et peu importent les moyens: l'aile de l'oiseau et celle de la chauve-souris n'ont rien de commun pas plus que l'aile du papillon et celle de la mouche; le vol "aérostatique" des araignées s'éloigne autant que possible du vol plané des poissons volants; ils n'ont rien de commun eux mêmes avec la propulsion par réaction de certains calmars qui arrivent à se lancer hors de l'eau et à percourir dans l'air des distances considérables. Mais tous ces animaux volent et peu importe le moyen; de même que notre volonté lève notre bras quand nous le désirons, et peu importe les commandes neuromuscolaires que nous ne connaissons pas forcément;

3) l'évolution veut certainement le développement du cerveau et cela dans toutes les lignées: chez les primates, comme nous le savons et cela jusqu'à l'homme; chez les dauphins qui avaient un énorme cerveau comparable par son poids à celui de l'homme, mais 45 millions d'années avant nous; chez les oiseaux dont les cellules cérébrales sont plus petites et plus nombreuses que celles des vertébrés, ce qui permet d'en loger davantage dans le petit volume du crâne des oiseaux; ceux-ci manifestent des comportements égaux par leur complexité, sinon supérieurs, à ce que les primates font de mieux ... N'oublions pas les pieuvres, sommet de l'évolution des mollusques, qui possèdent un énorme cerveau, des yeux très semblables à ceux des vertébrés, et des tentacules dont elles se servent comme de mains. Et enfin les insectes sociaux: ici la nature a tourné l'obstacle créé par la très petite taille en combinant, en quelque sorte, les petits cerveaux individuels;

---

[1] Ai-je besoin de dire que l'emploi du verbe vouloir est évidemment métaphorique?

l'opération a été réussie, car par certains côtés les insectes font ce que seuls les hommes savent faire, à l'exclusion de tous les autres animaux: par exemple l'agriculture n'a été inventée que deux fois, dont une par les hommes et la seconde par une fourmi du genre *Atta* qui cultive un champignon sur de la pulpe de feuilles, en culture pure, dans des jardins à champignons souterrains de 200 mètres carrés!

4) L'évolution veut la complexité, ce que j'ai appelé la "vis implicatrix naturae". Qu'un dispositif simple fonctionne parfaitement ne l'empêche pas d'en imaginer un autre infinement compliqué, qui fonctionne aussi, malgré un degré de complication poussé jusqu'à l'absurde (pour un cerveau humain) C'est une des tendances les plus surprenantes que nous ayions déchiffrée dans la nature; elle se manifestait sans doute au tout premier début, quand les premières molécules vivantes cherchaient obscurément à se réunir.

Je pourrais continuer ainsi, mais je devine que mon langage anthropomorphique peut choquer, bien que j'aie averti qu'il était métaphorique. L'est-il totalement? Je ne sais... On a parfois tellement l'impression qu'une volonté préside à l'édification de certains dispositifs, très souvent admirables par leur organisation et leur efficacité. Les réductionnistes le savent, mais leur religion leur interdit de les considérer; ils freinent autant qu'ils le peuvent la stupeur à laquelle on ne peut échapper, en oubliant que l'étonnement devant un fait est la condition sine qua non de la recherche scientifique. Ou bien ils trouvent des "explications" échafaudées sur des hypothèses plus échevelées les unes que les autres.

On voit par ce que je viens de dire que les problèmes sont aussi bien connus que bizarres. Encore n'en ai-je, bien entendu, énumérés que quelques uns. Supposer à l'origine une Intelligence Unificatrice ne résout pas le problème purement scientifique qui est de découvrir "comment fonctionne la nature" plutôt que de verser prématurément dans la métaphysique. Malheureusement, comme on l'a vu, son mode de fonctionnement confond bien souvent l'intelligence humaine.

Je ne perds pas espoir cependant: nous trouverons peut être l'hypothèse salvatrice, si nous nous décidons à observer la nature sans prejugés, en mettant sur le papier, une bonne fois, toutes nos ignorances.

Nous ne savons qu'une chose jusqu'à présent: le réductionnisme darwinien, après des heures de gloire, n'explique pas grand'chose en somme et probablement rien du tout.

DISCUSSION
(McConnell, chairman)

LICHNEROWICZ: J'ai apprecié votre exposé et ses thèses, je voudrais simplement prendre un point de vue diamétralement opposé au vôtre. Je suis
un mathématicien et il m'est arrivé d'étudier des petits modèles
mathématiques de génétique des populations. Pour les étudier et pour les
définir il faut un critère, il faut essayer de dire de manière précise, ce
qu'on veut. Rien ne marche si on prend des multi-critères. Il faut un seul
critère. Mais selon le choix de ce critère, un population va évoluer
differemment. Donc je veux aller dans votre sens, en ce sens que la
notion d'aptitude a besoin dans chaque cas d'être parfaitement définie,
mais que le résultat de ce choix d'aptitude permet d'expliquer chaque
chose et le contraire.

CHAUVIN: Chaque chose et le contraire c'est aussi la définition du darwinisme
que vous venez de donner.

MALU: J'ai fort apprécié votre contribution que j'ai lu avec beaucoup de
plaisir. Votre contribution devrait être publiée en français pour rendre de
façon correcte tout l'humour qui l'imprègne. Quand aux thèses qui y sont
exposées je ne m'aventure pas à les commenter n'étant pas biologiste.

LAMBO: After the question of aptitude, the question of progeny, you went on
to bigger brains. I could not really understand whether you were
thinking in terms of — just big in terms of size or big in terms of
differentiation of functions of the brain. I'm sure that when Prof. Eccles
comes on he may well touch upon this. Another point which I would
mention. In Africa, in Nigeria, where I come from, there is a particular
ethnic group that produces more twins than any other ethnic group in
the world. More multiple births and so on. Because again you mentioned
the question of progeny, the question of fertility, I may have to discuss
this with you in private.

SINGER: Vous attaquez les darwinistes en disant qu'ils prennent de mauvais
critères de sélection comme l'adaptabilité ou la fecondité. Je pense, si j'ai
bien compris, que vous courez le même danger avec les définitions que

vous donnez des aptitudes biologiques et que vous présentez comme établies. Vous prenez la taille du cerveau, par exemple, comme une mesure de l'efficacité cérébrale, ce qui est faux. Meme le nombre des neurones n'a qu'une importance limitée. Ce qui détermine l'intelligence d'un cerveau c'est l'architecture, la complexité et la sélectivité de ses connections. On peut réaliser les mêmes fonctions avec beaucoup ou peu de cellules, donc votre argument est incompatible avec des données scientifiques. Vous dites, et je pense que c'est inadmissible, que les darwinistes croient que la fécondité est la seule chose qui compte. Je ne partage pas cet avis, ce qui compte c'est l'aptitude à transmettre certains gènes de bonne qualité. Cette aptitude n'a rien à voir avec la fécondité propre, le nombre d'enfants qu'on peut produire au cours d'une vie. Votre argument selon lequel les chimpanzés ou les gorilles ne devraient pas exister doit justement être interprété dans le sens opposé: le fait que l'espèce ait survécu en ne produisant que peu de bébés par couple et periode de vie est un grand compliment que l'on-peut faire au cerveau et aux gènes de cette espèce et non l'inverse. Vous voyez combien il est simple de renverser les arguments.

CHAUVIN: Alors, permettez-moi de faire un double renversement. Pour nous autres il est absolument certain que les espèces animales dont le comportement est les plus développé, ont le plus gros cerveau, la plus grande taille de cerveau. C'est un critère très grossier, mais je vous assure qu'il est accepté de tous, il n'y a rien a discuter, une taupe a beaucoup moins de cerveau qu'un singe. Quant à l'aptitude à transmettre les gênes je ne vois pas en quoi cela combat ma thèse, pour transmettre beaucoup de gênes il faut avoir une grande progéniture: je vous assure aussi, que si vous n'avez qu'un enfant vous ne transmettrez pas statistiquement beaucoup de gênes, parce que vous les transmettrez à un seul individu. Alors, sur le plan de la biologie générale, vous n'aurez pas transmis beaucoup de gênes. Alors, je ne me rappelle plus l'argument, qu'est-ce que vous aviez dit des gênes?

SINGER: Je crois que c'etait un compliment aux chimpanzés et aux gorilles.

CHAUVIN: Mais non, Ce n'est pas un compliment à leur faire parce que leur nombre décroit à une vitesse vertigineuse! Leur cerveaux sont très développés, mais ils ont une autre cause d'affaiblissement, ce qui prouve que la taille du cerveau et la faculté à se reproduire ne s'accerclent pas si facilement.

THOM: Je voudrais faire une remarque peut-être un peu plus abstracte en ce qui concerne le darwinisme. Je crois que l'objection de Waddington que le darwinisme est une tautologie est perfectement exacte et le seul moyen de supposer le darwinisme c'est en effet d'essayer de perfectionner la definition de l'extremum à realiser de manière à pouvoir effectivement contrôler. Autrement dit, s'il y a quelque chose de valable dans le darwinisme, ce doit être dans la possibilité d'enoncer quelque chose comme un principe variationnel qui déterminerait le sens de l'evolution. Or, si on essaye de déterminer ce principe variationnel, on tombe tout de suite sur des difficultés comme celles que notre conférencier vient d'exposer. Mais il y a une difficulté encore plus subtile qui est due au fait justement que si on prend l'analogie d'un principe variationnel, il faut determiner la quantité à extremaliser. En admettant qu'il s'agit simplement de la fécondité comme quantité à extremaliser, il faudrait déterminer la variation de cette fécondité sur l'espace de variation virtuelle d'une espèce. Aussi, ce qui ruine le déterminisme, du point de vue théorique, c'est l'impossibilité de définir cet espace. Alors les généticiens ont dit "oh, mais il faut considérer les variations du génome. Il suffit de determiner les variations du génome". Malheureusement, comme l'a dit après Waddington, il l'a très bien dit en anglais — "the evolution acts on the phenotypes and not on the genotypes". Et même si on fait l'informalisation du génotype, on ne pourra pas atteindre l'informalisation des effets du milieu-ambient sur le phénotype. C'est le tournoir entre le phénotype et le génotype qui en effet ruine la capacité théorique du darwinisme.

CAUVIN: Naturellement je vous remercie d'avoir subtilement manifesté votre accord à mes thèses.

PULLMAN: Il reste une toute petite remarque en partie controversable, c'est à propos du sens de la vie. Je ne veux certainement pas ouvrir une discussion sur le sens de la vie, mais je voudrais seulement rapporter une phrase, dont je ne me rappelle pas l'auteur: "Même si la vie n'a pas de sens, ce n'est pas une raison de ne pas donner un sens à sa vie".

CHAUVIN: Cela me parait d'une élégance considérable, mais c'est un argument qui se mord la queue. Si la vie n'a pas de sens, comment voulez vous lui en donner un?

RUNCORN: Do you not think that it is relevant to discussions on evolution to take into account that the environment has changed from time to time rather violently and as a result things evolved as if an experiment were done on the evolutionary process? I'm particularly impressed by the evidence that the dynosaurs were wiped out by a meteorite impact. The evidence seems strong and the implication from the point of view of evolution is that the environment was strange for a while, particularly the climate was sufficiently cold to make it difficult for the dynosaurs to survive. But the process of evolution — darwinism — has produced such a great variety of types of creatures that the smaller ones, more adapted for this period of intense climatic change went ahead and mammals occupied the the ecological niches which were made available by the destruction of the dynosaurs. That is a very significant "experiment". It wouldn't of course imply direction to evolution, it would imply that the great variety of the mutations which developed — you know, the wonderful variety of species — made new developments possible through the elimination of one particular branching.

CHAUVIN: Mais l'extinction des dynosaures est quelque chose de très complexe. On avait imaginé comme cause la chute d'un météorite, mais ce n'est en fait pas du tout certain. Les dynosaures ne sont pas disparus d'un seul coup, ils ont mis plus d'un million d'années à disparaître complètement. On a évoqué l'argument d'un refroidissement du climat qui n'aurait pas permis a ces animaux de survivre. Mais vous savez aussi qu'il y avait de petits dynosaures pas plus gros qu'une poule, et qu'ils avaient sans doute des plumes et sont disparus aussi ou peut être se sont transformés en oiseaux. C'est possible, bien que la filiation des oiseaux à partir des dynosaures est peut être problématique. Mais en réalité qu'est-ce que je tire de l'histoire des dynosaures? Rien. Comment ont-ils disparu? Je ne sais rien. Il y a plusieurs causes qui ont été invoquées et si j'en crois Gould qui se dit darwinien il ne faut pas tellement parler d'évolution mais de révolution, c'est à dire qu'on revient au vieux catastrophisme de Cuvier; c'est quand même une "involution de l'évolution".

WHAT IS LIFE?

On the Irreducibility of Life to Chaotic and Non-Chaotic Physical Systems

JOSEF SEIFERT

Internationale Akademie für Philosophie
im Fürstentum Liechtenstein

1. *Of 'Life' we speak in many senses*

Aristotle says in his *Metaphysics*[1] that we speak of being in many senses. It is not less correct to say, 'Of life we speak in many senses'. The two Greek terms of *zoee* (life as such) and *bios* (organic life) suggest at least two of the many meanings of 'life'.

Plato and Aristotle saw one of the most prominent marks of life in a movement which has its source in the living being itself and not outside of it[2]. Such a self-movement is not restricted to *bios* but characterizes all forms of *zoee*. In view of such a 'self-movement' - which must be distinguished from the purely material 'self-movement' of sub-atomic particles, pendulums, or planets, Plato and many other philosophers see the highest forms of *zoee*-life linked to freedom. When Kant says: "For all life rests on the inner faculty to determine oneself at will",[3] he seems likewise to have in mind the highest form of such self-movement, namely freedom[4] which involves a spontaneous and rational self-determination that has most clearly its source in the agent.

Zoee designates that unique and irreducible actuality and internal activity which we call 'life'. This notion of *zoee*-life encompasses *all conceivable*

---

[1] 4.1003 a 33: *"To de òn légetai men pollachoos"*.
[2] ARISTOTLE, *De Anima* 404 a 9; 406 a 1-5.
[3] I. KANT, *Träume eines Geistersehers*, Akad. Ausg. Bd. IV, p. 544.
[4] Cf. AUGUSTINE, *De Civitate Dei*, V.

forms of life and also refers to the pure positivity of life which is a 'pure perfection' of which Anselm says that[5] to possess it is absolutely better than not
to possess it. One cannot surpass life in perfection - without living - whereas
one can be more perfect than gold, for example, or nobler even than man -
without being them. By their necessary limitations, mixed perfections differ
from pure perfections. Such qualities of which we gain the insight that they
are not surpassable, except by a being that possesses their form in a higher
mode, are the condition of the possibility of any knowledge of God. Duns
Scotus adds that each pure perfection is irreducibly simple (undefinable),
admits of infinity, is communicable to more than one subject, and is compatible with all other pure perfections. *Zoee,* which can be attributed also to the
highest imaginable forms of life[6] was, because of this its character as pure
perfection, attributed by Aristotle - and already by Plato - to God to the
'Demiurge', to the activity of the highest, intellectual contemplation, or
rather to its subject.[7] We remember here also the famous definition of the
eternity of God in terms of life offered by Boëthius: eternity is the wholly
simultaneous and perfect possession of interminable life' *(interminabilis
vitae tota simul et perfecta possessio).*

If we come down from such a transcendental[8] consideration of life to the
world of our immediate experience, we encounter life in three fundamental
kinds of being, get acquainted with it in two experiential ways, and can
understand it by means of two chief methods.

We know from experience three kinds of living beings: plants, animals,
and human beings. Human life, and partly also the life of plants and animals, involves again distinct phenomena: 1) the partial life-processes in individual cells and cell-cultures (which can continue after the death of the
multi-cellular organism); 2) the life of the organism as a whole; 3) the life of
the mind as such (which many hold to continue after death), and 4) the

---

[5] Cf. ANSELM OF CANTERBURY, *Monologion,* 15. Duns Scotus, *Oxon.* (or *Ordinatio*) 1, d. 1-3 ff.,
ALLAN WOLTER, *The Transcendentals and their Function in the Metaphysics of Duns Scotus* (St.
Bonaventure, New York: Franciscan Institute Publications, 1946). JOSEF SEIFERT, *Essere e persona. Verso una fondazione fenomenologica di una metafisica classica e personalistica* (Milano: Vita
e Pensiero, 1989), ch. v.
[6] Anselm of Canterbury (Aosta), *Monologion,* ch. xv, 29-31:
    Quare necesse est eam [summam naturam=God] esse viventem,
    sapientem, potentem et omnipotentem, veram, iustam, beatam, aeter-
    nam, et quidquid similiter absolute melius est quam non ipsum.
[7] ARISTOTLE, *Metaphysics,* XII, 1072b 27 ff. Aristotle says there that *zoee* belongs to God and
that "the actuality (*energeia*) of the *nous* is life", and "eternal life".
[8] Not in Kant's but in the Scholastic sense explained above.

union of mind and body.[9] (To the last concept of human life corresponds the concept of death as the separation of the soul from the body).

In the first experiential way in which we know life we do not immediately experience biological life itself but only its observable essential marks and signs such as nutrition, growth, propagation, or regeneration. Also death - or rather mortality - is an essential characteristic of *bios*.[10]

The experience of life also involves understanding of causality (for example when we speak of reproduction), of the unity of form of an organism and of its integrated wholeness, and of finality found in the purposes of genetic codes and organs. Any understanding of life also involves some grasp of further categories the explanation of which is a task of philosophy: for example of unity, being, time, motion, and of life itself.

Biological life is not directly observed by us. This thesis is even more true with respect to facts such as the highly complex proteins in living organisms DNA and RNA structures, etc. The life in animals and plants, nevertheless is not merely concluded by means of syllogism but manifests itself more immediately in the described phenomena, which give rise to the immediate common-sense experience of plants or dogs as living. We grasp biological life in some form of mediated immediacy: i.e., in and through its manifestations and effects.

Even our own vegetative life we know in a certain way from without, although we both live our body consciously and feel the signs of its life, such as breathing, from within our body.

The second experiential way in which we encounter life is the immediate inner experience of our own life in the living of our conscious life.

The two chief methods of theoretically cognizing life are:

1) Scientific observation (by means of instruments lab-tests, etc.), and theoretical explanations of the observed phenomena by means of: induction,[11] corroboration or falsification of theories by means of trial and error methods, etc.

---

[9] The first two apply also to the life of animals and - to a lesser degree - of plants, the fourth exclusively to the human being and only analogously to animals, the third to all personal beings.

[10] ARISTOTLE, *De Anima* 412 a 14 f.

[11] I tried to defend induction against Karl Popper's critique of it, in JOSEF SEIFERT "Objektivismus in der Wissenschaft und Grundlagen philosophischer Rationalität. Kritische Überlegungen zu Karl Poppers Wissenschafts-, Erkenntnis- und Wahrheitstheorie", in: N. LESER J. SEIFERT, K. PLITZNER (Hrsg.), *Die Gedankenwelt Sir Karl Poppers*: Kritischer Rationalismus im Dialog (Heidelberg: Universitätsverlag C. Winter, 1991), S. 31-74. See also J. SEIFERT "Wissen und Wahrheit in Naturwissenschaften, Philosophie und Glauben", in: *Naturwissenschaft und Weltbild* (Hrsg. H.-C. Reichel, E. Prat de la Riba), (Wien: Hölder-Pichler-Tempsky, 1992).

2) Philosophical methods: intuition into the essence of life and into its highly intelligible and necessary essential marks and into the intelligible causes of the observed phenomena; and deductive methods based on evident premises and laws of correct inference. Philosophical intuition, analysis, and deduction are not based on experience in the sense of mere observation of facts; nevertheless, such an a priori knowledge is not entirely independent of experience. Rather, it is based on a such-being experience in which we become acquainted with the intelligible essences of things.[12]

2. What is Biological Life?

But what then is organic life? The astonishing phenomenon of *bios* - just like all absolutely original and ultimate data, (for example being) defies our attempts to define it. In a certain sense, life allows us only, to say what G.E. Moore said about the good: Life is life.[13] In spite of its undefinability in terms of anything else, however, we can analyse its essential marks. Although we are unable to define it in terms of its generic features and its specific differences in such a way that anyone who does not already know it can be led to understand it through our explanation we still can unfold its different marks and express the results of such an analysis in an essential definition.[14] We can certainly describe life, in spite of its mysterious character, in terms of such phenomena as nutrition, growth, regeneration, propagation, complex DNA-RNA-structures, genetic codes, etc.[15] Life is precisely that power which brings forth the described effects.

That life is, in the last analysis undefinable, is not due to a weakness of our minds but rather to its originality and irreducible simplicity[16]. Even different colours cannot be defined in terms of anything else. Aristotle has well

---

[12] DIETRICH VON HILDEBRAND, *What is Philosophy?*, 3rd ed., with a new Introductory Essay by Josef Seifert (London: Routledge, 1991), especially ch. iv.

[13] G.E. MOORE, *Principia Ethica* (1903);14th ed. (London, 1971), ch. i ff.

[14] Cf. J. SEIFERT, "Essence and Existence. A New Foundation of Classical Metaphysics on the Basis of 'Phenomenological Realism,' and a Critical Investigation of 'Existentialist Thomism'," *Aletheia* I (1977), pp. 17-157; I,2 (1977), pp. 371-459, 56 ff.

[15] Cf. J. SEIFERT, "Genetischer Code und Teleologie. Information, Kausalität und Finalität. Kurzer Abriß naturwissenschaftlicher genetischer Erkenntnisse als Hintergrund neuer philosophischer Probleme", *Arzt und Christ* 34 (1988), 185-200.

[16] This term, introduced by Duns Scotus to designate the 'pure perfections', could be applied to all fundamental and ultimate (irreducible) data such as red or blue, numbers, being, knowledge, etc.

shown that the quest to define everything through something else is just as irrational and circular as the desire to prove everything. For all definitions and all syllogisms rest on foundations which are prior to them and must be understood in their own right and by means of some immediate understanding or perception.[17] For as each definition requires first undefinable essences, so each argument presupposes first premises and laws of correct inference which cannot be further proven without falling into circular argument or without begging the question.[18] Any effort to define such data as being or life through their parts, or by proximate genus and specific difference, not only presupposes preceding notions of other essences which cannot be endlessly defined but also requires the notion itself which one seeks to explain. This truth is also a central point of any phenomenological method.[19] But what then are the essential marks of life? Let us first turn to investigation of the empirical marks of life.[20]

The two chief methods of science in characterising life proceed either a) by an exhaustive enumeration of the main characteristics and biological functions of living organisms (with the problem that not every living organism shows all these functions); or b) by an attempt to identify among these functions those which are present in *all* living individuals: [21]

---

[17] ARISTOTLE, *Posterior Analytics*, II, 99 b 20 ff., 100 b 10 ff.

[18] Cf. JOSEF SEIFERT, "Essence and Existence. A New Foundation of Classical Metaphysics on the Basis of 'Phenomenological Realism' and a Critical Investigation of 'Existentialist Thomism'," *Aletheia* I (1977), pp. 17-157; I,2 (1977), pp. 371-459.

[19] ADOLF REINACH, 'Concerning Phenomenology,' transl. by Dallas Willard, *The Personalist* 50 (Spring 1969), pp. 194 221. Reprinted in *Perspectives in Philosophy*, ed. Robert N. Beck (New York: Holt, Reinhart, & Winston, 1961 and 1969); the same author, "Über Phänomenologie", in: ADOLF REINACH, *Sämtliche Werke*. Texktritische Ausgabe in zwei Bänden, Bd. 1: Die Werke, Teil 1: Kritische Neuausgabe (1905-1914), hrsg.v. KARL SCHUHMANN und BARRY SMITH (München und Wien: Philosophia Verlag, 1989) pp. 531-550, see also MAX SCHELER *Formalism in Ethics and Non-Formal Ethics of Values* transl. Manfred S. Frings and Roger L. Funk, (Evanston: Northwestern University Press, 1973), especially the part 'Formalism and Apriorism', DIETRICH VON HILDEBRAND, *What is Philosophy?* cit., especially pp. xii ff., and ch. iv FRITZ WENISCH, *Die Philosophie und ihre Methode* (Salzburg: A. Pustet, 1976), JOSEF SEIFERT *Back to Things in Themselves. A Phenomenological Foundation for Classical Realism* (London: Routledge, 1987).

[20] Some of them also constitute the experiential starting point for the following analysis of the essential marks of life and may, at least in their bare outline, partake in the necessary essence of *bios*.

[21] I wish to recognize the great help I received from Dr. med. Paulina Taboada.

a) Following the first method, one could give the following list of properties of living beings:[22] 1. Reproduction, 2. Adaptive capacity, 3 Irritability, 4. Endogen motility, 5. Nutrition (which, in its wide sense, includes a. ingestion, b. digestion, c. absorption, d. transport,[23] e. metabolism, f. exchange of gases,[24] g. excretion).

b) Following the second type of scientific 'definition' of life, one identifies some properties or even one property present in all living entities, for example 'nutrition' as an exchange of matter and/or energy which the organism establishes with the surrounding environment. Nutrition in this wide sense can be considered biologically as *the* basic function of the organism in three senses: 1. as empirical *"conditio sine qua non"* for the existence of living beings and of all their other functions, 2. as a 'total function' in the sense that it is exercised by each of the parts of an organism as well as by the organism as a whole; 3. as a 'permanent function' from, the very beginning until the end of the organism's existence.[25] Another attempt to identify one single property present in all living organisms is the theory that the possession and ability to transmit genetic information constitutes the essence of organic life. But this position can certainly not be defended successfully. For the same genetic 'code' is stored in the body cells immediately after the death of the organism and could be stored even on a computer disk. But this 'pure information' is entirely different from life.[26] A further problem with the scientific 'definition' through one central property is that it loses the fullness of the content and of the appearances of life.

The methods used in biology to 'define' life should be complemented by a philosophical analysis which penetrates into the intelligible essence of life. Is there a deeper unified essential core of biological life? This is suggested by the fact that none of the mentioned single features of life is absolutely necessary for biological life. Organisms live which no longer grow, which have no longer any capacity of regeneration, and which are sterile and unable to propagate, and even nutrition can, at least for an extended period of time stop, while life continues. This happens quite regularly in the seeds of plants and, under exceptional circumstances, also in animals and human beings,

---

[22] Cf. N. JESSOP in *Zoology. Theory and Problems*.

[23] Which includes the circulatory system.

[24] Which includes respiration.

[25] We shall see that the ideas contained in this attempt of a biological definition of life are partly erroneous.

[26] Cf. J. SEIFERT, "Genetischer Code und Teleologie", cit., pp. 187 ff.

for example, in a cryo-conserved state in which no nutrition and no gas-exchange through breathing take place.[27]: Philosophically speaking not only *zoee* but also *bios* can in principle continue after a total collapse of the nutritive system.

Also, the mere potentiality to produce all the effects of life *is not* life yet. If nourishment, growth, etc. were produced by mere physical processes, as de la Mettrie and Descartes assumed - none of these dead entities which could be imagined to show such signs as those which we observe in living things would actually *be* living organisms.[28] This is precisely the thesis of Descartes: that all of these entities *are complicated machines* and *therefore not living*.

Philosophers have approached the quest for an underlying core of the essence of life in five manners chiefly:

1) Life as *bios* is first of all inseparably linked to the physical sphere of some bodily being. Aristotle even defines *bios* as the *act* (essential form) *of an organic body*[29].

2) Some philosophers, including Kant,[30] identified life in terms of "the inner power to determine itself to act from an inner principle".[31]

3) The underlying essence of biological life also includes an essential

---

[27] KENNETH B. STOREY and JANET M. STOREY, "Frozen and Alive" *Scientific American (Dec., 1990)*, pp. 62-67. "While frozen, all these animals show no movement, respiration, heart beat or blood circulation, and ... barely detectable neurological activity." (ibid., p. 62-63).

[28] JOSEF SIMON: "Life is in this context a mere lingering concept from traditional metaphysics without objective meaning. Organisms are supposed to be understood in purely mechanistic terms fashioned after the model of machines". J. SIMON, "Leben", in: *Handbuch philosophischer Grundbegriffe* Bd. 3, ed. by H. Krings, H. M. Baumgartner and C. Wild, (München, Kösel, 1973), pp. 844-859, p. 847.

[29] ARISTOTLE, *De Anima* 412 b 11 ff. This definition is also somewhat reductionistic or circular, I believe. If it refers only to the described biological phenomena and identifies life with them, it is reductionistic; if it presupposes already an intuition that this 'act of the organic body' - as *psyche* or *entelechy* - is transcendent to matter, it is both confused and circular. For then it says: life is a principle of the act of an organic body which is transcendent to the latter's activities themselves. But then the formulation of this sense of soul by Aristotle is deficient and unclear, for then his definition amounts to saying: life is the transcendent cause of biological phenomena, or: life itself is distinct from the activation of the body in terms of which it is here defined.

[30] I. KANT, *Träume eines Geistersehers*, ibid., p. 544. This overlooks that there are activities essential to life which are more receptive than self-determining, for example cognition. Moreover, any such self-movement as it is essential to life already presupposes the life of that which can move from an internal principle of action.

[31] ARISTOTLE, *De Anima* 404 a 9; 406 a 1-5. See also Hedwig Conrad-Martius *Die Seele der Pflanze*. In: CONRAD-MARTIUS *Schriften zur Philosophie* (Hrsg. v. Eberhard Avé-Lallement), Bd. I (München: Kösel, 1963), pp. 276-362, p. 313 ff.

teleological direction towards an end residing inside the living organism.[32] This attempt to identify the central mark of life combines finality with *entelechy* in the sense of "having within itself its own end". 'End' has to be interpreted here as a meaningful form.

4) While this could also be applied to works of art, in living organisms we find also a dynamic structure aiming at the realization of its own form. Also such an active self-engendering can be called entelechy.

The organism builds itself up through nutrition and growth, preserving its own form through nourishment and oxygen-transfer, but also recreating and transmitting it through regeneration and propagation.[33] Conrad-Martius compares this feature of life to the case in which the most ingenious architect who produces extraordinary and beautiful works of art would succeed in putting his idea into the colours, stones and other materials so that the work could paint and build itself.[34] This feature of *bios*[35] constitutes an astonishing trait of life.

5) A fifth mark of life gives an ontological explanation to the preceding phenomena in terms of entelechy. *Entelechy* is here understood in a third sense, namely as an active principle of its own, distinct from matter, which generates from within the form, order and essence of the living entity:[36] That some soul *(psyche, entelechy)* is found in all forms of *bios*, was defended particularly by Conrad-Martius and Driesch.[37]

We grasp by a pure philosophical intuition that the observed signs of biological life and the actual power to produce them do not exhaust the essence of life. Organisms which do not grow are not nourished, do not regenerate damaged parts, and are sterile, may still live also the cryoconserved organ-

---

[32] R. SPAEMANN and R. LÖW, *Die Frage Wozu? Geschichte und Wederentdeckung des teleologischen Denkens* (München, 1981).

[33] I. KANT, *Kritik der Urteilskraff*, B 287 f.

[34] HEDWIG CONRAD-MARTIUS, *Die Seele der Pflanze*. In: CONRAD-MARTIUS, *Schriften* zur *Philosophie*, cit., p. 284.

[35] In an analogous and even more profound sense, the *eigengesetzliche Selbstgestaltung* is of course also a mark of the *zoee of free* beings who, through their value responses and free acts, in a certain way build themselves up and engender themselves. Still, even this feature would not be part of the essence of life as such: at least not of the perfectly actualized life theists attribute to God, in whom all life is eternally already in full being.

[36] Cf. ARISTOTLE, *De Anima*, See also FRANZ BRENTANO, *Psychology from an Empirical Standpoint* (1973), p. 179.

[37] HEDWIG CONRAD-MARTIUS, *Die Seele der Pflanze*. In: CONRAD-MARTIUS, *Schriften zur Philosophie* (Hrsg. v. Eberhard Avé-Lallement), Bd. I (München. Kösel, 1963), pp. 276-362 especially p. 337, and p. 327, note 51. See also J. SEIFERT, *Leib und Seele. Ein Beitrag zur philosophischen Anthropologie* (Salzburg: A. Pustet, 1973), pp. 71 ff.

ism still lives.[38] This possibility of lack of vital activity in a living organism is even clearer in the case of apparent death[39]. Thus we can identify *psyche* alone - an active principle of its own, distinct from matter, which actualizes and generates from within the form, order and essence of the living entity[40] - as life itself which can exist even without exercising its mentioned operations.

Living organisms are radically new systems of physical entities which are more complex[41] and obey other laws than inanimate obiects. In fact, scientists observed following the lead of Erwin Schrödinger - that organic sub-

---

[38] KENNETH B. STOREY and JANET M. STOREY, *Frozen and Alive*, cit., p. 62: "At the extreme, insects of the high Arctic such as wooly bear caterpillars... may spend 10 months of the year frozen solid at temperatures that descend to -50° C ... or even lower...."

[39] Also described by phenomena such as asphyxia, cataleptic trance, or state of suspended animation.

[40] Cf. ARISTOTLE, De *Anima;* See also FRANZ BRENTANO, *Psychology from an Empirical Standpoint* (1973), p. 179.

[41] See on this JAY ROTH, "The Piling of Coincidence on Coincidence", in. HENRY MARGENAU and ROY ABRAHAM VARGHESE (Ed.) *Cosmos, Bios Theos. Scientists Reflect on Science, God, and the Origins of the Universe, Life, and* Homo sapiens (La Salle, III.: Open Court, 1992), p. 199:
> "If one considers even a single protein, for example, glycogen phosphorylase, this displays such an immense complexity that it boggles the mind. Considering the processes of protein synthesis, DNA replication and repair, and hundreds of equally complicated processes, one is left with a feeling best described as awe.
> I have carefully studied molecular, biological, and chemical ideas of the origin of life and read all the books and papers I could find. Never have I found any explanation that was satisfactory to me.... Even reduced to the barest essentials, [the original template (be it DNA or RNA)] ... must have been very complex indeed. For this template and this template alone, it appears it is reasonable at present to suggest the possibility of a creator.
> ...the odds of such a template forming by chance are 1 in about 10^{300} or, possibly, a much larger number...".

Similar ideas are expressed by Henry Margenau, "The Laws of Nature are Created by God", in: *Cosmos*, cit., p. 63: "the British Astronomer Fred Hoyle is widely noted for the statement that believing the first cell originated by chance is like believing a tornado ripping through a junk yard full of airplane parts could produce a Boeing 747." A very similar thought was expressed before by the famous discoverer of the biological notion of *Umwelt* (surroundings), Johannes von Uexcüll, *Umwelt und Innenwelt der Tiere* (Berlin, 1921) with respect to the immense improbability that an amoeba could not only come to exist, but find a surrounding in which to live and then to propagate itself regularly. He compares this with the idea that by accident a car is formed out of different parts of metal a key falls - by accident - in the key-hole, gas flows in - by accident - in the tank, the key turns by accident, by accident there is a street on which the car can drive, etc. And when this car crashes and breaks, from its parts by accident it is restored or little cars emerge from the pieces of the preceding one. And this goes on all the time and quite regularly, millions of times, and not just 1 time per 10^{300} - all by chance.

stances are partly governed by principles which are antithetically opposed to those of inorganic matter.[42] Whereas material things, for example, underlie the thermodynamic laws of the inanimate cosmos and obey principles of (positive) entropy which involve a growing transition to lower forms of energy and less ordered material wholes, the living organisms develop in accordance with principles of 'negative entropy', i.e., they develop towards more ordered energetic systems.

Living organisms possess in themselves an order whereas, in inorganic matter order is produced from without.[43] While the principle of 'order from order' indicates well phenomena such as propagation, the living organism also creates higher order from disorder, whereas inorganic physical systems undergo a movement from higher organisation to chaos.

Though Schrödinger admits that classical mechanistic reductionistic explanations of life fail, he himself seeks to explain life through a physicalistic principle. And this concept finds ample nourishment in the recent advances of chaos-theory[44] And in the efforts to analyse chaotic orbits mathematically.[45] By non-chaotic physical systems, I mean systems which obey well-explored laws - for example those of classical physics - possibly including the statistical laws of quantum-mechanics. By chaotic physical systems, I refer to systems which appear to be chaotic but which can be proven to produce certain effects with higher frequency than would follow from the general laws of pro-bability. To this class belongs the weather or the changes of which one as-sumes occurred during the alleged Big Bang,[46] and other events.[47] There are

---

[42] Cf. ERWIN SCHRÖDINGER, *What is Life?* (Cambridge, London, New York Melbourne, 1944); *Was ist Leben?*, 3. Aufl. (München-Zürich: Piper, 1987).

[43] SCHRÖDINGER, *op. cit.*, ch. vi, 57; vi, 65.

[44] GREGORY J. CHAITIN, "Zahlen und Zufall - Algorithmische Informationstheorie. Neueste Resultate über die Grundlagen der Mathematik" in: *Naturwissenschaft und Weltbild* (H.-C. REICHEL, E. PRAT DE LA RIBA), (Wien-Hölder-Pichler-Tempsky, 1992, editors), pp. 30-44. See also Margenau-Varghese, *Cosmos*, cit., p. 29

[45] JAMES P. CRUTCHFIELD, "Chaos 1: Chaos and Complexity", in: HANS BURKHARDT and BARRY SMITH (Editors) *Handbook of Metaphysics and Ontology*, Vol. 1 (Munich-Philadelphia-Vienna: Philosophia, 1991), pp. 139-144: "What will last, then, is not so much the phenomenon of deter-ministic chaos, but rather the methodology, *experimental mathematics*, that has been developed to explore it." (*ibid*, p. 142).

[46] Some scientists (as Ilya Prigogine, who received the Nobel Prize of Chemistry in 1977 for his discoveries in the field of nonequilibrium thermodynamics) do not believe in the Big Bang theory. See the following exposition on the Big Bang in the main text.

[47] Such a modern chaos-theoretical explanation of life is proposed, for example, by llya Prigogine, in his "The Universe Started from an Instability in the Quantum Vacuum", in: H. MARGENAU and ROY A. VARGHESE, ed., *Cosmos*, cit., pp. 188-192; esp. p. 189.

limits of such chaotic physical explanations of life.[48]

The term 'chaos' means at times a macro-physical application of microphysical concepts from quantum-physics (indeterministic chaos which Schrödinger himself introduces to explain the arising of life).[49]

Another concept of (deterministic) chaos is based on the notion introduced in the last century by Maxwell, of unpredictable complex events[50] which can be calculated only in terms of non-linear equations and/or where the exact conditions found in the original starting point of a given process are not known.

A third meaning of chaos is the following: facts or processes which science considered for a long time as occurring just by chance, are governed by some laws which cannot be explained until now but where science can prove that determined processes or facts occur more frequently than expected just by the laws of probability.[51]

Before we investigate this issue in greater depth, we shall reflect critically on the idea and different meanings of the Big Bang (some forms of chaos-theory, as for example Andrej Linde's 'chaotic inflation', are used against the Big Bang theory):

1) First, we find a philosophically and metaphysically undetermined notion of the Big Bang as some event that marks the beginning of our cosmos and took place some 15 billion years ago.[52] In this sense, 'Big Bang' means only that our cosmos and particularly the motion of the astronomic bodies and systems away from an imaginary center had a beginning at a more or less calculable time of some 15 billion years ago. It is essential for this first meaning of 'Big Bang' that the nature of this event remains open. It

---

[48] The absence of any empirical proof of such chaotic events giving rise to life is expressed by a physicist who believes in such a materialist explanation of the origin of life. See ROBERT JASTROW, "What Forces filled the Universe with Energy Fifteen Billion Years Ago?", in: MARGENAU-VARGHESE, *Cosmos*, cit., pp. 45-49, esp. p. 47: "Nobody has demonstrated that life, even a simple bacterium, can evolve from a broth of molecules." See also EUGENE LINDEN, "Can we Really Understand Matter?", *Science*, April 16, 1990, p. 43.

[49] Heisenberg's own philosophical interpretation of quantum physics - as demonstrating events which are in themselves indetermined or uncaused contains philosophical confusions between various epistemological and ontological issues.

[50] ADAM MORTON, "Chaos II: Fractals and Chaos", in: HANS BURKHARDT and BARRY SMITH (Editors), *Handbook of Metaphysics and Ontology*, Vol. 1, cit., pp.

[51] JAMES P. CRUTCHFIELD, "Chaos 1: Chaos and Complexity", in: HANS BURKHARDT and BARRY SMITH (Editors), *Handbook of Metaphysics and Ontology*, Vol. 1 cit., pp. 139-144.

[52] ILYA PRIGOGINE and EDGARD GUNZIG, "Cosmology II: The Reasons for the Cosmos", in: HANS BURKHARDT and BARRY SMITH (Editors), *Handbook of Metaphysics and Ontology*, Vol. 1, pp. 186-189.

could even - according to the interpretation of some theistic scientists - be the moment of the creation of the world from nothing.

2) A second notion of 'Big Bang' refers to a far more specific event in the physical universe: namely to a massive explosion described in some detail by some physicists (in terms of the preceding temperatures, its duration, etc.) which marks the beginning phase of the cosmos and regarding which it remains open whether this state is some absolute beginning of the universe or only the initial event of a new 'cosmic age'. The metaphysical question remains open whether or not the static and dynamic beginning state of the world (and the Big Bang itself) required a personal Creator and whether the world stands in need of many further creative interventions of the supreme Being or is to be conceived in atheistic terms according to which the physical event called 'Big Bang' is the absolute cause of the universe.

3) A third notion of the Big Bang assumes that all later developments and forms of life developed 'out of this event' and were caused by it. Such a monistic evolutionary notion of the Big Bang can be theistic (assuming that the Big Bang had to-be planned and initiated by a transcendent intelligent God) or atheistic-materialistic-monistic (alleging that this event alone consti-tutes a sufficient cause of the universe). In both cases one attributes to the first material elements and events in the universe all the evolutionary terms which then give rise - *in an immanent development* to the coming-to-be of the whole universe.

4) There is also a fourth notion of the Big Bang as a purely physico-chemical event which - without any transcendent intelligent or divine causality - produced the whole universe by some chaotic and non-chaotic laws and events. This fourth understanding of the Big Bang is neodarwinistic and monistic and seeks to make quantum-mechanical indeterminacy and deter-ministic chaotic laws of chance and necessity the 'divinity' responsible for the origin of the world.[53]

What should we think about the described four meanings of 'Big Bang'?

1) The first sense of Big Bang corresponds to many facts of science and is a reasonable assumption fully compatible both with metaphysics and with present astronomical and cosmological knowledge. It is well-nigh universally accepted today. It is opposed by those who interpret the seven days of crea-tion literally and by those who deny any beginning of a universe which they

---

[53] One is reminded here of Benedetto Croce's view that natural science does not use authen-tic concepts but words without definite meanings, something like nonconcepts which perform functions in scientific systems but remain conceptually indetermined .

regard as eternal, or who hold the opinion that the cosmos moves in some cosmic cycles.

2) The second meaning of 'Big Bang' constitutes a scientific hypothesis - or even myth - which to demonstrate conclusively is impossible today but which can be introduced as a reasonable working hypothesis for cosmology and cosmogony. It was accepted widely by scientists over the last decades. But it is in no way uncontroversial. Some scientists today criticise it sharply and hold it to be an unsuccessful hypothesis - for various alleged scientific deficiencies.[54] Such scientists admit a temporal origin of the cosmos (and thus a Big Bang in the first sense) but reject the assumption of a 'Big Bang' in the second sense at the origin of the universe; instead, they introduce other cosmological hypotheses about the origin of the world.

A more philosophical criticism emphasises that the purposeful aspects of the universe cannot be explained by a concept of a Big Bang which seeks to identify only the efficient cause and here only the physical part thereof which produced the cosmos. Rather any proper cosmological explanation has to take into account the obvious teleological aspects of the universe. The so-called 'anthropic principle' which shows how many astonishingly delicate conditions in the universe had to be fulfilled in order to render human life on earth possible, emphasises this teleological aspect in modern cosmologi-cal reflection inasmuch as only - physically speaking - incredibly improbable combinations of material conditions in the universe as habitat of life make possible any terrestrial form of life, and not only human life, this principle could also be called a teleological 'biotic principle' which governs the rise of the universe. This teleological principle also demands an intelligent efficient cause behind any 'Big Bang', as Plato and Aristotle saw.[55]

The third meaning of the Big Bang can never be demonstrated by science and is a philosophical theory of evolution. It contains monistic and evolutio-nary conceptions of the structure of the universe which fail to recognise any of the significant essential and unbridgeable differences between various beings on earth, as we shall try to substantiate them in our arguments for the human soul.

---

[54] For example, Hannes Alfvén and Fred Hoyle. They think the Big Bang theory in the second sense has not been successful and does not account for elements heavier than helium. WILLEM B. DREES, *Beyond the Big Bang. Quantum Cosmologies and God* cit., pp. 22 ff. See also *ibid.*, pp. 18 ff., where a great variety of opinions about the nature and validity of the Big Bang theory are discussed .

[55] GIOVANNI REALE, *Platon*, cit., ch. xv ff.

If these arguments hold, the third meaning of the 'Big Bang' theory is definitely false because it contains the impossible assumption that entirely irreducible realities (life and lifeless things; matter and mind) could 'develop into each other' and explain each other in terms of an 'evolution'. Evolution is only possible where there are no radical necessary and essential differences between different species. Only in such cases is evolution through series of genetic mutations possible, similarly to the amazing possibilities of metamorphoses which we witness in nature.[57]

The fourth meaning of 'Big Bang' implies a materialist and atheist explanation of the coming into being of the world which is entirely unfounded on scientific facts and fails to do justice to countless teleological aspects of the development of the universe,[58] as well as to the anthropic principle.

Realist phenomenology[59] forces us to recognise: The essence of life is irreducible to physical systems of any sort. Therefore, *if* a reductionistic explanation of organisms in terms of chaotic or non-chaotic physical systems were successful in accounting for the phenomena we observe in organisms, all such a theory could possibly prove is that what appeared to be living substances are nothing but machines and do not live at all.[60] In the light of understanding biological life as the life of a living cell and as the life of an organism as a whole (a distinction which applies to all multicellular organisms and which plays a big role today in the definition of brain death)[61] - we may say that even granted that so-called living organisms were mere physical systems governed by chaotic and non-chaotic physical laws, this would only show that what we call life, is not really life.

The method to demonstrate this involves two steps: First, a simple intui-

---

[56] For a refutation of such metaphysics see CORNELIO FABRO, *L'uomo e il Rischio di Dio* (Roma: Editrice Studium, 1967), and JOSEF SEIFERT, *Essere e Persona*, cit., ch. x-xv, and my *Leib und Seele*, cit., ch. i-ii.

[57] On the many meanings of the term 'evolution' see JOSEF SEIFERT, *Leib und Seele*, cit., pp. 140-159.

[58] Cf. on this also WILLEM B. DREES, *Beyond the Big Bang. Quantum Cosmologies and God*, cit., pp. 24 ff., where the metaphysical and religious neutrality of the Big Bang theory is shown and the fact that some interpreted in the first sense, others in materialist and atheist senses.

[59] DIETRICH VON HILDEBRAND, *What is Philosophy?*, cit., especially pp. xii ff., and ch. iv; JOSEF SEIFERT, *Back to Things in Themselves*, cit., ch. i-ii.

[60] Cf. on this the Concluding Address of His Holiness Pope John Paul II to the Academicians of the Pontifical Academy of the Sciences, on October 31, 1992, on the Galilei Case, in which precisely this point (of the whole being irreducible to the sum of its parts) is emphasized.

[61] J. SEIFERT, "Hirntod", in *Ethik und Technik* (Zürich: M&T edition, 1988) and, "Is Brain Death Actually Death?" in the Proceedings of the Pontifical Academy of Sciences 1989 on Defining Death, and in the *Monist* (April 1993).

tion into the essence and marks of life reveals these to be distinct from physical systems of any chaotic or non-chaotic kind.

But there is a second and much less simple philosophical problem to solve: can one exclude philosophically what we said above: namely that what appears to live is in fact nothing but a complicated physical system? Science alone can attempt to explore the empirical side of this question. Apart from the fact that scientists have not succeeded in producing life from inorganic matter,[62] it is clear that empirical science can only investigate those phenomena which lie, as it were, 'before' the ultimate solution of this question because life qua life escapes the methods of observation and cognition of natural science.

3. The Irreducibility of Mental Life and of Its Subject (Soul) to Ordered Chaotic Physical Systems

Life found in the operations and in the being of the ego-subject of conscious acts is accessible to us in an experience much more immediate and intimate than introspection or inner perception, namely in a differentiated experience of consciousness inseparable from performing conscious acts.[63] Augustine has spoken of the indubitable evidence of the skeptic that he is and that he *lives*.[64] Therefore, when we deal with the experience of our own conscious life, we are in a far better position to reject any reductionistic interpretation of life, for we possess apodictic certainty and scientific-philosophic evidence about the irreducibility of the mind's life to chaotic or non-chaotic physical systems.

The arguments for the substantiality of the mind summarized in the following[65] were expounded elsewhere.[66]

---

[62] Not even viruses have been artificially produced. Some distinguished scientists, like J. Lejeune, deny life to them, while others insists that they are living organisms.

[63] Cf. KAROL WOJTYLA, *The Acting Person* (Boston: Reidel, 1979), cf. also the corrected text, authorized by the author (unpublished), Library of the International Academy of Philosophy in the Principality Liechtenstein, Schaan.

[64] AUGUSTINE, *De Trinitate*, X, x, 14:

> Who can doubt, however, that he lives, remembers, understands,
> wills, thinks, knows and judges? For even if he doubts, he lives...

Cf. also LUDGER HÖLSCHER, *The Reality of the Mind. St. Augustine's Arguments for the Human Soul as Spiritual Substance* (London: Routledge and Kegan Paul, 1986).

[65] We take the liberty here to use, in the following few pages, an adapted brief section of a paper entitled "Is Brain-Death Actually Death?", published in the *Monist* (Spring 1993).

[66] JOSEF SEIFERT, *Das Leib Seele Problem und die gegenwärtige philosophische Diskussion. Eine kritisch-systematische Analyse* (Darmstadt: Wissenschaftliche Buchgesellschaft, 2: 1989), pp. 111 ff., 180 ff.; Hölscher 1986, and others.

1. Our conscious life and acts require a subject which is not an attribute of another thing but stands by itself; this is what Aristotle means by substance. 'Substance' in this sense has nothing to do with a thing (*res*) as opposed to a person but indicates only the irreducible trait of standing on its own feet in being as opposed to inhering in another entity. Thought and any conscious life clearly require such a substance-subject: someone *who* thinks, wills, etc.

But this I or self, this person who wills or thinks, is clearly given in our experience as one single and indivisible subject. Material objects can carry many properties such as colour, beauty, etc. A material object such as a computer disk can store the physical substratum of information. But can a composite thing like our brain think? Evidently not, for the mind as subject of our thought-life is an absolutely indivisible, simple subject that cannot be observed by the senses. To say that my ego divides into two, etc. is absurd. The non-duplicability, any non-divisibility of the individual person manifests itself most clearly in self-consciousness, as Sir John Eccles has repeatedly noted. Now the brain possesses billions of non-identical cells and parts, into which, as Leibniz remarks in the *Monadology*, n. 17 - we could enter, if they were enlarged, and which we could observe through the senses, whereas we understand clearly the non-sensible nature, the simplicity and the absolute indivisibility of the I-subject of mental life and thought. Thus neither the mind nor its life can be reduced to physical systems.

2. Another argument for the reality of the mind is based on two evident premises: first, on Aristotle's intuition that the subject (substance), as that which supports in being every attribute inherent in it, surpasses in reality that which just exists in dependence on it; for it is the latter's supporting ground and the most real part of an entity. The second premise is this: Capable of rational acts man surpasses in reality the entire cosmos of physical substances, including the perennial mummified existence of our body. Conscious and rational life possesses an actuality and contains a self-possession of being, compared to which the universe of material substances is nothing. As a matter of fact, for the conscious life of the self to be reduced to the being of a material substance would be equivalent to its annihilation.

From these two premises it follows that the incomparably higher life of the spirit, which is more properly real and of greater dignity than all material substances of the universe, cannot possibly be a mere function or accident of one of them, the brain. If the substance is the most real entity in the world, then the conscious self whose reality and life surpasses that of all material bodies put together (which are mere 'nothings' compared with it), cannot coincide with the single material substance of the brain and even less with some accidents or epiphenomena of it.

3. A third argument in favour of the substantiality of the mind is based on the specific essence of certain acts. Knowledge, in order to be knowledge, must depend in its content on the nature of that of which it is knowledge. Its subject mentally grasps the nature and/or existence of that which he knows and in certain knowledge, such as that he himself exists, the subject can be absolutely certain of the object of his knowledge. Knowledge exists only to the extent that someone grasps that or how something is - because it is and is in a certain way. Cognition possesses an intentional relationship to its object and is characterised by a receptive transcendence of the subject in the cognitive contact with the object of cognition. Knowledge must not only be formed objectively by the nature of the object, as a piece of wax can we objectively formed by a mould, but it must mentally reach this object, cognitively touch it in such a way that it discovers that and how something is. Cognition would therefore be impossible if knowledge were just dependent in its content on brain-functions which are subject to causal laws and are entirely determined by external material causes which have no intrinsic relation to the true nature of the objects of knowledge. As it is impossible that a computer, whose output is entirely dependent on physical events, could know or check its own program in any real and ultimate way, so man could never know anything if his knowledge were causally dependent on material processes and their causal connections, and if he did not have a mind distinct from matter and its causal effects. But man knows with certainty some things, for example that he truly exists and is a subject. Therefore he has a mind. Moreover, even the most radical doubt is absolutely impossible without cognitions such as that I exist, that I doubt, that I am aware of objects, that I pose a question, that the two contradictory states of affairs which I consider in doubt cannot both be, etc. These and countless other such things not only *can* be known but they *are* known by all of us. We possess such knowledge inevitably in any assertion or thought. Hence any assertion of the identity of knowledge with brain-processes or of its causal dependence on the brain is self-contradictory.

The same could be shown regarding freedom, promising, etc. Jonas (1981) has demonstrated the absurdity of some materialists' promise to promote materialism. They pledged to promote a theory of the power of matter over mind but presupposed, in the very assumption that their promise could and would be kept, an original power of the mind over matter. Any free act is necessarily impossible, nay absurd, if such an act is determined by material or organic processes or by evolutionary developments. Moreover, every man presupposes some free acts such as searching for truth, asserting, promising, etc. Hence he contradicts his own theory in every moment in which he - inevitably presupposes his freedom. We deal here not with a mere inevitable subjective presupposition à la Kant but with an evident datum of the essence of freedom and of its real existence in us. And this datum of freedom refutes materialism, according to which free acts could not exist.[67] Thus the life of free acts and of their subject is irreducible to the brain and to any conceivable material system whether it is dominated by well-explored or by chaotic rules.[68]

Cognition and free actions contradict their being an epiphenomenon of, let alone their being identical with, the brain or its functions.

4. Possibly the most striking argument for the irreducibility of the mind's life to material systems is taken from a reflection on the immediately experienced selfconscious subject. This subject we do not have to *infer* as the cause of our acts but we experience it immediately as subject and understand it to stand in itself. Its predicates differ from those of any conscious experience as such: our experiences are not conscious of each other, but we as subjects are consciously living them; they can not reflect upon themselves and know themselves, but we as subjects can reflect both on them and on ourselves, etc.

This *substantial* human mind possesses each trait of substance in an incomparably superior way to the weak and analogous manner in which material entities can be substances. The mind is not only the ultimate subject of acts, as thing-substances are of accidents, but it is a consciously living subject; it is for this reason that today we reserve the term subject which originally designated all substances for Persons only. The subject is not an 'it' but a 'he' or 'she'. The person's acts do not merely 'inhere' in a mind-substance but are performed and lived by the mind-subject. As one single, non-composite self and as conscious of himself, the person stands in himself and is 'this single substance-subject' in a way which is quite incapa-

---

[67] Of course, also philosophies which recognize the existence of a soul can embrace determinism but materialistic philosophies must inevitably do so.

[68] This is overlooked by JAMES P. CRUTCHFIELD, "Chaos 1: Chaos and Complexity" in: HANS BURKHARDT and BARRY SMITH (Editors), *Handbook of Metaphysics and Ontology*, Vol. 1, cit., p. 143.

ble of being matched by the material world where accidents inhere in various parts and appearances of bodies and not in one single material subject. In sharp contradistinction to this, my identity as one simple (non-composite) 'I' - the subject of myriads of experiences - is given to me.[69] And this subject is different from the brain and from any totality that consists of non-identical distinct parts. Material things are neither indivisible nor do they possess an inner unity of substance comparable to that of the mind. The subject of personal life also possesses other traits of substance more properly speaking than any material substance could possess. Material things lack the absolutely indivisible, irreplaceable and incommunicable individual identity of the sort that is found in the personal self. While also material things or Rembrandt paintings possess these traits in some way, each material substance and each part thereof can be replaced by another, without this making a big difference, whereas persons cannot be replaced or substituted by others. Even masterful paintings could conceivably be replaced by a perfect copy. Moreover, in material things the different traits of substantiality are somewhat separate from each other. Their individual thisness and distinctness stems from an external form which does not coincide with their self-standing substantiality. Again, what is lasting in them is not the material substrate of body cells which are entirely renewed every seven years. In the subject of the conscious life immediately experienced by us from within the various traits of substantiality are united: the same subject lasts, is substance, is an individual self, etc.

From these considerations it becomes evident that the mind cannot be a supervenient attribute or epiphenomenon of the brain, and that it is necessarily different from the brain with its billions of cells and countless non-identical parts. The magic word of 'emergence' or of 'laws of chaos' cannot help here because chaos can never bridge intelligible essential differences or explain that which is intrinsically impossible: that the operations of the mind as a substantial entity distinct from matter could be produced by unexplored laws of matter.

The difference of the mind from the body also constitutes the only conceivable basis for doing justice to the unity of man and of human life, in whom body and soul are joined to form a much more unified being than two

---

[69] An investigation into memory, retention, and 'superactual consciousness' could help to elucidate the lasting identity of the subject of consciousness. See DIETRICH VON HILDEBRAND: *Ethics*, cit., ch. xvii, xxvi ff.; *Das Wesen der Liebe* (Regensburg-Stuttgart: Habbel/Kohlhammer, 1971), ch. ii.

sorts of matter - or a material thing and its accidents - could ever constitute. For only a mind distinct from the body can become 'form' and animating spiritual principle of the body.

The manifold dependence of the mind on the brain, too is by no means incompatible with the superior life of the mind. Since the union of two persons who love each other requires the duality of persons, so the unity of the word with the concept it expresses requires the radical distinctness of its meaning from its audible *Gestalt, and* so the unity of man requires the duality of body and mind.

Another significant discovery has to be stressed here. Every arising of accidents in a substance presupposes some real potentiality in that same subject. Therefore, acts of understanding cannot arise from nowhere but presuppose a faculty of intellect, etc. Some of the human potencies and faculties are inseparable from the substantial being of the mind, some are also 'transcendent potencies': they open the person to all beings.[70]

The individual human personal life on earth objectively begins when the spiritual human soul is present in the human body, it continues as long as the mind is united with the body, and our bodily mundane life objectively ends at the moment when the human mind definitively leaves the body.

A radical materialism and a new form of Cartesianism throw into question the unity of life in man. According to such a view, the embryonic life before brain-birth is not personal human life. And again, the medical-legal theory of brain death separates biological life of man from personal human life. If the mind coincides with, or totally depends in its existence on, higher brain functions, then of course the mind cannot precede or survive irreversible disfunction of the brain.[71]

Against these and similar theories we have insisted elsewhere on the union between body and soul and on the inseparability of biological human life from the life of the human soul.[72]

---

[70] On potencies, see SEIFERT, 1977, pp. 379-385.
[71] H.T. ENGELHARDT JR., *The Foundation of Bioethics*, (New York/Oxford: Oxford University Press, 1986).
[72] JOSEF SEIFERT, *Das Leib-Seele Problem und die gegenwärtige philosophische Diskussion*, cit.; the same author, "Is Brain Death Actually Man's Death?", in *Monist 76, 2* (1993),175-202.

DISCUSSION
(Chairman McCONNELL)

ARECCHI: I agree in large part with what you have said. Let me emphasize, however, that not only is life an ambiguous word which has different connotations in science and philosophy, but there is nothing exclusively reducible just for a scientific investigation. Now, you have built arguments which are on the borderline and you have trespassed on the field of science with considerations which I do not completely share. I would like to face the complete independence of science in investigating life from its point of view, even though this, of course, need not coincide and should not interfere with the freedom of a philosopher to investigate life from his own point of view. We never exhaust a reality. I tried to make reference to this in my talk, to a remark made by an historian of science, Eugenio Gaveni, about the conflict between Florentines and Paduans in the fifteenth century. If you want to assign in principle a limitation to scientific endeavours, as far as this problem is concerned, this would look like the old Galileo debate. On the contrary, usually science will have in principle limitations arising from its inner processes like, for instance, the Gödel theorem with respect to limitations of a formal theory.

SEIFERT: A great number of words have many meanings only for arbitrary linguistic reasons. For example the German word 'Schloß can mean a lock and a castle.[1] But it is hardly meaningful to say 'of Schlösser we speak in many senses' - precisely because there is no inner reason in the object for such a manifold meaning. Other terms have only slight differences in meaning, as when you apply the word 'red' to a light or a dark red colour. Only where two radically different meanings of the same term are easily confused is it necessary to distinguish the 'many senses' of a term. The terms 'being' or 'life' possess at the same time a profound difference in meaning and involve a deeply meaningful unity and similarity between their distinct referents. With respect to such analogous terms it is important to take into consideration the whole spectrum of

---

[1] Interestingly, a word with the same equivocal meanings exists in Polish: zamek.

their meaning and to become aware that and how they are used differently. In a discussion of the origin of life it is crucially important for the biologist to realize that the organic life he investigates is not the only kind of life and, if he could successfully reduce it to material systems (which I deem impossible), he should not therefore consider also the life of the mind reducible to matter. Mental life is much more evidently irreducible to matter than the life of bacteria or that of viruses (where we can have some doubts as to whether they actually live or whether their reproduction is something analogously mechanical or at least purely physical like the 'growth' of crystals). Keeping in mind the radically different forms of life, - the biologist will become also more cautious in his claims favoring reductionistic models of explaining life.

Concerning the idea of the autonomy of the methods of science and of philosophy, I fully agree with it if this 'autonomy' means that both disciplines and their methods are quite different and have different formal objects. Empirical science uses sense perception, observation experiments, hypotheses etc. in order to explore the factual sides of the world. Philosophers look for highly intelligible and necessary essences and essential features of the world open to evident cognitions and insights into necessary essences as well as to deductive demonstrations.[2] Mathematicians do the same when they really attend to the principles of numbers, geometric objects, etc. and do not only construe ideas, a procedure which also plays a legitimate role in mathematics. The reason for such a difference in method lies in differences in the formal objects and purposes of the respective sciences. Nonnecessary essences - the elements of which are held together by some factual, non-necessary although very meaningful bond - do not allow insights into necessary evident truths. Thus scientists who explore them must use empirical and experimental methods. But to apply such methods to questions of philosophy and mathematics would be absurd. The mathematician would laugh if he were asked to conduct years of experiments to know that 2x2 can never equal 5. Similarly, It would be absurd to solve philosophical questions of ethics and oughtness by means of empirical studies of human or animal behaviour.[3]

---

[2] Also *existence (esse)* plays a decisive role as object of philosophical knowledge.

[3] See DIETRICH VON HILDEBRAND, *What is Philosophy?*, 3rd edn, with a New Introductory Essay by Josef Seifert (London: Routledge, 1991), especially pp. xii ff., and ch. iv; see also BALDUIN SCHWARZ, "Dietrich von Hildebrands Lehre von der Soseinserfahrung in ihren philosophiegeschichtlichen Zusammenhängen", in B. SCHWARZ (Editors): *Wahrheit, Wert und Sein. Festgabe für Dietrich von Hildebrand zum 80. Geburtstag,* (Regensburg: Habbel, 1970), pp. 33-51.

Philosophers cannot explore, by means of their methods, for example the different species of insects or the cellular structure ot living organisms. This notwithstanding some philosophers especially those who also conducted empirical investigations, tended to exert an illegitimate influence on science, barring experimental sciences for centuries. Although Aristotle himself had conducted empirical investigation, the Aristotelean idea that all universal natures are necessary impeded the rise of experimental sciences for a long time.[4]

But what applies to philosophy, applies even more to science. It is a disaster if ethologists who observe animal behaviour deduce ethical conclusions from this, as Konrad Lorenz, Wolfgang Wickler and many others do. And scientists frequently trespass over the limits of their discipline. This is legitimate only as long as they realize that now they speak as philosophers and not as scientists. For example, Einstein's theory of the relativity of time has purely scientific aspects resulting in the important formula regarding mass, acceleration, and energy: $E=m.c^2$. But he also develops philosophical theories that the essence of time and simultaneity are relative which in no way follow from his physics. The same is true about Heisenberg who deduces from the physical aspects of his discoveries regarding the uncertainty relation far-reaching and mistaken philosophical consequences (which in no way follow from scientific results) regarding indeterminacy freedom, causality, the principle of excluded middle, etc. The same is even more obviously true about the theory of evolution, of the Big Bang, etc. which are, in most of their forms, chiefly philosophical theories for which scientific research provides at best some starting point. This applies still more to J. Monod's metaphysical claims about chance and necessity. As soon as scientists hold false philosophical theses as if they were empirically demonstrated, this becomes the source of many errors.

The danger of mutual intrusion must not be construed to imply a complete divorce between scientific and philosophical investigations.

---

[4] Professor S. Jaki has examined this phenomenon equally deeply and sharply in many books and has shown convincingly that only a creationist metaphysics which sees the origin of nature in a free divine act and therefore recognizes contingency in nature, could provide the proper metaphysical basis for empirical sciences. This knowledge must not lead us, however, to believe that there are not *also* absolutely necessary and immutable essences which precisely a priori sciences such as mathematics or philosophy investigate, nor should we believe that in the field of objects of such sciences it would be possible, for example, to establish the morally good by statistics or by behavioural studies of animals.

Therefore one has to recognize both the described autonomy and the many mutual relations between philosophy and science. For example, biologists presuppose many philosophical categories such as reality, existence proof, argument, logical laws, matter, space, time, indeterminacy, determinism, finality, etc., many aspects and the general essence of which they cannot explore by means of their methods but which are objects of philosophical analysis. The same is true about truth, the scope and purpose of each science, the value and limits of scientific knowledge. All these are philosophical problems, and the scientist presupposes implicitly some answers to them which only philosophy can give. Professor Jaki showed convincingly that even a metaphysics and theology of contingency and creation was necessary in order to make empirical sciences possible and to provide their proper philosophical foundation.

On the other hand, philosophical questions are posed by science; scientific results and experiences about the brain can widen the scope of philosophy and corroborate or inspire new philosophical discoveries of objective essential necessities and intelligible truths which become accessible only through a special type of experience. However, the role which experience plays for philosophy and for biology is totally different. The experiment as such has no relevance for philosophy except for corroborating or calling into question from the outside the results of philosophy. Thus the philosopher can be pleased when the results of his philosophical studies are corroborated by empirical sciences[5] and he is forced to check the adequacy of his methods when their results do not harmonize with the results of empirical science. But the philosophical method is never the experiment but an experience of intelligible essences which allow insight (intuition) into highly intelligible and evident essences and states of affairs, and deductive demonstrations.

HELLER: I think that large parts of your talk were based on the tacit assumption that two mutually excluding standpoints are possible in the philosophy of biology: either you can reduce the phenomenon of life to physics, or you need something more than the set of physical laws to explain

---

[5] I tried to show this in detail with respect to brain-science and with reference to the work of SIR JOHN ECCLES in JOSEF SEIFERT, *Das Leib-Seele Problem und die gegenwärtige philosophische Diskussion. Eine kritisch-systematische Analyse* (Darmstadt: Wissenschaftliche Buchgesellschaft, ²1989).

this phenomenon. I think that the main philosophical merit of all we have discussed during this Session: the complexity problem, the dynamical chaos theory, non-linear thermodynamics, and so on, is that the third possibility has been discovered. On the one hand, life is not reducible to physics in the sense that it cannot be entirely explained by dissolving a living system into its physical components. On the other hand, however, in explaining life one needs nothing more but the laws of physics. The third possibility is that from the purely physical laws, on the strength of their non-linear character, a new level of organisation can emerge. In this process, the laws of physics operate in such a manner that life cannot be deduced from them in any predictable way. I stress this but, even if we assume that the laws of physics are deterministic, the process of the emergence of life is non-predictable. I think this should be taken into account in philosophical analyses.

SEIFERT: Any theory which opens up to new, verifiable or plausible principles of explanation is certainly less primitively reductionistic than older mechanistic reductionisms which did not include reference to teleological (finalistic) principles and purposes, such as the 'anthropic principle', or to dynamic chaos-theories, nonlinear thermodynamical etc. Certainly, today most scientists will agree that life cannot be reduced to predictable patterns of mechanics. And to allow for purposes which are not reducible to predictable or even to unpredictable chance such as the anthropic principle in the cosmos, is less reductionistic than theories which did not admit such principles of the emergence of order. Matter has become far more mysterious and explanations of matter which allow for the unpredictable emergence of phenomena certainly are far less reductionistic than purely mechanical models of explanation. But any form of explanation which does not do full justice to the essences of things is reductionistic. The explanation of a chess-game in terms of a pure theory of wood of which the pieces consist, and of mechanical causality by means of which the pieces are moved by hands or computers etc. would not explain the moves because it would ignore the principles and rules of chess, including the timeless mathematical, logical and chess-specific principles. It would ignore the chess-conventions, the intelligible systems of end-games or of openings, etc. Even to introduce into the theory of chess the occurrence of nonmechanical, unpredictable patterns of behaviour not explicable in terms of classical mechanics would not suffice as an explanation because they would not take into consideration the entirely new sphere of rules geometrical and logical and chess-specific necessities and the sphere of their purposeful action, intelligent understanding, etc.[6] Therefore even the most high-level physical theory of chess in terms of chaotic and non-chaotic physical systems would be radically reductionistic.

But to explain life in terms of those same principles alleged to lie beyond reductionism would be much more reductionistic. For in chess it would indeed be enough to admit rules, logical and mathematical necessities, etc.; and then to say that the substantial being in which all these are stored, the computer, is a purely material thing. But such an explanation would not be enough for a non-reductionist explanation of life or of mental life. For mental life requires minds.

Heisenberg could indeed discover certain analogies between the alleged indeterminacy of quantum-physical micro-processes and freedom but his philosophical interpretation of his ingenious discoveries in physics involves - in my opinion - both a confusion between freedom and chance and another one between our inability to measure with precision both speed and location of particles and our resulting ignorance about certain aspects of physical causes of micro-physical events, and their alleged ontological indeterminacy or uncausedness. Matter and material events necessarily require an efficient cause through the power of which they are or happen, whether the efficient cause of material events is immanent in nature itself or transcendent to it (as in the case of free actions upon matter). For inanimate matter can neither be free (or act from an inner principle of spontaneous movement) nor is absolute chance possible or a viable substitute of a causal principle of explanation. Only the intelligible natures of things themselves dictate whether or not a given explanation of life is reductionistic. If life requires a *soul*, then even the most sophisticated materialist models of explaining life, using non-linear equations, chaos-theory, quantum physics, etc. remain basically just as reductionistic as the old reductionisms. Life, especially mental life, just cannot be thus reduced. If biological life and mental life cannot be reduced to complex brain-events, brain-processes, cells, modules, etc., any attempt to explain them in such terms remains reductionistic.

---

[6] JOSEF SEIFERT *Schachphilosophie* (Darmstadt: Wissenschaftliche Buchgesellschaft, 1989), ch. ii ff.

NICOLESCU: I would just like to make a remark about terminology. I think it would be useful to replace the word interdisciplinary, in your case, by trans-disciplinary, because inter-disciplinary still supposes the autonomy of disciplines, it is just a transfer of methods from one discipline to another and it is in some sense against what you want to say. So I will reformulate what you said, by saying that life is a highly trans-disciplinary concept.

SEIFERT: Interdisciplinary studies should not just mean a sum-total of studies which remain each enclosed in one discipline only. Their goal is precisely to arrive at a more comprehensive understanding of the issues at stake. In this respect it is legitimate to speak of transdisciplinary studies. But 'transdisciplinary' must not mean that either scientists propose dilettantic or false philosophy which they pass for 'transdisciplinary' or that philosophers put forward dilettantic scientific theories. Jacques Monod for example philosophizes in the most hair-raisingly dilettantic way about the highest matters of metaphysics. His knowledge in micro-biology did not prepare him in the least for philosophical theories. If one is unprepared to pursue philosophical investigations, one should leave those to the philosopher.
It is obvious that philosophy does not turn a philosopher into a scientist but many people think that anyone is prepared to philosophize. This is true in the sense that every man makes some philosophical assumptions and can in principle know philosophical truths. Whereas Socrates knows quite well that he does not understand the different crafts, poetry, and the intricacies of political life, the craftsmen, poets and politicians think they know all philosophy. Socrates knows his limits, whereas the representatives of other fields think, because they know some things which the philosopher does not understand, that they know all others and the most important objects of philosophy as well, even if it can easily be shown that they know nothing about them. This does not exclude, of course, that some individual scientists were also great philosophers and that many scientists who have not been great philosophers nevertheless gained many philosophical insights, as I tried to show of Sir John Eccles with respect to the 'self-conscious mind'.[7]

---

[7] JOSEF SEIFERT, *Das Leib-Seele Problem und die gegenwärtige philosophische Diskussion*, cit.

SINGER: I would like to take up Dr. Heller's point and apply it to your statement that consciousness and matters of mind cannot in principle be reduced to anything else. I think that when we ask for stepping-down from consciousness or matters of mind to individual brains, we are asking too much in a single step, we tend to forget that phenomena such as consciousness may not be emergent properties of individual single brains. If individual brains grew up in isolation from other brains, they would most likely not develop any of these phenomena. Certainly, they would have no names and concepts for matters of mind. The reason why we are able to distinguish such phenomena is, that there have been many brains evolving together in a cultural environment, mutually observing and imageing one another and inventing descriptions for the observed states of other brains. The whole phenomenology of "matters of the mind" emerged only as a collective property from a society of brains. This suggests that it should be impossible to establish a simple one-step down path from matters of mind to individual brains. There are a few steps in between the individual brains and the phenomenon of consciousness and I think that needs to be considered.

SEIFERT: I fully agree that the 'programming' of the brain as well as the complex spatio-temporal patterns in the modules of the brain include manifold changes of brain-activity and 'brain-programs' which are consequences of experience and of manifold external influences. Memory in all its forms gives a splendid example of this, and so does any form of learning.
Nevertheless, if the indivisible, simple, and spiritual *subject* of thought and will can necessarily not be identical with one brain which - in its physical complexity of cells and of parts thereof - is utterly incapable of producing even the tiniest thought or conscious act, then it is equally impossible that the mind be reduced to one such brain modified by other brains. And programs, spatio-temporal patterns of brain-activity, etc. themselves are even less capable of being the ontological subjects of intellectual or free acts than a brain. For they are not substantial entities such as persons. The fundamental brain-physiological reductionism does not cease to be reductionistic if one adds to our own brain other brains and their interaction with our brain events. None of the phenomena discussed in my paper in the context of the proofs for a spiritual soul are more explainable in terms of many brains than in terms of one brain, or in terms of a brain programmed by external influences as opposed to an

isolated brain. I tried to show this in various books and articles on the body-soul problem. Consciousness simply cannot be an emergent or a supervenient property of a brain or of a 'society of brains' any more than conscious life can be an epiphenomenon of a single brain.

LOVASZ: I would like to make a remark, just a little bit with the eye of a complexity theorist on the use of the verbs "can" and "cannot" in the definition of reduction. I had a slide in my talk where I put the word "cannot" in quotation marks, because there are many different senses in which you can use it, and many different senses in which you might want to reduce the behavior of a complex structure to its constituents. I think there are many examples of a complex structure where, although its behavior is completely determined in some sense by its constituents, you need entirely new notions, and a new approach and a new phenomenology to describe the behaviour of the structure. There can be complexity theoretic reasons behind this (simply, say, to write up and solve the equations would be impossible, and I think this is a very serious impossibility in any sense). There are nice examples like in statistical mechanics where the behavior of certain systems depends a lot more on the structure of it than on the particular properties of the constituents. Now "can" or "cannot" the behavior of such a structure be reduced to its constituents? So I would like to see that philosophers deal with this "can" a little bit more carefully — or maybe give various interpretations.

SEIFERT: I fully agree with your opinion about the dangers of saying 'can' or 'cannot' too quickly. In the object of natural science the most astonishing forms of 'reductions' are possible, even where our experience seems to refute the possibility of such reductions: as for example the simple example of some 'reduction' of diamond to coal. And philosophers must be extremely cautious not to put illegitimate obstacles to the freedom of experimental research. They should abstain from assertions of impossibility wherever only empirical methods can determine what is possible or impossible. But where we meet intelligible and necessary data such as the question as to whether the number 11 is prime or whether or not minds can emerge from brains, it would be utterly untenable to say that not enough experiments have been made to verify this assumption.[8] Here

---

[8] Similar empirical methods and guess-work could also apply to certain aspects of number theory - but only where immediate insight is impossible regarding a priori questions, such as in the theory of prime numbers or of chess-openings. In such cases, however, the experiment

no experiments are needed, not because philosophy makes blind and unverifiable as well as unfalsifiable assumptions, but much rather because the intrinsic objective evidence of objects makes empirical experimental methods and their objects superfluous. When we understand such essential necessities, we can indeed say: never can minds be reduced to brains, never can justice be reduced to *ressentiments* of the weak against the strong, never can moral values be reduced to conventions never promises to declarations of wills or to their social acceptances, etc.[9] Due to the intelligibility of the object we can indeed reject any such reduction.[10] While obstacles erected by philosophers blocked the progress of science for many centuries,[11] there are also properly philosophical methods and truths which can never be overthrown by empirical research. If it is evident that a free act can never be produced by a brain but that the brain can only be its condition or be in other similar ways related to it, then the necessity of such a fact holds for the real and for any possible world. Then it is as impossible to prove the materiality of the subject of thought as it is impossible that mathematical or philosophical, or any empirical results would ever show that 2x2=12 or that envy is morally good or that a being can at the same time not be, and innumerably many other such necessary falsities. And here we are able to determine parameters of what can and cannot be.

---

has a role relative to our incapacity of understanding such matters *by* a priori insights, mathematical inductions or deductive arguments. GREGORY J. CHAITIN, "Zahlen und Zufall: Algorithmische Informationstheorie. Neueste Resultate über die Grundlagen der Mathematik", in: *Naturwissenschaft und Weltbild* (H.-C. Reichel, E. Prat de la Riba, Editors), (Wien- Hölder-Pichler-Tempsky, 1992), S. 30-44. JOSEF SEIFERT *Schachphilosophie*, cit., ch. ii ff.

[9] A. REINACH, 'The Apriori Foundations of the Civil Law', transl. by J. F. Crosby, *Aletheia III* (1983), pp. xxxiii-xxxv; 1-142.

[10] LUDGER HÖLSCHER, *The Reality of the Mind: St. Augustine's Philosophical Arguments for the Human Soul as Spiritual Substance* (London-New York: Routledge and Kegan Paul, 1986); second edition prepared by C. Winter (Universitätsverlag Heidelberg, in the series *Philosophy and Realist Phenomenology*).

[11] Karl Popper holds a falsificationistic concept of science which calls into radical doubt the method of induction and has many other problematic implications. Also T. Kuhn and L. Laudan have proposed theories of science in which the rationality of inductive methods is denied. Regarding a critique of this elimination of induction, cf. JOSEF SEIFERT, "Objektivismus In der Wissenschaft und Grundlagen philosophischer Rationalität. Kritische Uberlegungen zu Karl Poppers Wissenschafts-, Erkenntnis- und Wahrheitstheorie", in: N. LESER, J. SEIFERT, K PLITZNER Editors), *Die Gedankenwelt Sir Karl Poppers:* Kritischer Rationalismus im Dialog (Heidelberg: Universitätsverlag C. Winter, 1991), pp. 31-74.

EVOLUTION OF COMPLEXITY OF THE BRAIN WITH THE EMERGENCE OF CONSCIOUSNESS*

JOHN C. ECCLES

Max-Planck-Institut für Hirnforschung

Frankfurt/M - Noederrat 71, Germany

Synopsis

It has recently been proposed that before the evolution of the mammalian brain the animal world was literally mindless.

Even in the most primitive mammals, the basal insectivores, there came to exist a neocortex with a higher level of neural complexity, particularly in its pyramidal cell structure. Their apical dendrites have an enormous synaptic input and they form bundles as they ascend through the cortical laminae. There are hundreds of thousands of synaptic inputs, through *boutons*, onto a dendritic bundle, which is the reception unit, named *dendron* of the cerebral cortex. The axons of the pyramidal cells have a wide distribution in the brain. In this simplified conventional account of the structure of the cerebral cortex the story of the feelings that may be generated by the brain activity is completely missing.

In developing that story it is necessary to move into a higher level of complexity, the ultramicrosite structure and function of the synapse, as discovered particularly by Akert and associates of Zürich. The boutons of chemical transmitting synapses have a presynaptic ultrastructure of a para-crystalline arrangment of dense projections and synaptic vesicles, a *presynaptic vesicular grid*. Its manner of operation in controlling chemical transmission opens up an important field of neural complexity that is still at its

---

* The Editors thank professor Danilo Gherardi, of the Dept. of Neurosciences of the University of Rome "La Sapienza", for help with this contribution.

inception. The key activity of a synapse concerns a synaptic vesicle that liber-ates into the synaptic cleft its content of transmitter substance, an exocyto-sis. There are about 50 synaptic vesicles in the presynaptic vesicular grid. A nerve impulse invading a bouton causes an input of thousands of Ca^{2+} ions, 4 being necessary to trigger an exocytosis. The fundamental discovery is that at all types of chemical synapses an impulse invading a single presynaptic vesicular grid causes at the most a single exocytosis. There is conservation of the synaptic transmitter by an as yet unknown process of higher complexity.

The conservation challenge becomes intense when it is recognized that synapses of the cerebral cortex have an exceptionally effective conservation with a probability of exocytosis as low as 1 in 5 to 1 in 4 in response to an impulse invading a hippocampal bouton.

Because of the conservation laws of physics, it has been generally believ-ed that non-material mental events can have no effective action on neuronal events in the brain. On the contrary it has been proposed that all mental experiences have a unitary composition, the units, being unique for each type of experience and being called *psychons*. It has been further proposed that each psychon is linked in a unitary manner to a specific dendron, which is the basis of mind-brain interaction.

Quantum physics gives a new understanding of the mode of operation of the presynaptic vesicular gird and of the probability of exocytosis. Changes in this probability are brought about without an energy input, so the mind could achieve effective action on the brain merely by increasing the probabi-lity of exocytosis, for example from 1 in 5 to 1 in 3. That would give a large neuronal response when the mind through its psychons causes this incre-ment in the hundreds of thousands of presynaptic vesicular grid on specific dendrons. A higher level of neural complexity is thus envisaged in order to lead to an understanding of how the mind can effectively influence the brain in conscious volition without infringing the conservation laws.

I. Introduction

The key concept of my lecture is that in the evolution of the mammalian brain there have come to exist in the neocortex levels of complexity in its ultramicrostructure, that we may literally call transcendent because[1] they

---

[1] ECCLES, J.C. *The evolution of Consciousness*. Proc. Nat. Acad. Science, vol. 89, 1992; ECCLES, J.C. *Evolution of the brain: Creation of the Self,* London, New York: Routledge, 1989.

opened the brain to the world of conscious feeling. Before that the living world was *mindless*, as we would now recognize for bacteria, plants and lower animals. We may ask how low? The usual answer would be that all mammals, dogs, cats, monkeys, horses, rats experience feelings and pain and possibly also birds, but not invertebrates and lower vertebrates such as fish and even amphibia and reptiles that have instinctual and learned responses. However, much more experimental testing may be possible in the light of concepts of how animals could use their consciousness as I will mention at the end of the lecture.

We can assume that the mammalian neocortex evolved for the purpose of integrating the greatly increased complexity of the sensory inputs: visual, auditory, tactile, olfactory, gustatory, proprioceptive, so as to give effective behaviour. We can now try to understand how the functional structure of the mammalian brain could have properties mediating consciousness of another world from that of matter-energy: the world of feelings, thoughts, memories, intentions, emotions.

We have to concentrate on the neocortex because all other parts of the brain such as the striatum, the diencephalon, the cerebellum, the pons exist in lower vertebrates, reptiles, amphibia and fish that do not exhibit evidence of conscious feelings[2], [3].

2. The mammalian neocortex

In this enquiry we come to the mammalian neocortex that is qualitatively similar to ours, though usually much smaller. It has the same neuronal structure. The cortex has six layers (Fig.1B) and all true pyramidal cells are in laminae V,III,II, each with an apical dendrite and many side branches projecting towards the surface to end as a terminal tuft (Fig. 1A). A pyramidal cell has a nerve fibre or axon for transmitting information. It projects downwards from the cell and leaves the cerebral cortex. It ends eventually in many branches either elsewhere in the cortex or in more distant sites in the brain.

It is necessary to make some further statements with illustrations on the mammalian neocortex before we come to its transcendent properties. It is

[2] THORPE, W.H. *Animal Nature and human Nature*. London: Methuen, 1974.
[3] THORPE, W.H. *Purpose in a world of chance*. Oxford: Oxford University Press, 1978.

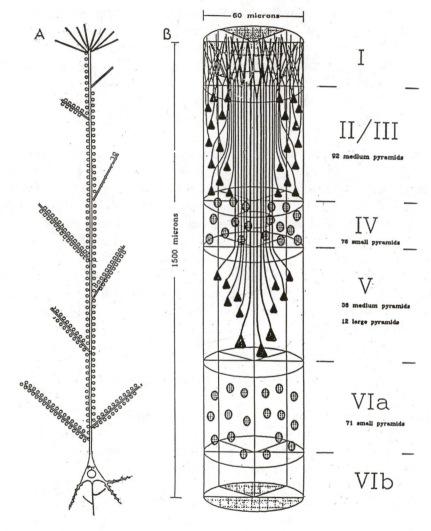

Fig. 1 - A. Drawing of a lamina V pyramidal cell with its apical dendrite showing the side branches and the terminal tuft all studded with spine synapses (not all shown). The soma with its basal dendrites has an axon with axon collateral before leaving the cortex. B. Drawing of the 6 laminae of the cerebral cortex with the apical dendrites of pyramidal cells of laminae II, III and V, showing the manner in which they bunch in ascending to lamina I, where they end in tufts. The small pyramids of laminae IV and VI do not partecipate in this apical bunching (A. Peters, personal communication).

composed of an immense number, even thousands of millions, of individual nerve cells and each is the recipient of information from other nerve cells by means of the fine axonal branches that terminate as synaptic knobs or boutons. There are thousands of excitatory spine synapses on each pyramidal cell, as are partly illustrated by the spines on the pyramidal cell apical dendrites and by the clustered boutons on the apical dendrite and its branches. Even a drawing is inadequate to show the thousands of boutons on the synaptic spines of the apical dendrite of each pyramidal cell and of course also on other cell dendrites and the cell body.

Nerve impulses are brief signals of about 1 millisecond depolarization passing along nerve fibres to finish in the terminal bouton and it is greatly enlarged in Figs 2 and 3. Transmission across a synapse occurs when an

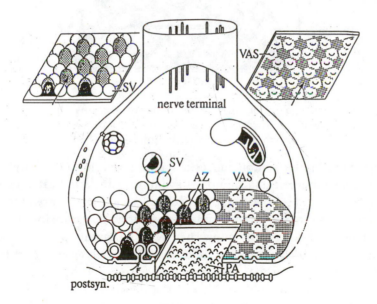

Fig. 2 - Schema of nerve terminal (bouton) of mammalian central synapse showing active zone, the presynaptic vesicular grid with geometrical design of dense projections (AZ) in triangular array and of synaptic vesicles (SV) in hexagonal array. One vesicle is shown in exocytosis indicated by arrow in the synaptic cleft. Below is the postsynaptic membrane with particles PA below the cut out. The presynaptic vesicular grid is stripped off to right to display hexagonal arrangement of presynaptic attachments (VAS). Insets to left show presynaptic vesicular grid and to right the vesicle attachment sites (VAS). Modified from Akert et al., (1975).

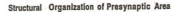

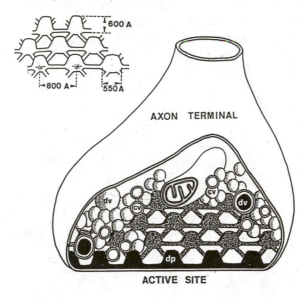

Fig. 3 - Axon terminal or bouton showing dense projections (dp) projecting from the active site with cross linkages forming the presynaptic vesicular grid, PVG, that is drawn in the inset with dimensions (Pfenninger et al., 1969).

impulse invading the bouton causes a synaptic vesicle to discharge its contents of transmitter substance into the synaptic cleft, as indicated by the curved arrow in F, and so to act on the specific receptor sites across the synapse and postsynapse in Fig. 2. For excitatory synaptic transmission in the cerebral cortex we are specially concerned with glutamate as the transmitter with its action in briefly opening ionic channels to decrease momentarily the electric potential across the postsynaptic membrane, so causing a mini-excitatory postsynaptic potential (EPSP) of the dendrite. By electrotonic transmission along the dendrite there is summation of the mini-EPSPs generated by each bouton activated at about the same time. When this occurs for a multitude of boutons (cf. Fig. 1A) the summed mini-EPSPs could result in a membrane depolarization of 10 to 20 mV, which could be enough to generate an impulse in the pyramidal cell that would travel down its axon eventually to the many synapses of the cerebral cortex on dendrites of neurons or to other regions of the brain.

This is the conventional macro-operation of a pyramidal cell of the neocortex (Figs 1A) and it can be satisfactorily described by classical physics and neuroscience, even in the most complex design of network theory and neuronal group selection[4], [5], [6], [7].

It may seem that this generally accepted simplified account of the neuronal mechanism of the neocortex indicates already a high level of complexity in its design. However, this account neglects the conscious feelings that may be generated by the brain activity. In order to move into this field it is necessary to consider in detail the manner of operation of synapses on the pyramidal cells, which is a new level of complexity.

Furthermore these complex neural structures have been postulated to have mental properties[8]. For example Changeux speaks of "consciousness being born". However, Stap[9] asserts that the origin of consciousness cannot be explained by classical physics; quantum physics is necessary. Classical physics is dedicated to matter-energy at all levels of complexity, but is not concerned with the mental world. By contrast quantal physics is closely related to the mental world. So our enquiry into the manner of operation of synapses on pyramidal cells moves into a higher level of complexity in the quest for mind.

3. Organization of the neocortex

There is agreement by Peters of Boston and Fleischhauer of Bonn and their associates that the apical dendrites of the pyramidal cells in laminae V,

---

[4] SZENTAGOTHAI, J. *The neuron network of the cerebral cortex: a functional interpretation.* Proc. Roy. Soc. B 201: 219-248, 1978; SZENTAGOTHAI, J. *Local Neuron Circuits of the Neocortex.* In: *The Neurosciences Fourth Study Program.* Eds F.O. SCHMITT and F.G. WORDEN. Cambridge, Mass.: MIT Press, pp. 339-415.

[5] MOUNTCASTLE, V.B. *An organizing principle for cerebral function: the unit module and the distributed system:* In: *The Mindful Brain.* Cambridge, Mass.:MIT Press 1978; PETERS, A. and KARA, D.A. *The neuronal composition of area 17 of the rat visual cortex. IV. The organization of pyramidal cells.* J. comp. Neurol. 260, 573-590, 1987; PFENNINGER, K., SANDRI, C., AKERT, K. and EUGSTER, C.H. *Contribution to the problem of structural organization of the presynaptic area.* Brain Res. 12. 10-18, 1989.

[6] EDELMAN, G.M. *The remembered present: A biological theory of consciousness.* New York. Basic Books, 1989; GRAY E.G. *Rehabilitating the dendritic spine.* Trends Beurosci 5, 5-6, 1982.

[7] CHANGEUX, J.P. *Neuronal Man.* Paris, Fayard, 1985.

[8] CRICK, F. and KOCH, C. *Towards a neurobiological theory of consciousness.* The Neurosciences, Seminars in 2, pp. 263-275, 1990; STAPP, H.P. *Brain-Mind Connection.* Foundations of Physics 21: Nr. 12, New York: Plenum Press, 1991.

[9] STAPP, H.P. In: *Nature, Cognition and System.* Ed. MARC CARVALLO. Dordrecht: Kluver Academic Publishers, 1992.

III and II bundle together as they ascend to lamina I (Figs 6B and 6). So there are neural receptor units of the cerebral cortex composed of about 100 apical dendrites plus their branches that is called a *dendron*. The enormous synaptic input into the 70 to 100 apical dendrites bundled into a dendron (Fig. 1B) can be calculated to be much more than 100,000 synapses, if there be on the average about 2000 on each apical-dendrite (Fig. 1A).

4. Ultrastructure of synapses

It is now necessary to describe the ultrastructure as revealed to Akert and associates by the techniques of freeze fracture, electronmicroscopy and selective staining. It is here that we enter into a higher level of complexity. Fig. 2 and 3 are a key diagram showing as a central feature a nerve terminal or bouton confronting the synaptic cleft. The inner surface of a bouton has been recognized as an assemblage of synaptic vesicles, but now is shown as a beautiful structure with dense protein projections (DP) (Fig. 3) in triangular array, AZ in Fig. 2, forming the presynaptic vesicular grid (PVG) (Figs 2, 3). The spherical SV's, 50 Å to 60 Å, with their content of transmitter molecules can be seen in the idealized drawings of the PVG (Fig. 4 and inset to left) with the triangularly arranged DP's, AZ and the hexagonal of the SV's. The SV's are so intimately related to the presynaptic membrane that it dimples outwards to meet them (Fig 2 to left), and when the SV's are stripped off, these dimples reveal the hexagonal pattern of the presynaptic attachment sites (VAS in Fig. 2 and right inset). The usual number is 40 to 60 SV's in the single PVG of a bouton (Figs 2 and 3).

The exquisite design of the PVG can be recognized as having an evolutionary origin for chemically transmitting synapses. In a more primitive form it can be seen in synapses of the mollusc Aplysia[10] and the fish Mathner cell[11]. Its essential rationale would appear to be conservation of transmitter molecules during intense synaptic usage.

---

[10] KANDEL, E.R. KLEIN, M., HOCHNER, B., SCHUSTER, M., SIEGELBAUM, S.A., HAWKINS, R.D., GLANZMAN, D.L., CASTELLUCCI, V.P. and ABRAMS, T.W. *Synaptic modulation and learning: New insights into synaptic transmission from the study of behaviour.* In: *Synaptic Function.* Eds. G.M. EDELMAN, W.E. GALL, W.M. COWAN. A Neurosciences Institute Publications. John Wiley, New York, 1987.; KELLY, R.B., DEUTSCH, J.W., CARLSON, S.S. and WAGNER, J.A. *Biochemistry of neurotransmitter release.* Ann. Rev. Neurosci: 2, 399-446, 1979.

[11] KORN, H. and FABER, D.S. *Regulation and Significance of Probabilistic release mechanisms at Central Synapses.* In: *New Insights into Synaptic Function.* Eds G.M. EDELMAN, W.E. GALL, W.M. COWAN. Neurosci. Res. Foundation. New York: John Wiley & Sons, 57-108, 1987.

With chemical synaptic transmission not only was there the problem of manufacturing the transmitter and transporting it to the synaptic site of action, where it was packaged in vesicles, but there was also the necessity for conservation. As stated above, there are no more than 40 to 60 vesicles assembled in the PVG ready for liberation in exocytosis. Yet the demand may be caused by presynaptic impulses invading the bouton at about 40/sec. So the necessity for conservation is evident.

5. Exocytosis

A nerve impulse propagating into a bouton causes a large influx of Ca^{2+} ions (Fig. 4A). The input of 4 Ca^{2+} ions activates via calmodulin a synaptic vesicle and may cause it momentarily to open a channel (Fig. 4C) through the contacting presynaptic membrane, as indicated by the curved arrow in Fig. 4, so that its total transmitter content is liberated into the synaptic cleft in a process called *exocytosis*.

At most, a nerve impulse evokes a single exocytosis from a PVG (Fig. 2). This limitation must involve organized complexity of the paracrystalline PVG. It has not yet been explained how exocytosis can be controlled when the nerve impulse causes an influx of Ca^{2+} ions into a bouton that is thousands of times in excess of the 4 required for the calmodulin that generates one exocytosis[12].

Exocytosis is the basic unitary activity of the cerebral cortex. Each all-or-nothing exocytosis of synaptic transmitter results in a brief excitatory postsynaptic depolarization, the EPSP. As already described, summation by electrotonic transmission of many hundreds of these mini-EPSP's is required for an EPSP large enough (10-20 mv) to generate the discharge of an impulse by a pyramidal cell. This impulse will travel along its axon to make effective excitation at its many synapses.

Exocytosis has been intensively studied in the mammalian central nervous system where it is now possible to move to a new level of complexity by utilizing a single excitatory impulse to generate EPSP's in single neurons that were being studied by intracellular recording. Immense difficulties are

---

[12] McGEER, P.L., ECCLES, J.C. and McGEER, E. *The molecular neurobiology of the mammalian brain*. 2nd Edition New York: Plenum Press, 1987. MARQUIZE-POUEY, B., WISDEN, W., MALOSIO, M.L. and BETZ, H. *Differential expression of Synaptophysin and Synaptoporin in RNA's in the post-natal rat central nervous system*. J. of Neuroscience 11, 3388-3397, 1991.

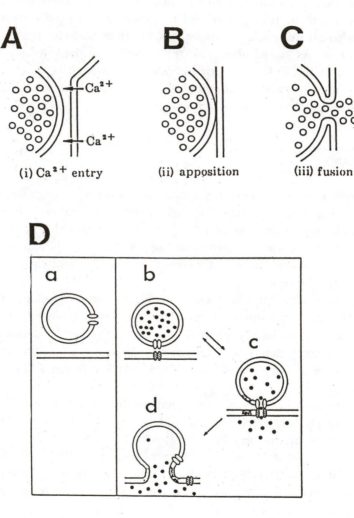

A

(i) Ca²⁺ entry

B

(ii) apposition

C

(iii) fusion

D

a

b

c

d

Fig. 4 - A,B,C Schematic drawing of a synaptic vesicle filled with transmitter molecules in stages of exocytoses (Kelly et al., 1979). D. Possible functions of synaptophysin. a. Synaptophysin may serve as a transmembrane channel connecting cytoplasm and vesicle interior. b. Synaptophysin participates in fusion pore formation during transmitter release. Docking of synaptophysin (O) to a channel protein (☉) in the presynaptic membrane (b) causes formation of a gap junction-like pore structure (c) followed by fusion of vesicle and plasma membranes (a). Membrane areas fusing incidentally are dotted. Note that membrane fusion is not required for release (Thomas et al., 1989).

presented by the background noise that was even as large as the signals being studied (Fig. 5). Fortunately the signal can be repeated many thousand times for effective averaging above the background noise, and special statistical procedures of deconvolution analysis have been devised to extract the probabilities for exocytoses[13].

The initial studies were on the spinal cord, the monosynaptic action on motoneurones by single impulses in the large Ia afferent fibre from muscle[14]. Recently[15] it was found that the signal to noise ratio: was much better for the neurons projecting up the dorso-spino-cerebellar tract (DSCT) to the cerebellum, and many quantal responses generated by exocytosis on DSCT neurons were studied. The quantal EPSP's had a mean probability of 0,76.

6. Probability of exocytosis

Fig. 5 shows diagrammatically the experimental arrangement for making the most important study on the probability of exocytosis in neurons of the hippocampus, which is a special type of cerebral cortex[16]. Advantage was taken of a unique neuronal connection. The axon of a CA3 neuron gives off a branch, a Schaffer collateral (Sch), that makes synapses on the apical dendrite of another neuron, type CA1. A microelectrode inserted into a CA3 neurone (A) can set up the discharge of an impulse that goes by the Schaffer collateral to the CA1 neuron to end there and generate an EPSP that is recorded intracellularly (C). This meticulous technique ensures that the CA1 EPSP is generated by an impulse in a single axon, but as shown on E and F the EPSP's set up by a stimulus at the arrow are superimposed on noise that is even larger than the EPSP. Nevertheless superposition of many 1000 impulses virtually eliminates the random noise so that smooth EPSP's can be recorded (F and H) and measured to be about 160 μV[17]. Moreover the statistical

[13] REDMAN, S. Quantal Analysis of synaptic potentials in Neurons of the Central nervous system. Physiolog Rev. 70: 165-198, 1990.

[14] JACK, J.J.B., REDMAN, S.J., WONG. The components of synaptic potentials evoked in cat spinal motoneurones by impulses in single group Ia afferents. J. Physiol., 321, 65-96, 1981.

[15] WALMSLEY, B., EDWARDS, F.R. and TRACEY, D.J. The probabilistic nature of synaptic transmission at a mammalian excitatory central synapse. J. Neurosci. 7: 1037-1046, 1987.

[16] SAYER, R.J., REDMAN, S.J. and ANDERSEN, P. Amplitude fluctuations in small EPSP's recorded from CA1 pyramidal cells in the guinea pig hippocampal slice. J. Neurosci: 9: 845-850, 1989.

[17] SAYER, R.J., FRIEDLANDER, M.J., REDMAN, S.J. The time-course and amplitude of EPSP's evoked at synapses between pairs of CA3/CA1 neurons in the hippocampal slice. J. Neurosci: 10 (3) 626-636, 1990.

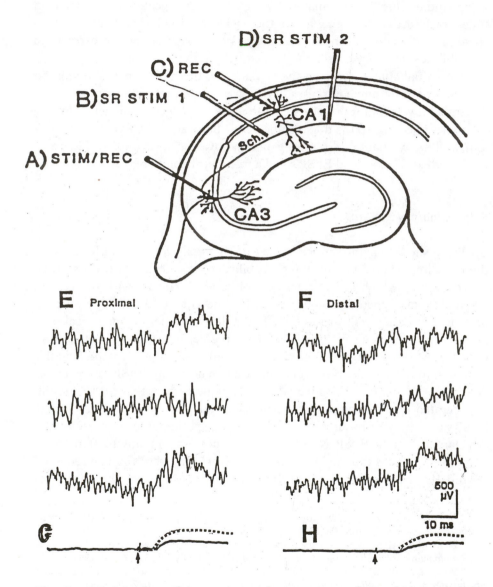

Fig. 5 - A,B,C,D show the experimental arrangement for making probability study on exocytosis of neurons in the hippocampus as described in the text. E are three records recorded as in C in response to a stimulus of a single Sch fibre as in D, F being the average of many thousands responses. G,H as in E,F but for a Sch fibre distributed more distally to the CA1.

technique of deconvolution analysis enables the determination of the probability of release by a nerve impulse of a single synaptic vesicle, an exocytosis. As in simpler situations this probability of release is always less than one, in fact very low for the hippocampus with average mean values of 0,27, 0,24 and 0,16 for the three completely reliable experiments on the hippocampus (Redman, personal communication 1992). So an impulse invading a bouton induces an exocytosis with a probability as low as 1 in 4 to 1 in 6. This is a fundamentally important finding, introducing a new level of complexity.

The control of exocytosis has been investigated by Professor Heinrich Betz and his associates, who have for several years being making an intensive study of the proteins of synaptic vesicles in the hope of understanding the quantal release mechanism. The two highly significant proteins are rather similar, synaptophysin and synaptoporin. However, in their last paper they state that "the function of these two homologous proteins in the vesicle membrane is at present unknown". Fig. 4D with legend gives their suggested explanation of exocytosis[18].

In the dynamic structure of the PVG the dense projections (DP in Fig. 2) are at least as important as the synaptic vesicles but there seems to be no study of them complementary to that of synaptic vesicles by Betz and associates. Elucidation of the paracrystalline structure of the PVG requires detailed knowledge of both components. The ultimate goal is to account by quantum physics for the low probability of quantal emissions (exocytoses) in response to nerve impulses invading the bouton.

7. Psychons

The hypothesis that the dendron is the neural unit of the neocortex leads to the attempt to discover the complementary mental units which interact with the dendron, for example in intention and attention.

Mental experiences, such as feelings, may not be vague nebulous happenings, but may be microgranular and precisely organized in their immense variety so as to bring about accurate description of the type of feeling. For example, the experience could be a special sensation from a spot on the big right toe.

---

[18] THOMAS, L., KNAUS, P. and BETZ, H. *Comparison of the presynaptic vesicle component synaptophysin and gap junction proteins: A clue for neurotransmitter release?* In: *Molecular biology of neuroreceptors and ion channels.* Ed. A. MAELICHE. Berlin, Heidelberg: Springer Verlag, 1989.

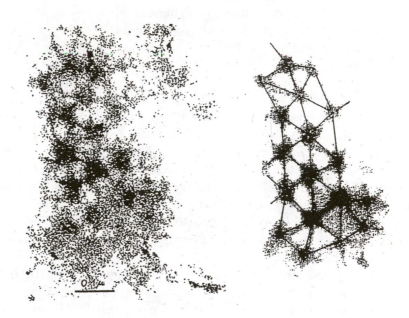

Fig. 6a - Drawing of three dendrons showing manner in which the apical dendrites of large and medium pyramidal cells bunch together in lamina IV and more superficially, so forming a neural unit. A small portion of apical dendrites do not join the bunches. The apical dendrites are shown terminating in lamina I. This termination is in tufts (cf. Fig. 1a). The other feature of the diagram (fig. 6b) is the superposition on each neural unit or dendron, of a mental unit of psychon, that has a characteristic marking (solid squares, open squares, solid circles). Each dendron is linked with a psychon giving its own characteristic unitary experience.

The hypothesis has been made[19] that all mental events and experiences, in fact that the whole of the outer and inner sensory experiences are a composite of elemental or unitary mental events at all levels of intensity. Each of these mental units is reciprocally linked in some unitary manner to a dendron, as is illustrated ideally in Fig. 6 for three dendrons. The three associated mental units are represented as an ensheathing of the three dendrons by designs of solid squares, open squares and dots. Appropriately we can name

---

[19] ECCLES, J.C. *A unitary hypothesis of mind-brain interaction in the cerebral cortex.* Proc. Roy. Soc. Lond. B 240: 433-451, 1990.

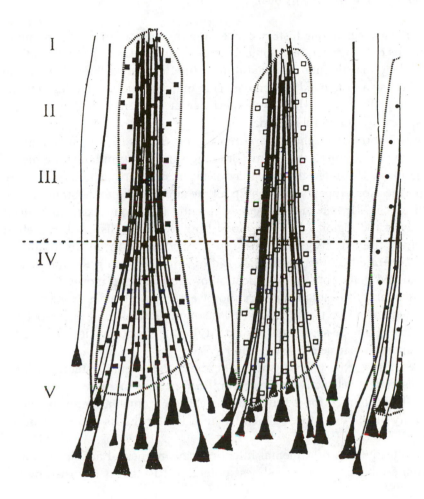

Figura 6b - See figure 6a.

these proposed mental units as "psychons". According to the unitary hypothesis there is a unique linkage of each psychon with its dendron in brain-mind interaction, for example the special feeling in the big right toe.

Psychons are not perceptual paths to experiences. They are the experiences in all their diversity and uniqueness. There could be millions of psychons each linked uniquely to the millions of dendrons. It is the very nature of psychons to link-together in providing a unified experience.

8. Generation of neural events by mental events

There has been a long history concerning the manner in which voluntary movements can be generated. Some neuroscientists, taking up a materialistic dogma, deny that the non-material mind can effectively influence the brain so as to cause an intended movement. This materialistic dogma neglects the conscious performance which we experience at every moment, even in the linguistic expression of this dogma!

Ingvar[20] introduced the term *pure ideation* that is defined as cognitive events which are unrelated to any ongoing sensory stimulation or motor performances. He stated that "a study of brain structures activated by pure ideation therefore appears to open up a new approach to understanding the human psyche". Ingvar and associates at Lund (1965 onwards) introduced the study of the regional cerebral blood flow (rCBF) to display by cerebral ideography the activity of the brain in pure ideation in all the immense variety generated by the psyche. By radio Xenon mapping Roland et al.[21] demonstrated that in pure motor ideation of complex hand movements there was activation of the supplementary motor area on both sides (SMA, Fig.7). By the more accurate technique of PET scanning Raichle and his associates demonstrated a wide-spread patchy activity of the neocortex during specific mental operations in selective attention[22].

So extensive experimental studies establish that mental intentions (psychons) can effectively activate the cerebral cortex. This increased neural activity can be accounted for if the psychons caused momentarily an increased probability of the exocytoses generated in a bouton by its incoming nerve impulses[23].

The effectiveness of a mental intention causing neural activity has also been well established in the readiness potential (RP) that is recorded by the

---

[20] INGVAR, D. *On ideation and "ideography" In: The Principles of design and operation of the brain.* Eds. J.C. ECCLES and D.D. CREUTZFELDT. Exp. Brain Res. Series 21, 433-453, 1990. SCHMOLKE, C. and FLEISCHHAUER, K. *Morphological characteristics of neocortical laminae when studied in tangential semi-thin sections through the visual cortex in the rabbit.* Anat. Embryol. 169, 125-132, 1984.

[21] ROLAND, P.E., LARSEN, B., LASSEN, N.A. and SKINHOJ, E. *Supplemental motor area and other cortical areas in organization of voluntary movements in man.* J. Neurophysiol. 43: 118-136, 1980.

[22] POSNER, M.I., PETERSEN, S.E., FOX, P.T., RAICHLE, M.E. *Localization of Cognitive Operations.* Science 240: 1627-1631, 1988. RAMON y CAJAL, S.R. *Histologie du Systeme Nerveux.* Paris: Maloine 1911.

[23] BECK, F. and ECCLES, J.C. *Quantum Aspects of Brain Activity and the Role of Consciousness.* Proc. Nat. Acad. Sci. submitted 1992.

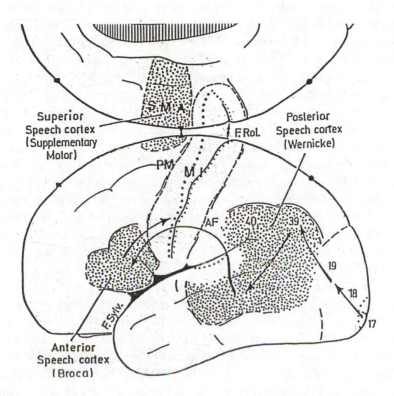

Fig. 7 - The left hemisphere from the lateral side with frontal lobe to the left. The medial side of the hemisphere is shown as if reflected upwards. F.Rol. is the fissure of Rolando or the central fissure; F. Sylv. is the fissure of Sylvius. The primary motor cortex, M1, is shown in the precentral cortex just anterior to the central sulcus and extending deeply into it. Anterior to M1 is shown the premotor cortex, PM, with the supplementary motor area, SMA, largely on the medial side of the hemisphere.

averaging technique as a slow negative potential over the scalp[24]. It is largest over the supplementary motor area (SMA, Fig. 7) anterior to the motor cor-

---

[24] DEEKE, L. and LANG, V. *Movement-related potentials and complex actions: Coordinating role of the supplementary motor area.* In: *The Principles of design and operation of the brain.* Eds. J.C. Eccles and O.D. CREUTZFELDT. Exp. Brain Res. Series 21, 303-341, 1990.

tex (MI in Fig.7). By exquisitely designed experiments Libet[25] has discovered that the readiness potential (RP) begins at least 0,5 s before the subject is conscious (W) of willing the movement, which is at the earliest only 0,2 sec before the onset of the movement. So it has been concluded that the brain is active about 0,3 sec before the movement is consciously willed (W) by the subject. However, the earlier part of the RP is probably artefactual (Libet,1990, General Discussion). So it seems that the effective *willing* (W) of the movement does not occur until about 0,2 sec before the movement. The mental event of willing (W) can be regarded as preceding the neural events in the brain, particularly in the SMA (Fig.7).

The presynaptic vesicular grid provides a unique structure in the attempt to account for the effective action of mental events on the brain by a process that does not infringe the conservation laws of physics (Beck and Eccles, 1992). A nerve impulse induces an exocytosis in a bouton and an effective mini-EPSP with a mean probability as low as 1 in 5 for the cerebral cortex (Sayer et al., 1989,1990). This probability requires an explanation by quantum physics. If a mental intention momentarily increased that probability to an average value of 1 in 3, it would have almost doubled the EPSP's for the whole dendron. Thus there would be an *effective mind-brain action* without infringing the conservation laws (Beck and Eccles,1992, op. cit.). The very low probabilities of quantal release in the cortex give excellent opportunities for mind-brain interaction in the cerebral cortex. It could be very effective if the psychon's influence was distributed widely to the hundreds of thousands of synapses on a dendron (Fig. 6). So the low probability of quantal emission in the cerebral cortex was of fundamental significance in the origin of consciousness. The complexity of the ultra-design of cortical synapses presents the ultimate scientific challenge.

As has been described (Eccles 1990, 1992 op. cit.) consciousness gives global experiences from moment to moment of the diverse complexities of cerebral performances, e.g. it would give a mammal global experience of a visual world or a tactual world for guiding its behaviour, far beyond what is given by the robotic operations of the visual or tactile cortical areas *per se*. Thus conscious experiences such as feelings would give evolutionary advantage. This opens up a field of behavioural psychology in which the consciou-

---

[25] LIBET. B. *Cerebral processes that distinguish conscious experience from Unconscious mental functions.* In: *The Principles of design and operation of the brain.* Eds J.C. ECCLES and O.D. CREUTZFELDT. Exp. Brain Res., Series 21 (1990) Berlin, Heidelberg, New York: Springer Verlag, pp. 185-205 plus general discussion pp. 207-211.

sness of animals could be tested. For example a reptile such as a tortoise and an insectivore (mammal) such as a hedgehog with its neocortex can be tested to see if the hedgehog displays intelligence when compared to the tortoise,which has no neocortex that could give it consciousness (Eccles, 1992, op. cit.).

It may seem to be a poor evolutionary design for the fundamental performance of brain-mind interaction because of its dependence on a rather indirect action via quantal probability. It would have been expected that mental intentions would directly excite the neurons of the SMA that are precisely concerned in the intended movement. The indirect action by increase of quantal probability would seem to lack precision and speed, which are of paramount importance in motor control. However, once the movement is initiated, it is subject to all the subtle controls of the complex neural machinery of the brain in the conventional neuroscience of motor control.

The hypothesis is that, because the vertebrate brain evolved in the matter energy world of classical physics, it was mindless, deterministic and subject to the conservation laws. Then, because of the complex design of the mammalian neocortex with its operation of quantal probability, there came to exist experiences of another world, that of the conscious mind, presumably most primitive and fleeting (Eccles, 1992). However, with hominid evolution there eventually came higher levels of conscious experiences as expressed eventually in human culture (Eccles, 1989), and ultimately, in Homo Sapiens Sapiens, self-consciousness - which is the unique life-long experience of each human self, and which we must regard as a miracle beyond Darwinian evolution.

DISCUSSION
(J. SZENTAGOTHAI, chairman)

LAMBO: First of all I would like to show my gratitude to Sir John Eccles whose work is well known to most us. My major interest has always been in the area of mind and brain interaction and the controversy over the effect of mind on the brain. I was extremely pleased indeed this afternoon to see his acceptance of a long-standing controversy. About fifteen years ago, I was privileged to read a letter from Freud to Einstein in which this particular subject was discussed. It is historical and memorable to see that they both, after so many arguments, arrived at the same conclusion: that mind also has an effect on the brain.

NICOLESCU: What I am saying is not meant to be an objection but a genuine question. If I understand it correctly, the way in which you introduce the psychon is completely parallel with what we call in quantum physics, the quanton, which is at the same time wave and particle. Now, there is a problem. It is well known that if we perform experiments at the macrophysical level, the quantum conditions are abolished. In other words, there is a kind of tautology in the sense that we introduce classical objects and we recover classical objects. So, my question is: if you associate the psychon with a macrophysical structure I think that quantum conditions have to be abolished.

ECCLES: We are using quantum physics in a strict way, it is done by the quantum physicist Federick Beck, the Professor of Quantum Physics at Darmstadt.
An impulse invading a synaptic bouton causes the liberation of a single synaptic vesicle with a probability of much less than 1.0, as low as 0.3 to 0.2. Beck has made a very strict probability study of exocytosis, and it shows that mental intentions could increase the frequency of exocytosis without infringing conservation laws. This is shown in the mathematical treatment in my paper. The reference is Beck and Eccles, Proceedings National Academy of Sciences, U.S.A., vol. 89.

ARECCHI: Can you draw a comparison between your psychons and the correlated space patterns of electrical activity that Freeman is associating with different sensations of smell in the rabbit? This appeared in Brain and Behavioural Science in 1987, Physica D in 1989 and Scientific American in February 1992.

ECCLES: The psychons can be related to any activity of the neocortex including the electrical activity produced by detection of odours. Let's get back to the question of what a psychon is. It is nothing mysterious, it is just the unit of experiences of every experience. We go through life with patterns of experiences that are all made up of psychons, all the visual world, all the hearing world, all the olfaction world, all the thinking world, all the performing and behaving world, all that is happening to us from moment to moment is the immense lot of experiences or psychons that we are experiencing as a unit. That is the strange thing, the mystery is the unity of the mind stressed by William James and it is accepted by philosophers even today, but this unity contains, immense numbers of elementary components, psychons, made up of every kind and variety of perception you can think of. Analogical unitary concepts are photons and light, electrons and electricity. Physichons are the units that interact with the neural units, dendrons in the mind-brain problem.

ARECCHI: I haven't read the Beck paper, so I have the following question; how does this fit into conventional quantum mechanics? What is the probability amplitude? Schrödinger's wave function always develops according to the Schrödinger equation, except for one moment when the probabilities are manipulated and this is the process of measurement. There, according to the present ideas, the wave function suddenly collapses if the observer takes note of what the result of the measurement was. Now this has always been a somewhat unclear point in quantum mechanics. Who is really qualified to be the observer? What one could show is that it really doesn't matter so much where you make the cut between the observer and the object, if you shift a little bit you get the same result. So people were satisfied with this, although it still remained a somewhat unclear point. I was wondering if this is the point where your interpretation sets in, that the mind somehow manipulates the quantum mechanical probabilities?

ECCLES: There are many ways of doing quantum mechanics, Beck and I are doing it in a very special way, which Beck has defined quite clearly in the paper in relation to the exocytoses. It involves a tunnelling process. The way in which a synaptic vesicle goes through and is liberated involves tunnelling mathematics. This is all in the paper and the probability comes out from that. There is a drawing that Beck made of the tunnelling problem in fig. 4 of our joint paper quoted above.

THIRRING: I just wanted to say that these different ways of doing quantum mechanics are mathematically equivalent. I know Beck and his work well, so I am quite satisfied.

WHITE: Concerning the displays you presented of Professor Ingvar's work, particularly the areas in the frontal lobe in reference to cerebral blood flow, performing various tasks, mathematics, imageing and so forth, should we read anything into the fact that those are multiple sites and not single sites in terms of the areas of increase in cerebral blood flow? The other thing is, as you know, there is some difficulty in providing exactly the same template using PET scanning (Positron Emission Tomography) the centers are not superimposable in their entirety, now we know that these are two different modes of measuring, one is blood flow and the other is metabolism, but the multiplicity of sites, that's my question, I mean when we talk about mathematics or a number of areas, what does that mean to us in terms of consciousness, in terms of mind function, the multiplicity of sites?

ECCLES: The ones I showed you were by Roland. There is a paper by Raichle on PET scan work. It shows again multiple foci. That's because you see the psychon world picks it up from all of these patterns of operation relating to the world we experience. It makes it into a unity, and that's what we are doing all the time. I am here seeing you all as an immense number of inputs, of very various kinds, into my visual cortex,but I make a global picture of it seeing it as unity, and I can add in the sound as well. We are not just psychological experiments. We are living in a world of incredible complexity, and yet we manage to deal with it. Take this conference, all the different ideas and people you've met, what it all meant to you. At the same time we are enormously adding, amplifying and simplifying our experience too. It's our life and what we see and hear now is dependent on our experiences of our whole lifetime, our background, we're not just starting from zero, we are all immensely experienced people and that's how we can handle it at all. If you took a man off the street and put him in this conference chamber, he wouldn't know what was going on.

WHITE: I notice in your two worlds that you have now located the human soul in the brain, is that correct?

ECCLES: Not located in the brain, that's not true. What I've shown is that what goes on in the brain is giving us mental experiences, the whole world of the self. I use the word self because each of us is a self, a unitary experiencing being that is of course intimately related to its brain, but it isn't the brain.
The brain is a material structure with laws of physics etc., in Popper's World and the self is what you know. We know it more directly than the brain. Probably none of you has seen its brain, but your self is with you all the time, from the time you wake up in the morning or when you are asleep even with dreams. It's all your self. You can call it your Soul, if you are religiously minded, but I like to use the philosophical word, Self. I have a book in press: *How the Self controls its brain*.

LAMBO: I just want to mention that of course Sir John had no time this afternoon to elaborate on the whole areas of memory, of consciousness and self, which are also vital. You can see a good deal of this not only in hypnotism but in trances and primitive cultures where the self image is not in physical terms, but in terms of the inner self and I feel that this again in something which should be developed in a very elaborate form.

LEVELS OF COMPLEXITY AND LEVELS OF REALITY: NATURE AS TRANS-NATURE

BASARAB NICOLESCU

Division de Physique Théorique, Institut de Physique Nucléaire, 91406 Orsay, France
and LPTPE, Université Pierre et Marie Curie, Paris, France

How can one study the emergence of complexity in hard sciences without a rigorous understanding of what we mean by "Nature"? In fact, complexity and Nature are intimately interconnected. However, if complexity is widely recognized as a valuable notion, Nature finds itself in the paradoxical situation of being expelled from the realm of hard sciences.

Our century has invented all kinds of "deaths" and "ends": the death of God, the death of man, the end of ideologies, the end of history. However, there is one death about which one speaks much less, probably from ignorance: the death of Nature. In my opinion, this death of Nature is in fact the origin of all the other deadly concepts we have just invoked. In any case, the word Nature itself has disappeared from the scientific vocabulary. Of course, the layman and even the scientist (in his popular books or in his non-specialized talks) still uses this word, but with a confused and sentimental meaning, as a reminiscence of magic. In our times, it is sufficient to pronounce the word "Nature" in order to be immediately qualified as "ecologist", which is, to say the least a huge over simplification. How have we arrived at such a situation?

Man constantly modifies his vision of Nature. The historians of science agree that in spite of what one might superficially believe, there is not just one Nature intersecting all historical periods[1]. What can there be in common between the Nature of so-called "primitive" man, the Nature of the Greeks,

---

[1] ROBERT LENOBLE, *Histoire de l'idée de Nature*, Albin Michel, Paris, coll. "L'évolution de l'humanité", 1990.

the Nature of the age of Galileo, of the Marquis de Sade, of Laplace or of Novalis? Nothing besides man himself. The vision of Nature in a given age depends on the leading imaginal at this age[2] which, in its turn, depends on many parameters: the degree of growth of sciences and techniques, social organization, art, religion etc.. The image of Nature once formed acts on all fields of knowledge. The passage from one vision to the other is not continuous: it occurs through sudden, radical, discontinuous breakdowns. Several contradictory visions can even coexist. The extraordinary diversity of visions of Nature clearly demonstrates why one cannot speak about *Nature*, but just about a certain nature in agreement with the imaginal of the given age.

It is important to underline that the privileged and even exclusive relation between Nature and science is just a recent prejudice, based on the scientistic ideology of the XIXth century. The historic reality is much more complex. The image of Nature always had a multiple action: it influenced not only science but also art, religion, social life. This fact explains a lot of strange synchronisms. I'll just give two examples: the simultaneous occurrence, at the beginning of this century, of abstract art and of quantum mechanics and, at the end of this century, of the theory of the end of history and of unified theories in particle physics. The first example is relatively well-known while the second has not been mentioned till now. The unified theories in physics have as their aim to formulate a complete approach, founded on a single interaction and which will predict everything we would like to know (hence the name "Theory of Everything" - TOE). It is obvious that if in future such a theory is formulated, this will mean the end of physics, because there will be nothing else to look for. It is interesting to note that the ideas of the end of history and the end of physics arose simultaneously from our "end of century" imagination. Is that a simple and genuine coincidence?

In spite of the abundant and fascinating diversity of images of Nature one can nevertheless distinguish three main stages: magic Nature, Nature as a machine and the death of Nature. Obviously I cannot enter into details of the description of these three stages. I can at most deal briefly with the ques-

---

[2] The French term "imaginaire" is usually translated "the imaginary", which in English has the same wrong connotations of dreaminess and non-substantiality as "imagination". I therfore adopt the term popularized by Henry Corbin, "the imaginal" which Gilbert Durand describes in *The Encyclopedia of Religion* as a way of presenting images of the higher, the ultimate, the divine without slipping into the trap of idolatry: it is clearly a creative imagination or inspiration of the highest order. See e.g. HENRY CORBIN, *Mundus imaginalis ou l'imaginaire et l'imaginal*, Talk at the Colloque de Symbolisme, Paris, 1964; this text is published in HENRY CORBIN, *Face de Dieu, face de l'homme, "herméneutique et soufisme"*, Flammarion, Paris, 1983.

tion of the death of Nature, which is the starting point of my own research.

Magic thinking views Nature as a living organism, endowed with intelligence and consciousness. The fundamental postulate of magic thinking is that of universal relationship: Nature cannot be conceived without its relations with man. Everything is sign, trace, signature, symbol. Science, in the modern understanding of this word, is unnecessary.

At the other extremity, the mechanist thinking of the XVIIIth and specially of the XIXth century (which still dominates today) views Nature not as an organism but as a machine. In order to possess it completely it is sufficient to dismantle this machine bit by bit. The fundamental postulate of mechanist thinking is that Nature can be known and conquered by scientific methodology, defined in a completely independent way from what man really is. This triumphal vision of the "conquest of Nature" has its root in the redoubtable technical and technological efficiency of this postulate. In spite of a persistent presumption, one can demonstrate that the origin of mechanist thinking is not the invention of the methodology of modern science by Galileo. The mechanist vision has in fact its origin in the thinking of Aristotle and Democritos. One can show that the new vision of Nature introduced by Aristotle - Nature constituted from elements external to man - is intimately related to his binary logic (of identity, of non-contradiction and of the excluded middle). As to the atomist doctrine, it is well-known that, at that period of time, it had no scientific foundation. This doctrine arose from the visceral fear of a vacuum. The process of decomposing the substance had to stop somewhere in order that the universe and life should not vanish for ever into the vacuum. The "atoms", in the different meanings of this word, later became the fundamental building blocks of matter (quarks, leptons, etc.). One hoped that, starting from these building blocks, one could completely disentangle the code of the universe - machine of the contemporary neomechanist ideology. Hence, due to Aristotle, Democritos and their disciples everything was in place to engender the Nature-machine. However the vision of the founding fathers of modern science is, paradoxically, infinitely more complex, by a subtle mixture of magic, religious and scientific features[3]. Nevertheless it is the mechanist vision which was predominant, at least till the beginning of our century, for reasons which are too complicated to be made explicit here.

---

[3] GEORGES MINOIS, *L'Eglise et la science - Histoire d'un malentendu*, vol. I - *De Saint Augustin à Galilée*, vol. II - *De Galilée à Jean-Paul II*, Fayard, Paris, 1990-1991.

Some scientists, artists or philosophers felt the deadly danger of mechanist thinking fully. Thus the antagonist current of the German *Naturphilosophie* appeared[4]. One can quote important names like Schelling, Schlegel, Novalis, Ritter, without forgetting Goethe. The *Naturphilosophie* has its roots in the visionary work of Jakob Boehme. From the point of view of our times the *Naturphilosophie* can be considered as a grotesque deformation and a crude manipulation of science, as a dead-end in a ridiculous attempt at a return to magic thinking and to living Nature. However, how can one hide the fact that this Philosophy of Nature generated at least two major scientific discoveries: the cellular theory and electromagnetism (Oersted, 1820)? Perhaps the true fault of *Naturphilosophie* was to be born two centuries ahead of time, before the quantum, technological and informational revolution.

The logical outcome of the mechanist vision, which completely eliminated the *Naturphilosophie*, is the death of Nature, the vanishing of the concept of Nature from the scientific framework. The Nature - machine, with or without a watch-maker God, falls into a collection of attendant parts. From that moment, there is no more need of a coherent whole, of a living organism or even of a machine, which retained, after all, the musty smell of finalism. Nature is dead. The complexity remains. An astonishing complexity which penetrates all the domains of knowledge, from the infinitely small to the infinitely large. This complexity is perceived as being accidental, man himself being considered as an accident of complexity, a joyful vision, which brings us to our own world today.

We arrive therefore at an etymological paradox. In fact "living Nature" is a pleonasm, because the word "Nature" is intimately related to that of "birth". I quote Robert Lenoble: "... the latin word *natura* is attached to the root *nasci* (to be born, to come into the world) and signifies first: the action of giving birth, growing ...; ... *natura* signifies also the organs of engendering, and first of all the feminine organs. We can note that the form *natio-onis* has also, as primordial meaning, birth; as a consequent meaning, by personification and deification, it signifies the nation or one's native land, one's fatherland" (op. cit., [1], pp. 229-230). In the light of these etymological considera-

---

    [4] GEORGES GUSDORF, *Le savoir romantique de la Nature*, Payot, Paris, 1985; PH. LACOUE - LABARTHE et J.-L. NANCY, *L'absolu littéraire - Théorie de la littérature du romantisme allemand*, Seuil, Paris, 1978; PIERRE THUILLIER, *De la philosophie à l'électromagnétisme: Ie cas Oersted*, La Recherche, Paris, n° 215, pp. 344-351 (mars 1990); ANTOINE FAIVRE, *Franz von Baader et les Philosophes de la Nature*, in *Epochen der Naturmystik*, Erich Schmidt Verlag, Stuttgart, 1979, pp. 381-425.

tions, what could be the fate of a civilization that accepts the death of Nature?

This long introduction was necessary in order to put my approach into perspective.

First I have to make a methodological remark. I do not claim to propose a unique and "scientific" model of Nature even if I am continuously guided by quantum physics. In fact all visions of Nature in the past were many-sided or, in order to use a contemporary word, transdisciplinary[5]. The equation hard science = Nature was just a phantasm - the scientistic phantasm. My approach will therefore be resolutely transdisciplinary. Hence, as a specialist in a well-defined discipline (quantum physics), I take a non-negligible risk: to go out of my disciple by presenting you with a rough sketch of a long-term trans-disciplinary research project which of course in the future will involve a large number of eminent specialists. However, I know that I can take this risk in this high place. The true risk is elsewhere, in the vertiginous jump towards the unknown, represented by the metamorphosis of the death of Nature into the rebirth of Nature.

My starting point is the experience I have lived as a practician of quantum physics, namely noticing in my everyday life as a physicist the incompatibility between the results of quantum physics and cosmology and the neo-reductionistic attitude which concentrates exclusively upon the fundamental building blocks of matter and upon the four known physical interactions. According to this neo-reductionistic attitude every appeal to Nature is unnecessary and is even meaningless.

My aim is to formulate a definition of Nature in agreement with modern scientific knowledge. The two pillars of my definition of Nature are: the notion of *levels of Reality* and *the logic of the included middle*.

By the word *reality* I mean everything which *resists* experiments, our representations, descriptions, images. I mean by the word *level* a group of systems which is invariant under the action of certain laws. Finally I say that two *levels of Reality* are different if the passage from one to the other involves a breakdown of laws and a breakdown of fundamental concepts (such as

---

[5] The word "transdisciplinarity", as carrying a different meaning from "interdisciplinarity", was probably first introduced in 1970 by Jean Piaget, André Lichnerowicz and Eric Jantsch; see their contributions in *L'interdisciplinarité - Problèmes d'enseignement et de recherche dans les universités*, OCDE, Paris, 1972. From an etymological point of view "trans" means "across, beyond". We mean by "transdisciplinarity" that which crosses all disciplines and finds itself between (beyond) all disciplines. Therefore transdisciplinarity is clearly not a new discipline.

causality, for example). The obvious example is that of the couple microphysical level macrophysical level[6]. The breakdown between classical physics
and quantum physics is really radical. This is why the interpretation of quantum phenomena in macrophysical language generates an endless series of
paradoxes. Up to now nobody has succeeded in finding a mathematical formalism allowing the rigorous passage from the quantum world to our macroscopic world. Nevertheless these two worlds coexist. The proof: our own existence.

The second aspect concerns logic. Modern science was born from a
methodological breakdown, but not necessarily by a breakdown of logic, as
compared with the previous science. This explains why it generated both
classical and quantum physics. However, the quantum revolution concerns
first of all the problem of logic to be used. More precisely, the quantum revolution consists in the possible emergence of the included middle thanks to
scientific methodology.

Quantum mechanics and, later, quantum physics caused couples of
mutually exclusive contradictories (A and Ā, where Ā denotes not-A) to suddenly appear. I'll give just a few examples: wave *and* particle, continuity *and*
discontinuity, separability *and* non-separability, local causality *and* global
causality, autonomy *and* constraint, visible *and* invisible, manifested *and*
nonmanifested, symmetry *and* symmetry breaking, reversibility *and* irreversibility of time[6].

There are two possible solutions to this astonishing situation: either one
refuses a status of reality to the quantum scale (which is equivalent to asserting, once again, the death of Nature) or one has to change the kind of logic
which is used (as was proposed towards 1930-1940 by Alfred Korzybski and
Stéphane Lupasco). It is this second solution that I adopt here.

The fundamental logical change concerns the third axiom of the binary
Aristotelian logic the axiom of the excluded middle: there is not a third term
T (T from "Third" included) which is at the same time A and Ā. One has to
replace this axiom by the axiom of the included middle "there is a third term
T which is at the same time A and Ā - a formulation which might seem paradoxical but which becomes completely transparent if one takes into account
the first notion I discussed - levels of Reality. The three terms of the new
logic - A, Ā, T - and their associated dynamics could be represented by a

---

[6] BASARAB NICOLESCU, *Nous, la particule et le monde*, Le Mail, Paris, 1985; BASARAB
NICOLESCU, *Science, meaning and evolution - The cosmology of Jakob Boehme*, Parabola Books,
New York, 1991.

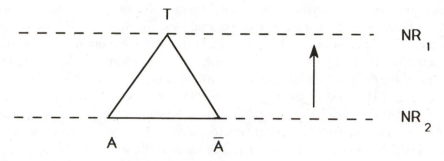

Fig. 1 - Schematic representation of the action of the logic of the included middle. NR_1 and NR_2 denote two different levels of Reality.

triangle, one of the vertices being located on one given level of Reality and the other two vertices - at another level of Reality (see Fig. 1). If one considers just one level of Reality everything appears as a fight between two contradictory elements (say, wave and particle). The third dynamism, that of the T-state, occurs on a different level of Reality, from where that which seems separated (wave and particle) is in fact unified (quanton) and that which seems to be contradictory is perceived as non-contradictory.

In spite of the distrust motivated by the fear of a return to magic thinking, the logic of the included middle was seriously studied, at least in France, by thinkers like Stéphane Lupasco, who proposed a possible formalization of this logic[7], Edgar Morin[8], Jacques Wunenburger[9], Gilbert Durand[10] and Antoine Faivre[11]. I, myself, in collaboration with Thierry Magnin, tried recently to show that the method of analysis of complexity used by modern scientists, in relation to the logic of the included middle, turns out to be fruitful also for theologians, without any confusion or concordism[12]. Interesting research into this approach has been carried out by Xavier Sallantin[13].

---

[7] STÉPHANE LUPASCO, *Le principe d'antagonisme et la logique de l'énergie*, Rocher, Paris, 1987.

[8] EDGAR MORIN, *La Méthode-4. Les idées*, Seuil, Paris, 1991.

[9] JEAN-JACQUES WUNENBURGER, *La raison contradictoire*, Albin Michel, Paris, 1989.

[10] GILBERT DURAND, *L'imagination symbolique*, Quadrige/Presses Universitaires de France, Paris, 1964.

[11] ANTOINE FAIVRE, *Accès de l'ésotérisme occidental*, Gallimard, Paris, 1986.

[12] THIERRY MAGNIN and BASARAB NICOLESCU, *The analysis of complexity in science and in theology: towards a common method?*, Proceedings of the 4th European Conference on Science and Theology, Villa Mondo Migliore on Lake Albano, March 23-29, 1992 (to be published).

[13] XAVIER SALLANTIN, *Le monde n'est pas malade, il enfante - Vers l'unité de la foi et de la connaissance*, O.E.I.L., Paris, 1989.

In any case, the logic of the included middle has revealed itself as the ideal tool for the analysis of complexity. I would like to quote in this context the epistemology of complexity elaborated over the years by Edgar Morin, which gives us a framework for the study of a large variety of phenomena. Edgar Morin rightly underlines that there is no question of abolishing the logic of the excluded middle but just about recognizing it as valid only in simple cases: "The field of the excluded middle is valid for simple cases... One cannot abolish the excluded middle; one has to inflect it in terms of complexity... The dialogics is precisely the included middle, two contrary propositions which are necessarily linked and opposed at the same time" (op. cit. [8], pp. 200-201). We can also note that the logic of the included middle is implicit in systems theory thinking[14] [15]. In my opinion, the notion of "level of Reality" formulated by us several years ago (op. cit. [6]) will in the future strengthen the empirical scientific foundation of the logic of the included middle.

There are many other landmarks for a new Philosophy of Nature (which is different from, but complementary to natural philosophy) quantum discontinuity, indeterminism, constructive random processes, chaos creating order, quantum non-separability, bootstrap dynamics, the unification of all physical interactions, supplementary space dimensions, big-bang dynamics, the anthropic principle, the quantum vacuum. To study the implications of all these ideas and phenomena I would need not a conference but a book. I can at most shortly formulate several questions. From where does the non-separability come when everything in our macrophysical world seems to be separated? From where the bewildering coherence between the infinitely small and the infinitely large? Why does the evolution of this universe seem to require, for its own existence and for the engendering of life, extremely narrow windows for the fundamental physical parameters? What is the "reality" of the supplementary space dimensions? What is the meaning of a single and unique physical interaction, unified at an energy which will never be reached in our accelerators? Such questions, solidly rooted in scientific ground, can guide our steps in our search for a new definition of Nature.

As a starting point I am taking *two axioms: the existence of several levels of Reality (their number being finite or infinite) and the universal action of the logic of the included middle.* These two axioms have, as we have seen, a cer-

---

[14] ERVIN LASZLO, *The systems view of the world*, George Braziller, Inc., New York, 1972.
[15] ERICH JANTSCH, *The self-organizing universe*, Pergamon, Elmsford, NY, 1980.

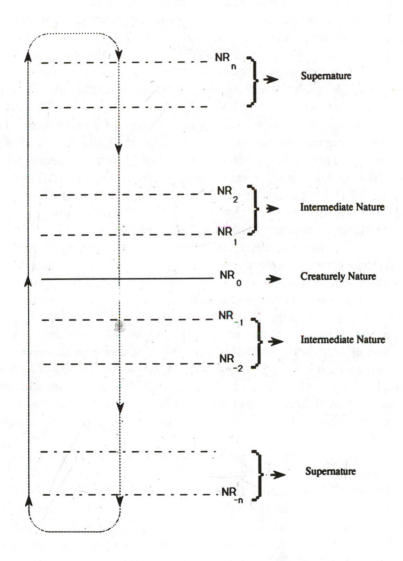

Fig. 2 - Schematic representation of the ternary (creaturely Nature, intermediate Nature, Supernature).

tain experimental scientific basis.

These axioms lead me first to introduce a ternary which depends on a certain level of Reality (NR), considered as a reference frame:

(creaturely Nature, intermediate Nature, Supernature).

The obvious reference frame to be chosen is of course the level of Reality to which we belong - the macrophysical level (NR_0).

The *creaturely Nature* is defined by the cosmic processes taken in their entirety, as they appear to us without the intervention of human activity: man, animals, plants, earth, planets, galaxies etc. This aspect of Nature reveals itself at the macrophysical level NR_0.

The *intermediate Nature* can be also explored by experiments, but the laws governing the corresponding levels of Reality (NR_1, NR_2, ...; NR_1, NR_2, ... - see Fig. 2) are radically different from those governing the macrophysical level NR_0. One can give as examples the quantum level (NR_1) and techno-Nature (NR_1). I have already described elsewhere the quantum level (op. cit. [6]). Here I would like to explain why I consider techno-Nature as a different level of Reality, belonging to the intermediate Nature.

By *techno-Nature* I mean the material external projection of the mind, generating results which are not produced by the so-called "natural" cosmic processes. These results have nevertheless their own self-movement, beginning with the system of highways in the United States and finishing with the troubling virtual Reality. Techno-Nature seems to obey a maximality principle: everything which can be made will be made. This principle is a particular case of a general maximality principle which seems to govern Nature: *everything which can happen, according to the existing laws, will happen.* The virtual Reality (which, for philosophical terminological reasons I would like to call "potential Reality") has a central place in techno-Nature, because it shows that abstraction is a component of Reality, a phenomenon already present in the quantum world.

Finally, *Supernature* concerns all levels of Reality inaccessible to scientific experiments.

Therefore, Supernature is not really "supernatural", because it simply translates a double limitation: the limitation of our sense organs and of their extensions and the limitation of the Galilean scientific methodology.

In Fig. 2, I represented a closed and oriented loop intersecting all levels of Reality. It signifies the existence of an informational flux crossing these levels. The closing of the loop is the symbol of a self-consistent bootstrap-type dynamics: *each level of Reality is what it is because all other levels of Reality exist at the same time.* The orientation of the loop signifies that the passage of the informational flux is operating in a coherent manner, either by loss or by gain of information. This orientation could be an interesting way of understanding and even of leading to a mathematical formulation of an outstanding scientific contemporary puzzle: that of time irreversibility. In our representation the irreversibility is the result of a loss of information, the passage from one level of Reality to another being necessarily associated with a modification of information. For example, the passage from the microphysical level NR_{-1} to the macrophysical level NR_0 occurs by a loss of information. The reversibility is present at the level NR_{-1} (it is governed by the theorem of CPT invariance), while the irreversibility occurs at the level NR_0.

In our representation, the levels of Reality are considered as energy levels, the passage from one level to another being, by definition, discontinuous. The energy is a unification concept: it appears in a coded form as information or in a concrete form as substance.

One can say that our representation of Nature corresponds to a *fibering*[16], leading to a *Gödel-like* structure of our knowledge.

We can define a local causality, concerning a well-defined NR and a global causality, concerning two, several or all levels of Reality. The global causality must not be confused with ordinary finality. The movement springs from the interaction between the local causality and the global causality. The "objectivity" depends on the considered NR. In the presence of several NR the binary partition (subject, object) has to be replaced by the ternary (subject, object, included middle). We are therefore able to describe the notion of *contradiction:* it signifies the ceaseless change from one quality of energy to another. The definition of *meaning,* as proposed by the philosopher Raymond Ledrut - the contradictory relationship between a presence and an absence - is integrated into our approach[17].

These few considerations allow us to introduce another ternary of *the living Nature:*

(Nature, anti-Nature, trans-Nature).

This ternary does not depend on a particular level of Reality, i.e. is *inva-*

---

[16] The name of "fibering" was suggested to me by Prof. René Thom. I find that this mathematical notion perfectly describes the self-consistent dynamics of Nature associated with the entirety of its structure in terms of levels of Reality.

[17] RAYMOND LEDRUT, *Situation de l'imaginaire dans la dialectique du rationnel et de l'irrationnel,* in *Cahiers de l'imaginaire,* n° 1 - *L'imaginaire dans les sciences et les ans,* Privat, Toulouse, 1988, pp. 43-50.

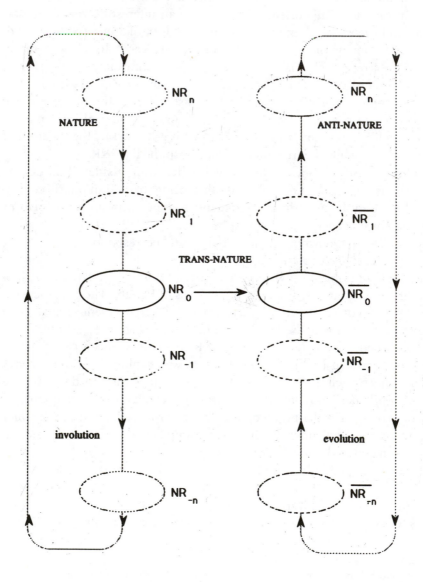

Fig. 3 - The ternary of the living Nature: (Nature, anti-Nature, trans-Nature).

riant as regards all levels of Reality.

Nature is the entirety of phenomena which appear to us as results of cosmic processes or of the mind. One can characterize it by increasing entropy, by a tendency to fragmentation, by a depletion of energy (see Fig. 3). *For us*, this Nature is the place of involution and death. Going against this movement signifies to go "against Nature", which does not yet mean the conversion to anti-Nature. Using the terminology of Lupasco, Nature can be defined as *the actualization of all the actualizations* .

It is therefore clear what *anti-Nature* means: the decreasing of entropy, the tendency towards unity in diversity, the growing in density of energy. Anti-Nature is therefore the place of evolution and life. Using the terminology of Lupasco, one can assert that anti-Nature is *the actualization of all the potentializations* .

Going from Nature to anti-Nature involves a discontinuity. This discontinuity forces us to introduce a third term - trans-Nature.

Trans-Nature is both what crosses and what is beyond Nature and anti-Nature. Trans-Nature is *the actualization of all T-states and of affectivity.* In particular, trans-Nature engenders *transculture.* Transculture appears as the incarnation of trans-Nature, as the experience of life and of the imaginal of all nations of the world. Trans-Nature induces a true *trans-presence* at all levels of Reality.

Living Nature is therefore defined by the two ternaries discussed above. These two ternaries allow me to consider, as concrete application to our life a multitude of ternaries such as:

(energy, movement, relation)
(unification, unity, uniqueness)
(levels of Reality, levels of knowing, levels of understanding)
(meaning, anti-meaning, nonsense)
(hard sciences, soft sciences, sciences of the included middle)
(Science, Art, Religion)
(techno-Nature, technoculture, technoscience)
(humanism, technohumanism, transhumanism)
(subject, object, included middle).

The last ternary (subject, object, included middle) is the generator not of post-modernity but of what one can call *cosmodernity* . I have no time to comment on all these ternaries. I am convinced that they represent tracks full of promise for transdisciplinary research. I would like to add that the confrontation of the new definition of Nature with the vision of Nature of other thinkers of threefoldness will turn out to be extremely rich. I am thinking especially of Charles Sanders Peirce[18], of his theory of three universes and of his ternary of fundamental categories (Firstness, Secondness, Thirdness).

It is also important to note that our definition of Nature is perfectly compatible with the three theses concerning the laws of Nature as formulated by Walter Thirring[19]: "(i) The laws of any lower level in the pyramid mentioned above are not completely determined by the laws of the upper level though they do not contradict them. However, what looks like fundamental fact at some level may seem purely accidental when looked at from the upper level; (ii) The laws of a lower level depend more on the circumstances they refer to, than on the laws above. However, they may need the latter to solve some internal ambiguities; (iii) The hierarchy of laws has evolved together with the evolution of the universe. The newly created laws did not exist at the beginning as laws but only as possibilities." In my opinion, the logic of the included middle offers a rational foundation for these three theses.

Before concluding, I would like to say a few words precisely on the logic which governs the interaction of the levels of Reality.

At a well-defined level of Reality (say, NR_0) Aristotelian logic is valid. In particular the *identity principle* (A is A) is true at this level (see Fig. 4). The influence of another level of Reality, revealed by scientific theory and experiment, manifests itself, at the considered level, by contradictory, mutually exclusive phenomena. Taking into account the quantum paradoxes, Alfred Korzybski introduced in 1933[20] a *contradiction principle* (A is Ā). The abrupt formulation of this principle explains why it inspired science-fiction authors. However, as we have seen, there is a rational solution of quantum paradoxes: the adoption of the logic of the included middle and the introduction of the notion of levels of Reality. The principle of contradiction appears as the projection of threefold dynamics on a given level of Reality.

Two contiguous levels are connected by the logic of the included middle,

---

[18] CHARLES S. PEIRCE, *Selected writings (Values in a universe of chance)*, edited with an introduction and notes by Philip P. Wiener, Dover, New York, 1966; ISABEL STEARNS, *Firstness, Secondness and Thirdness*, in *Studies in the philosophy of Charles Sanders Peirce*, Harvard University Press, Harvard, 1952.

[19] WALTER THIRRING, *Do the laws of Nature evolve?*, in Proceedings of the Study Week of the Pontificia Academia Scientiarum on *Understanding Reality: the role of culture and science*, 1991 (to be published).

[20] ALFRED KORZYBSKI, *Science and Sanity*, The International Non-Aristotelian Publishing Company, Lakeville, Connecticut, 1958 (first edition: 1933).

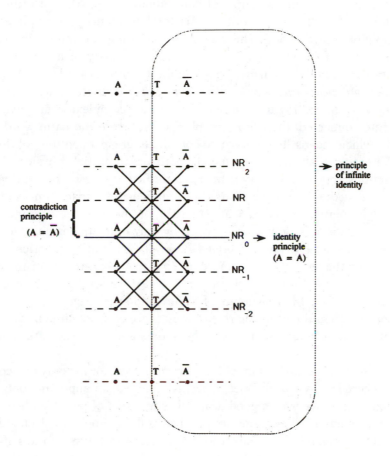

Fig. 4 - Schematic representation of the action of the principle of infinite identity.

namely the T-state present at a certain level is linked to a couple of contradictories (A, Ā) of the immediately contiguous level. The iterative action of
the logic of the included middle, represented by the triangulation shown in
Fig. 4, implies the imbrication of levels and the coherence of Nature as a
whole. A particular role is played by the topological envelope of all T-states,

represented by the closed loop in Fig. 4. This envelope represents the action of a new principle, having its root in trans-Nature. I call it *the principle of infinite identity*, because there is a perpetual, iterative and cyclical transmutation of a T-state into a couple of contradictories (A, Ā). The principle of infinite identity embodies both the Aristotelian identity principle and the principle of contradiction. This new principle brings an interesting light to the problem of complexity: the terrifying complexity of a single level of Reality can mean the harmonious simplicity of another level of Reality. *Complexity appears therefore as a measure of the distance between man and Nature.* It is interesting to note that "level" and "complexity" are etymologically interconnected: the word "complex" comes from the Latin word "complecti", which means "to embrace, to contain, to grasp an entirety of things."

For example, according to superstring theory in particle physics, physical interactions appear to be very simple and unified as a result of a few general principles, if they are depicted in a multidimensional space-time and at an ultra-high energy. Complexity arises at the moment of the passage to our world, which is inevitably characterized by only four dimensions and by the fact that considerably lower energies are available. Unified theories are very powerful on the level of general principles but they are rather poor in describing the complexity of our own level (op. cit. [19]). This seems to be one of the aspects of the Gödel-like structure of Nature and knowledge.

Let me stress the probable role of the nature of space-time in the definition of a level of Reality and thus in the understanding of the nature of complexity.

Our space-time continuum òf four dimensions is not the only one conceivable. In certain physical theories, it seems more like an approximation, like a "section" of a space-time a good deal richer in terms of possible phenomena. The supplementary dimensions are not the result of simple intellectual speculation. On one hand, these dimensions are necessary to assure the self-consistency of the theory and the elimination of certain undesirable aspects. On the other hand, they do not have a purely formal character - they have physical consequences on our own scale. For example, according to certain cosmological theories, if the universe was associated with a multidimensional space-time at the "beginning" of the Big Bang, the supplementary dimensions will remain hidden and unobservable forever, but their vestiges would be precisely the known physical interactions. In generalizing the example given by particle physics, it is not absurd to think that certain levels of Reality correspond to a specific space-time distinct from that of our own level. Complexity itself will depend on the nature of the space-time. The metaphysical relevance of the supplementary space-time dimensions was already stressed by Abdus Salam[21].

The principle of the infinite identity allows us to approach the difficult problem of *Reason*. Every level of Reality corresponds to a certain *degree of Reason* . The Reason is the entirety of the degrees of Reason. The imbrication and the coherence of the degrees of Reason translates the imbrication and the coherence of the levels of Reality. One can distinguish "degree of Reason" from "the small reason". The small reason results from the contraction of all levels of Reality to just one single level of Reality. The small reason is therefore, at least in the framework of our model, a delirious, irrational and even dangerous reason. Many historical events could be explained as results of the action of the small reason.

Our notion of *degrees of Reason* has to be confronted with the notion of *degrés du savoir* introduced by Jacques Maritain[22]. In any case we can distinguish *reason of knowing* from *reason of understanding*: the reason of knowing settles in us merely as information, whereas the reason of understanding fuses organically with a man's being. It is the reason of understanding in one form or another which could help in developing the dialogue between science and meaning[23].

The definition of Nature I propose does not signify a return to magic thinking or mechanist thinking, because it is founded on the double contradictory assertion: 1) man can study Nature via science; 2) Nature cannot be conceived as independent of man.

I am fully conscious that my model can be criticized, ameliorated and even disputed. However I think that the time has come for beginning true transdisciplinary research, which requires a new definition of Nature. As René Berger says "... nothing is more urgent than to save our imagination from a total surrender, by breathing into it the meaning of respect and wonder"[24].

In conclusion, I have tried, by my definition of Nature, to bring a contribution to the methodological foundations of transdisciplinarity. Transdisciplinarity, without being a new discipline, appears as *the science and art of living Nature*. I tied to show how the foundation of a world without

[21] ABDUS SALAM, *Islam and science*, in *Proceedings of the Venice Symposium on science and the boundaries of knowledge: the prologue of our cultural past*, UNESCO, Paris, 1986, pp. 163-169.

[22] JACQUES MARITAIN, *Distinguer pour unir ou Les degrés du savoir*, Desclée de Brouwer, Paris, 1982.

[23] MICHEL CAZENAVE, *La science et l'âme du monde*, Imago, Paris, 1983.

[24] RENÉ BERGER, *L'effet des changements technologiques*, Pierre-Marcel Favre, Lausanne, 1983, p. 229.

foundations is precisely the absence of foundations. Knowledge appears as a common generation of man and Nature. The famous "conquest of Nature" leads to the vanishing of our own nature. A true cooperation between man and Nature has to replace the murderous folly of the "conquest of Nature". An evolutive future seems to be intimately linked to the formulation of a new Philosophy of Nature, founded on Nature as trans-Nature.

Galileo had the vision of Nature as a text in mathematical language: it is sufficient to puzzle out and read it. This vision, which traversed several centuries, turned out to be of a redoubtable efficiency. However we know today that the situation is much more complex. Nature appears more as a pre-text: the book of Nature has not to be read but written.

A great physicist like Steven Weinberg does not hesitate to pose the problem of the *absurdity* of the universe: "... It is even harder to realize that this present universe has evolved from an unspeakably unfamiliar early condition, and faces a future extinction of endless cold or intolerable heat. The more the universe seems comprehensible, the more it also seems pointless"[25]. For his part, Edgar Morin raises the question of the tragic character of the universe: "Isn't the growing complexity only a detour in the generalized disaster of a universe that is intrinsically and definitively tragic?"[26].

Is the universe absurd? Tragic? Maybe, if one ignores the role of life, of man and his consciousness, of living Nature.

Acknowledgements

I would like to express my warm thanks to the Pontifical Academy of Sciences for its kind invitation to give this general talk.

I thank Michel Camus for his stimulating and pertinent questions and remarks. The interaction with Prof. René Berger, Prof. Antoine Faivre, Prof. Thierry Magnin, Prof. Edgar Morin and Xavier Sallantin was also extremely useful for the formulation of the present ideas. The very stimulating remarks made by Prof. Walter Thirring and Prof. René Thom during this Plenary Session are also gratefully acknowledged. Finally I would like to thank Prof. Elliot Leader for a careful reading of the manuscript and for precious suggestions concerning both the content and the style of this study.

[25] STEVEN WEINBERG, *The first three minutes*, Bantam Books, New York, 1979.
[26] EDGAR MORIN, *La relation anthropo-bio-cosmique*, in *Encyclopédie philosophique*, Presses Universitaires de France, Paris (to be published).

DISCUSSION
(J. Szentagothai, chairman)

Fondi: I admired the profoundness of your exposition and remain very grateful to you for it. The rediscovery of nature in the sense you use the
expression is revolutionary, opens new horizons to science and gives a
solid foundation to the holistic organismic or systemic paradigm which
is gradually substituting the Darwinian evolution paradigm which has
dominated until now. I agree on every point and in particular on the
necessity of adopting the principle of the included third, but as I am an
admirer of Aristotle I see his ideas of actuality and potentiality a good
key in which to read the phenomena taking place at the subatomic level.
From your last slides it would appear that your circle is very similar to
the apogee or the apogeic circle of consciousness or knowledge of
Aristotle. I regret that at this moment an expert on Aristotle's thought
such as Professor Berti is not present. Otherwise I wonder if it is possible
to reconcile the position of the great philosopher on the principle of the
excluded third, with that of Nicolescu.

Pullman: Je dois dire que je suis en désaccord avec une chose que vous avez
dit, à savoir que la doctrine atomique est née d'une peur viscérale du
vide. Non, le vide est une partie volontairement intrinsèque de la théorie
atomique. Les atomes y sont à égalité de signification avec le vide. Tous
les doxographes Grecs qui en parlent précisent explicitement que
Démocrite et Epicure basaient leur conception du monde sur *deux principes* qui sont les atomes et le vide. La théorie atomique devrait s'appeler,
en fait, la théorie des atomes et du vide. La raison principale pour laquelle Aristote a rejeté la théorie atomique, était parce qu'il a rejeté le vide. Il
a trouvé une façon bien meilleure d'éliminer le vide que de faire appel
pour cela aux atomes. Il a tout simplement déclaré que la matière est
continue. C'est une façon bien plus efficace d'éliminer le vide que d'y
introduire des atomes. Si Démocrite avait eu tellement peur du vide il
aurait mieux fait de faire comme Aristote. Le vide est une partie intégra-
le, intellectuellement fondamentale, de la théorie atomique.

Sela: You have enriched us with many notions and opened the "Pandora's
Box" of semantics. I'm not so familiar with how to deal with theories in

philosophy. I know that in science I like a theory if it is, first of all, something which I do not have to take too seriously and secondly, if it can provoke me to plan new experiments and find some parts of science, where I can make new observations through which I can see whether the theory is correct or incorrect. Now I don't know whether this is valid in your case. My question would be: can you, on the basis of the world you have constructed, see a scientific experiment or something of the same kind that could prove or disprove your theories?

GERMAIN: J'ai bien reconnu, effectivement, les concepts qu'Edgar Morin a introduits, mais je dois avouer que je vois les choses assez différemment. Les niveaux de réalité qui ont été mentionnés ici m'apparaissent plutõt comme les modèles mathématiques que nous faisons, pour essayer de comprendre cette réalité, a divers niveaux d'échelle, et les difficultés qui font introduire le tiers inclus me paraissent être, personnellement, la difficulté que l'on a de traduire une situation mathématique dont l'expression est complètement en dehors de notre langage ordinaire, euclidien ou newtonien, et qui alors conduit effectivement à d'apparentes contradictions. Autrement dit, je ne suis pas prêt, sauf évidence plus grande, à accepter ce tiers inclus qui me paraît introduire dans l'esprit des lecteurs des confusions très regrettables. C'est peut-être manque de réflexion, mais je dois dire que j'ai abandonné assez rapidement les voies dans lesquelles Edgard Morin voulait nous entraîner. Je n'ai pas vu effectivement ce que nous pouvions y gagner pour une description scientifique du monde ou pour une approche scientifique de notre univers.

NICOLESCU: Monsieur Pullman, je suis tout à fait d'accord avec vous. Comment pourrais-je dire le contraire de ce que vous dites? Mais, il est bien évident que l'ensemble "vide plus les atomes" est différent du "vide" tout seul. Or ce dont j'ai parlé c'est le vide tout seul, le vide sans atomes, le néant. C'est par peur du néant que les anciens ont inventé l'idee d'atomes. Je donnerai un seul exemple, en citant Epicure: "Parmi les corps il y en a qui sont composés et d'autres dont les composés sont constitués. Ceux-ci sont indivisibles, *si l'on ne veut pas que toutes choses soient réduites au non-être*, mais qu'il reste, après les dislocations des composés, des éléments résistants d'une nature compacte et ne pouvant d'aucune manière être dissous. Donc, les principaux indivisibles sont de *toute nécessité* les substances du corps". (Réf. 1, pp. 98-99). On pourrait citer beaucoup d'autres exemples dans ce sens. L'historien des sciences Robert Lenoble conclut à juste titre que "les 'principes'" de la physique épicurienne ont été conçus pour exorciser le monde des terreurs" (Réf. 1, p. 102). Cette attitude atomiste n'a aucun fondement scientifique, mais plutôt un fondement ontologique. Mon affirmation n'a rien de péjoratif. D'ailleurs les grandes constructions scientifiques ont souvent une motivation culturelle ou sociale — elles n'apparaissent pas d'une manière complètement indépendante de l'imaginaire de l'époqie.

Monsieur Sela, vous posez une question fondamentale concernant l'essence même de la recherche transdisciplinaire. Par définition, la transdisciplinarité n'est pas une nouvelle discipline, car elle étudie le flux d'information qui traverse et va au delà de toutes les disciplines. Son efficacité est d'une nature différente de l'efficacité disciplinaire. Pourtant elle présuppose un ensemble de méthodes et une méthodologie propre ce qui pose effectivement le problème, que vous soulevez, de sa vérification (tout de moins dans le sens de Popper). L'observation capitale est que la transdisciplinarité est une approche à la fois scientifique et culturelle, une approche de type nouveau, qui réclame des moyens de vérification appropriés. La validité de l'approche transdisciplinaire peut être de deux manières différentes. La première consiste dans des découvertes scientifiques majeures, qui sont inspirées par des idées transdisciplinaires. Dans le passé il y a eu plusieurs exemples, dont le plus spectaculaire est la découverte de l'électromagnétisme par Oersted, grâce à l'idée de polarité de la *Naturphilosophie*. Mais, a mon avis, ces découvertes seront extrêmement rares et ne pourront pas, par elles-mêmes, constituer une vérification convaincante de la transdisciplinarité. En revanche, la deuxième manière de vérification — par des études transdisciplinaires à partir d'une discipline bien déterminée — ouvre un fabuleux espace de recherche et de vérification. Le spectre de telles études est très large: de l'information des décideurs à l'accompagnement de mourants, de la formation de formateurs aux problèmes neurophysiologiques qui mettent en jeu l'identité humaine individuelle. L'importance des études transreligieuses et transpolitiques devrait être bien évidente dans la vie de nos sociétés. Car, voyez-vous, l'efficacité transdisciplinaire concerne le problème de la compréhension par rapport au savoir. On peut avoir un savoir extraordinaire et une compréhension très réduite. Aussi, on peut avoir une compréhension extraordinaire et un savoir très réduit. Je crois que la décadence des sociétés est reliée essentiellement au déséquilibre entre savoir et compréhension. Autrement dit, la transdisciplinarité pose le problème de l'unification entre le savoir et l'être. L'efficacité disciplinaire peut être dangereuse si elle n'est pas accompagnée par la compréhension transdisciplinaire. Je dois d'ailleurs avouer que l'un des aspects qui m'ont beaucoup intéressé à cette Session Plénière consacrée à l'étude de la complexité est précisément l'amorce d'un dialogue transdisciplinaire.

Monsieur Germain, je me permets de préciser que j'ai exposé ici mes propres idées. Mais, vous savez bien qu'un vrai scientifique doit toujours citer ses sources. J'ai cité Edgar Morin tout simplement parce qu'il me semble impossible de discuter le sujet de la complexité sans évoquer ses travaux, quelle que soit l'opinion personnelle qu'on puisse avoir sur ce même sujet. En ce qui concerne le tiers inclus, il est dangereux d'interdire l'usage d'une notion parce qu'il peut y avoir malentendu et distorsion de la part d'un esprit sans rigueur. Il faut définir d'une manière rigoureuse cette notion et alors il n'y aura aucun danger. Je dois ajouter que la logique du tiers inclus est formalisable et elle a été même formalisée. Elle nous permet une lecture des phénomènes quantiques qui élimine tous les soit-disant "paradoxes" engendrés, comme vous le dites avec justesse, par une difficulté de traduction en notre langage ordinaire. La logique du tiers inclus se situe pleinement dans le domaine des "logiques quantiques".

LICHNEROWICZ: Les exemples donnés ici ne me convainquent pas et correspondent à quelque chose qui m'ennuie. Ondes et particules, les deux concepts tels qu'ils sont, ils sont pour nous heuristiques, et sont inadaptés à notre analyse et même à la traduction. Nous savons ce que c'est qu'un champ, nous pouvons en tirer, il n'y a en réalité, vous le savez aussi bien que moi, il n'y a ni vraiment onde, ni vraiment particule, il n'y a pas de difficulté mathématique autre que dans notre concept de continuité et de discontinuité. Il n'y a pas de difficulté de séparabilité ou non-séparabilité, un système est séparable ou ne l'est pas, je vois mal les exemples donnés qui nécessiteraient un tiers inclus, dans les exemples donnés. Nous avons la vertu d'avoir un discours imaginatif qui nous porte dans la création qui est de niveau heuristique, où nous pouvons utiliser des particules comme de vagues images et des ondes. Nous avons un niveau vrai, de description mathématique qui en général marche bien dans un certain nombre de cas, et nous essayons de le traduire avec des mots charnels, mais la véritable pensée de la physique, c'est les équations.

NICOLESCU: Une réponse très brève. Je suis complètement d'accord avec vous. Le formalisme mathématique ne pose aucun problème. Mais le problème que nous venons d'évoquer est engendré par la traduction qu'on fait des équations dans le langage de tous les jours.

LICHNEROWICZ: Elle est impossible.

NICOLESCU: Effectivement, les équations mathématiques sont intraduisibles en langage ordinaire, nous le savons tous. Pourtant le physicien essaye et même doit communiquer avec le non-physicien. La preuve: notre présence ici. La vulgarisation est impossible, mais la communication est possible car elle concerne les résultats les plus généraux de la physique, qui se situent à l'interface connu-inconnu. La véritable pensée de la physique est, certes, de ses équations, mais aussi celle des résultats les plus généraux de l'ensemble de ces équations qui font intervenir l'intuition et l'imaginaire. En d'autres termes, c'est la Nature elle-même qui nous oblige de communiquer, car elle n'est pas la propriété exclusive du physicien mais elle s'addresse aussi au biologiste, au poète, à l'artiste, à l'homme religieux ou à l'homme de la rue. L'exemple des pères-foundateurs de la mécanique quantique — Planck, Bohr, Schrödinger, Pauli ou Einstein — doit nous servir de guide sur cette voie. Ils étaient tous de grands communicateurs.

ARECCHI: I was fascinated by your talk and, as you may remember, I myself tried to depict different levels of reality. However, I am perplexed. It is again the same problem. You use a logical tool to make a bridge between two different levels of reality, I've been speaking of metaphoric bridges, where a metaphor is the equivalent of analogy in medieval philosophy, relying on everyday language. You have formalized it into a logic. Now, the main question is the following: if you look at an extrapolation of the two-value logic, you can connect terms provided they belong to the same level. This is familiar for instance in the mathematics of Brouwer. How can you apply this new logic in order to connect concepts which belong to different levels of formalization?

NICOLESCU: Thank you for this question, which is a fundamental one. Here is a quick reply. Let me consider a simple image. Suppose that I represent, say physics, by a sphere, philosophy by a sphere, religion by a sphere, art by a sphere, etc. In this oversimplified picture, the volume of the sphere represents what is known at a given time from the point of view of a given discipline, while the surface of the sphere represents what is unknown. Let us consider now some bridges between spheres (representing isomorphisms, metaphors, etc.). Topologically speaking, it is absolutely clear that if we don't have something which prevents these bridges from encompassing more and more as time goes on, we will finally get a single new discipline which absorbs all the others. Therefore we will have a problem because we know that this process cannot possibly occur. It can be shown that the logic of the included middle precisely prevents this process towards a single hyper-discipline. Nevertheless there are bridges, i.e. a metadiscourse, of course not at the level of equations or of formalizations, but at the level of understanding. This metadiscourse implies the elaboration of a new language possessing a symbolic nature, which will coexist with the mathematical language of the hard sciences.

DEL RE: I completely accept the notion that reality is made up of levels, which are also levels of complexity. Personally, I tend to see the scale of increasing complexity as going from the submicroscopic to the macroscopic. Namely, we can think about: a) a low complexity level, corresponding to many elementary parts, each with few properties; b) a medium complexity level, corresponding to few parts, each with many properties; c) a maximum complexity level, corresponding to unity, associated with many more properties.
I have the impression that for you the direction of increasing complexity is just the other way around. Could you comment on this point?

NICOLESCU: You raise an important question which unfortunately I had no time to cover in my talk. The closed loop intersecting all levels of Reality has two possible orientations. I chose one of these orientations as being connected with the process of going from one level of Reality, characterized by a certain number of general laws, to the neighbouring level governed by a *smaller* number of general laws. It is natural to think that such a process is linked to "evolution" and "freedom". The opposite process, "involution", will correspond to the universe orientation of the arrow on the closed loop. There is rich experimental and theoretical evidence that the number of general laws *governing the quantum world is smaller than the number of general laws governing our* macrophysical world. Therefore the "evolutionary" arrow will point from the macrophysical level towards the microphysical one. This fact may seem paradoxical but is a consequence of the very restricted definition of the word "evolution" I adopted here. Of course, the word "evolution" is one of the richest and most ambiguous words in our language. To try to clarify this ambiguity would require not only a few minutes, but an entire workshop. By its very nature, "evolution", which is a way of expressing ordered complexity, is a transdisciplinary concept.

FIRST GENERAL COMMENTARY

MICHAEL SELA

The Weizmann Institute of Science, Rehovot, Israel

I am delighted that it is another life scientist, Professor Lejeune, who has the task of summarizing the meeting, whereas I have the liberty of just reflecting on it.

First of all, it has been a true intellectual challenge. I have the feeling that I have learned a lot, but I have been made clearly aware of the limitations of our knowledge, and I have found myself also clearly disagreeing with some conclusions.

The wondeful thing about scientific conquests — in contrast to territorial conquests — is that in a battle, when one side gains territory, the other side loses it, whereas in science, when the "known" is expanded, the predictable "unknown" also increases.

Definition of notions is important, and semantic discussions surfaced again and again. Actually, in a dictionary, the definition of complexity does not reflect the complexity of our deliberations. It says: "Complexity is the state of being complex", and "Complex" means: composed of two or more related parts; not simple.

We have also heard about systems, and theory of systems. Indeed, one way to try to cope with complexity is to systematize it. But here I must recollect what Ivan Turgueniev wrote in a letter to Leo Tolstoy in January 1857. I quote:

"Systems attract those who do not succeed in capturing truth in its totality and who want to catch it by the tail. The system is like the tail of the truth itself, it is like a lizard: it leaves its tail in your hands and runs away knowing that rapidly it will grow a new tail".

This reminds me of the famous expression of Niels Bohr, who said: "The opposite of a correct statement is a false statement. But the opposite of a profound truth may well be another profound truth". Of course, it is sometimes difficult to decide whether we deal with a statement or a truth.

I enjoyed Professor Berti's lecture and the discussion that followed very much. I would like to reiterate my remark that I found fault with his nomenclature of "ordre mécanique" and "ordre biologique", both because his use of "biological" had little to do with biology, and because all the phenomena are subject to the same laws of physics.

We discussed the life sciences relatively little, which might give the impression that they are less complex. It is true that I consider the problem of the origin of life - and this was discussed explicit here - more amenable to ultimate understanding than the problem of the origin of the universe, but I assure you that all the processes of life are extremely complex, and I shall allude to a few examples in a short while.

We all want to understand complexity in science, and to some extent to simplify it. But in some cases we learned about how complexity can be useful. Adi Shamir said that for cryptology, complexity is essential, and we also heard from Professor Rao how he puts complexity to good use in material science.

I enormously appreciated the fascinating lecture of Professor Peter Raven on the history and diversity of life on earth. We - in biology - do not call it "complexity", we call it "diversity", and we cherish it and we want to defend it.

Sir John Eccles has already discussed the complexity of the brain, and this is by far the most complex biological system. I would now like to mention the immune system, probably the most complex after the brain, and at this stage, definitely the best known and understood. In a man it contains 10^{12} cells and 10^{20} molecules (antibodies). The antibodies have a capacity for tremendous diversity, each class of immunoglobulins (Ig) being composed of four segments: V (variable), J (joining), D and C (constant). The generation of diversity occurs both through germ line and somatic mutations, and leads to a huge "Ig family".

The specific receptors on the T cell are as important as antibodies. Like the antibodies they are composed of two chains. Like antibodies, they contain variable, joining and constant segments, and thus the T cell receptor family belongs to the "Ig superfamily". Other examples of this superfamily include the so-called class I and class II antigens of the major histo-compatibility complex, which are crucial in the recognition of the processed

T cell epitopes, and their presentation to the T cell receptor.

There are many more members of the Ig superfamily, as well as other families of macromolecules crucial for the immune response, and for the biological phenomena. In the last decade, the world of cytokines, including lymphokines, interleukins, and interferons, as well as the world of growth factors, oncogenes, etc. has been successfully investigated, and may provide crucial help for our health in the future. It is my feeling that in the nineties the main excitement in this area of research will move to the adhesion molecules, molecules that permit comunication between cells.

I would now like to discuss another aspect of biological complexity. In identical twins diabetes occurs only in 33% of the cases in both siblings. Thus, environmental facts increase diversity orders of magnitude above the genetic variability and complexity.

If one can make a comparison with learning and books, I would say that, if one compares the size of all genetic information to the Bible, all learning due to environment may be compared to the contents of the U.S. Library of Congress. Environment increases complexity but it may also decrease it, e.g., by regulation. What I mean is that not all the information (genetic information) is expressed, thanks to controls (regulation). And, of course, the breakdown of regulation leads to cancer.

"It is now apparent that the development of cancer is a highly complex process with many possible causes, which include dietary and environmental factors and inherited and somatic mutations. The proposal that mutations may underlie cancer is not entirely novel; a genetic basis for cancer has been hypothesized for nearly a century, and support for this hypothesis has been obtained from familial, epidemiological, and cytogenetic studies. Only in the past decade, however, have powerful molecular biological techniques been used to provide direct evidence for a genetic basis for cancer through identification of both inherited and somatic (arising in non-germ cells during a patient's lifetime) mutations in human cancers. A preliminary view is that cancer results, at least in part, from mutations in two different types of genes - oncogenes and tumor suppressor genes. In general, wild-type alleles of cellular oncogenes function in a positive fashion in the normal growth regulatory circuits of the cell. Cellular oncogenes can act in a dominant fashion to promote tumor development when mutated or deregulated in their expression. In contrast, tumor suppressor genes normally function as negative regulators of cell growth, or perhaps transduce negative growth regulatory signals. Inactivation of tumor suppressor genes by either inherited or somatic mutations can result in loss of growth control and subsequent tumor formation"[1].

I would like to conclude these reflections by stating that complexity often increases in parallel with knowledge. This is apparent in our progress in physics and certainly in the developments since breaking the genetic code; and these are just two examples.

With this I would like to express my appreciation to all the speakers for their presentations, and for the depth of their convictions, and to end my commentaries here.

---

[1] E.R. FEARON and P.A. JONES: *Progressing toward a molecular description of colorectal cancer development*. FASEB Journal, 6, 2783-2790 (1992).

SECOND GENERAL COMMENTARY

STANLEY L. JAKI

Seton Hall University, Princeton, N.J., U.S.A.

Let me begin with some reflections on the title of this plenary session. A cursory look should make it very obvious what is meant by physics, chemistry, biology, and mathematics. But to some of their most illustrious practitioners those venerable fields of inquiry proved to be time and again a source of profound perplexity.

As to mathematics, it should be enough to think of Bertrand Russell's mathematics with the impression that it rested on foundations too complex to be solid. Kurt Gödel was still to come. Since his famous paper of 1931, mathematics seems to have lost its highest virtue, certainty, in the measure in which more and more complex branches have been growing on its ancient trunk. The trunk, arithmetic, was found to carry in it the virus of undecidability, a virus the more pervasive the more room given to it by the growth of new branches.

An even more serious loss began to threaten physics as it came to grips with the rich complexity of atoms, nuclei and of particles, all dubiously called fundamental or elementary. The threat looms large in a remark of Niels Bohr that physics is merely the talk of physicists of what other physicists say of physics. What then remains of that very matter, control of which is the uppermost ambition of physicists? This problem is more far-reaching than it may appear to be. Again, let me refer to Bertrand Russell, who in his *Autobiography* blamed modern physics for his loss of faith in physical reality. Without agreeing with his reasoning, I merely register the fact that the complexity, indeed paradoxical complexity of modern physics is all too often taken as a justification for doubting physical reality. That chemistry may be threatened in its very identity may be gathered from Max

Born's remark made in 1931, in the concluding page of his *Atomic Physics*, that with the conquest of the atom physics had taken over chemistry.

Biology too may face the possibility of losing its principal objective or subject matter, life, as it gains control of increasingly complex units or parts of living systems. To be sure, biochemistry and biophysics are far from fully accounting for the working of a single cell, whose complexity may surpass a vast chain of advanced industrial plants.

And what about the complexity of the interaction of millions and billions of such cells making up a non-primitive living organism? But, for the sake of argument, let as assume that all the process taking place in a worm, together with all its wriggling, will be fully described in terms of biophysics and biochemistry. The description would consist of rows upon rows of chemical chains and mathematical functions, say Fourier series, comprising data corresponding to molecular energy levels. A marvellous accomplishment which, however, will fail to convey one fact: the immediately perceived simple reality of life.

With all this in mind, one may look with some trepidation at the word, complexity, so innocuous-looking either by itself or in the title of this Plenary Session. It is tempting to think that complexity is an all too obvious notion or phenomenon. At the end of this most instructive and stimulating Session I am still to be enlightened by a broadly, perhaps universally valid definition of complexity. This is not to be taken for slighting, however indirect, of some efforts to give a limited definition of complexity. But in each case clarity was achieved at the price of making the complex appear deceptively simple. More than the need of economy of thought may be at work whenever a useful definition is given. Definition comes from *de-finire* or from setting limits. Therefore, by economizing on thought one may lose, intentionally or not, part of one's hold on a reality which is invariaby complex as we deal with it.

Unlike the word "complexity", the word "emergence" did not prompt even some limited reflection on what it really stands for. Possibly this happened because we scientists too are part of a world which thrives on words whose significance or meaning is taken for granted. When this happens to a word, it may cause a blindfold rather than a guide.

A brief recall of the history, relatively short history, of the word emergence may put this somber alternative into focus. Certainly the word "emergent" did not emerge in at least the English language in that intangible way which is usually seen nowadays as the meaning of the word. Rather, the word "emergent" was put on the scene by a sudden and almost imperious act. The one responsible for this was George H. Lewes, a now largely forgotten champion of Comte's positivism in England in the 1860s and 1870s[1].

It was a non-positive touch in Lewes' thinking that he was struck by elementary evidence. Evidence is the startling difference between the properties of given constituents and the properties of their compounds. The properties of hydrogen and oxygen are very different indeed from the properties of the result, water, obtained by mixing the two. The same can be said about the difference between nitrogen and oxygen and the result of their mixing, nitric acid.

Had Lewes been duly impressed by these and countless other differences between constituents and compounds, he might have extricated himself from the shackles of positivism. But those differences merely served him as a means of restating his positivist faith. He declared that ultimately science would fully account for all the properties that are novel in compounds with respect to the properties of their constituents. Then, again in a vein reminiscent of his hero, Auguste Comte, Lewes thought that by offering a new word the novelty would cease to be the kind of philosophical problem which is a perpetual thorn in the side of still sensitive positivists. We need not worry about insensitive positivists. The new word was "emergent". Lewes called emergent all compounds, or results of changes, whose coming about science had until then failed to trace with exactness but would certainly do so eventually. Lewes called "resultant" all changes that had been cleared up by science. This meant that emergence was, for its first user, no different from a purely mechanical interaction which, whatever it may do, certainly cannot account for a genuinely novel outcome. But is not the enjoyment of a cup of fresh water novel compared to whatever delight there may be in tasting a thimbleful of mere hydrogen and oxygen?

Within a generation or so the word "emergent" was assigned the role of overcoming reductionism. Careful readers of Morgan's Gifford Lectures, *Emergent Evolution*, will not miss his deliberate reluctance to see anything really over and above biochemical processes. To Morgan's studied vagueness, followed by many who began to celebrate emergent holism, I prefer the blunt stance of Lewes. The latter can at least be sized up for what it is, the former allows only for verbal shadowboxing.

---

[1] For a detailed discussion of Lewes' use of the word "emergent", see my Farmington Institute Lectures (Oxford), *The Purpose of It All*, Edinburgh: Scottish Academic Press, 1990, pp. 132-35.

The same is also true of unreflecting reliance on such words as probability, randomness, chance, to say nothing of its neo-Greek variant,
stochastic, let alone the word chaos. And if this claim may seem sacrilegious.
I might add in defense in all appearance a very blasphemous list compiled
not by me but by no less a theoretician of quantum mechanics than the late
J.S. Bell: "... however legitimate and necessary in application, the following
words have no place in a formulation with any claim to physical precision:
system, apparatus, environment, microscopic, macroscopic, reversible,
irreversible, observable, information, measurement. ... On this list of bad
words ... the worst of all is 'measurement'."[2]

A strange world of ghosts seems indeed to thrive behind these
specifically hallowed words. It would be dissipated by stressing the question:
What *is* chance? Is it anything, that is, anything more than a label for our
ignorance of what really takes place? Is it really more than a mere label put
on our inability to accurately describe the parameters of an interaction? In
all the scientific contexts I have been able to peruse so far I have found many
vague definitions of chance or probability. I am not alone. Twenty years ago,
P.G. Bergmann, a well-known student of probability in mathematics and
physics, made this confession: "Often we use the term probability in a very
loose and perhaps indefensible fashion"[3].

Vague definitions, precisely because they are such, do not constitute a
threat to the claim that chance is really a word for ignorance. An old
observation indeed. Hippocrates is said to have observed: "Chance, when a
close look is taken at it, is found to be nothing... There is no evidence that
chance could exist in nature, it is only a word". That chance was something
more, was the claim of Henri Poincaré: "You ask me to predict the
phenomena that will be produced. If I had the misfortune to know the laws
of these phenomena, I could not succeed except by inextricable calculations,
and I should have to give up the attempt to answer you; but since I am
fortunate to be ignorant of them, I will give you an answer at once. And,
what is more extraordinary still, my answer will be right. Chance, then, must
be something more than the name we give to ignorance"[4].

One should not be overimpressed by that prospect. It is the phantom of
confusion. Poincaré confused two outcomes. Had he been able to carry out
our first set of calculations, it would have contained information not only

---

[2] J.S. BELL, *Against Measurement*, Physics World 3, 34 (Aug. 1990).
[3] P.G. BERGMANN, *The Nature of Time*, Cornell University Press, 1967, p. 109.
[4] H. POINCARÉ, *Science and Method*, 1908; tr. F. Maitland; New York: Dover, n.d., p. 66.

about the ensemble, say, of molecules, but also about the position, speed, and energy of each molecule at any moment. It was that latter result that he could not obtain by resorting to probability methods. In other words, instead of yielding something different, or more, those methods yielded less and left one in total ignorance about the position and speed of individual molecules. That a great mind like Poincaré could ignore this big difference, relating to reality, may have its explanation in his being an idealist in the extreme as revealed by his saying: "Tout ce qui n'est pas pensée est le pur néant"[5].

Undue fondness for probability, randomness, and chaos may appear in its true nature when set against some obvious and 'hard' details of statistical or probabilistic physics, either classical or quantum mechanical. By 'hard' details I mean some very hard grounds to which, it seems to me, no commensurate attention is being paid. Let us take, for instance, Boltzmann's statistical gas theory. Its predictive strength concerning the behavior of the ensemble of a large number of molecules depends on the postulate that there is nothing statistical, not even variable, in the physical parameters of individual molecules. Boltzmann supposed them to be invariably perfectly spherical, invariably perfectly elastic, invariably subject to invariable laws of motion. Boltzmann, however, admitted this only in a cursory way in the opening pages of his great classic work. Had he made that admission with proper emphasis, far fewer of his readers would have become overimpressed by the picture which brings his discussion to a close, There one finds randomness engulfing the entire universe[6].

Boltzmann's particular aim was to explain the apparent irreversibility of time over a long period of time. He did this by turning "the entire universe" into an aggregate of individual worlds, all with different times. As to the specter of total cosmic confusion thus arising, he tried to dissipate it by letting ignorance separate those worlds from one another: "In the entire universe, the aggregate of all individual worlds, there will in fact however occur processes going in the opposite direction. But the beings who observe such processes will simply reckon time as proceeding from the less probable to the more probable states and it will never be discovered whether they reckon time differently from us, since they are separated from us by eons of time and spatial distances 10^{100} times the distance of Sirius — and moreover

---

[5] A phrase from the eulogy of Sully Prudhomme, delivered by Poincaré on taking his chair in the Académie Française in 1888.

[6] L. BOLTZMANN, *Lectures on Gas Theory*, tr. S.G. Brush, Berkeley: University of California Press, 1964, PP. 446-48.

their language has no relation to ours". But if all those universes had mutually unrelated languages, why could Boltzmann assume that his language (be it mathematical) was valid throughout all such universes?

Having been accustomed for over half a century or so to seeing the size of the universe in terms of 5 to 15 billion (10^9) light years, we may be stunned by Boltzmann's suggestion that the mere separation among universes should be of an incredibly larger unit of light years, or 10^{100} times the distance of Sirius, which is seven light years. And that is merely the separation between two of those universes. What is the total measure of all those separations? Boltzmann left us in the dark whether he had in mind an infinite Euclidean universe which is, of course, the worst kind of conceptual chaos, a contradiction in terms.

Those who propose a statistical universe on a quantum mechanical basis are equally cagey on a crucial point: Is the number of universes, out of which our universe with its laws is supposedly but one, infinite or not? If the number is infinite the question arises whether it is actually possible to have realized infinite quantity. If the number is not infinite, there has to be a cut off point which determines the actual ensemble of pre-universes. Unless conceptual chaos should engulf scientific cosmology, there too has to be a desk (like in President Truman's White House) where the buck must stop. Or should we say that cosmology, which is the most encompassing form of physical science, is not subject to the rule that warns against the pitfalls of infinite regress?

But statistical cosmology has other problems where one does not have to contend with objection against the strict conclusiveness of the impossibility of an actually realized infinite quantity or number. Brave talk about a huge ensemble of universes, all with statistically different physical properties, has still to show some measure of consistency. Is it possible to assume that the speed of light is absolutely constant only in our universe but in one or many of the countless other universes implied in the inflationary scenario? If indeed one is to be consistently statistical why not engage in similar, possibly very repulsive speculation about Planck's quantum of action and the electric charge? Why not speculate about a large number of different kinds of electricities, or at least electric forces that obey laws other than the Coulomb force? And why not speak of countless different statistical theories, one for each particular universe?

Respect for consistency may demand some other, less painful reflections. In some types of relativistic cosmologies universes are imagined as floating across the horizon into our universe. But if they can "float" into *our* universe, must not they be governed by exactly the same laws as our universe? And if such is the case, should or can they be considered *other* universes?

Unfortunately, it is becoming increasingly fashionable to subject even the universe, the totality of physical realities, to statistics, landing it thereby in a conceptual chaos. The fashion is a scientifically coated philosophical fallacy. It is the ultimate price paid for a fallacy which might not have been so readily accepted had it not first been proposed in one of the most important scientific papers published in this century. I mean Heisenberg's paper of 1927 which contains his indeterminacy formula, a most beautiful scientific result. However, the paper ends on a strictly philosophical note, the ultimate significance of the formula, so Heisenberg stated, that "the principle of causality has thereby been definitively disproved".

To attach such a significance to the formula is the kind of error which the Greeks of old called *metabasis eis allo genos*, a very elementary error in logic. This error has been pointed out in the pages of Nature, Dec. 29, 1930, by a well-known British philosopher, J.E. Turner, Professor of Philosophy at the University of Liverpool. The error he specified may be phrased as follows: "An interaction that cannot be measured exactly, cannot take place exactly"[7]. The error is the simultaneous use of "exactly" in two very different senses: One is operational (cannot be measured exactly), the other is ontological (cannot take place exactly).

Reference has been made in one of the papers to the butterfly effect in phenomena studied in meteorology. It is now illustrated in the philosophical history of quantum mechanics. The ontological uncertainty originally implied in Heisenberg's philosophical error is now enveloping the entire universe. Why, one may ask, do prominent scientific cosmologists claim that they can produce entire universes *literally* out of nothing and why, to make the farce complete, do they emphasize the word *literally*? Do they mean that they should not be taken, literally? Why cannot they say that they merely refer to the 0 level of energy? It should therefore be most satisfying that Prof. Moshinsky's insightful discussion of that level certainly warned us against taking it for nothing.

The essence of Heisenberg's philosophical error was that he took the

---

[7] I have used this rephrasing of Turner's argument for the past twenty years or so. See my *Chance or Reality and Other Essays* Lanham Md.: University Press of America, 1986.

principle of causality (every material event must have an adequate cause) for something that can be observed and measured exactly, in theory at least. Heisenberg overlooked the fact that man can observe only sequences and never causalities. David Hume was perfectly right in emphasizing this. To his credit he also admitted that therefore his philosophy amounted to a heap of bricks with no causal connection among them. He then went on to cultivating mere probabilities.

Once more, and to his credit, he drew all the consequences by touching on the whole that is the universe. It turned into a heap of bricks under the impact of Hume's sceptical probabilism. But this came only some thirty years after Hume had jotted down in his *Treatise on Human Nature* a phrase most relevant to all our discussions. In many of them much was made of random or spontaneous, that is, probabilistic processes giving rise to ever more complex forms and entities with *novel* properties. But it seems to me that the complex, or probable product, has been emphasized at the expense of the simple. It is against this background that I would like to quote David Hume, hardly a reincarnation of Thomas Aquinas, although Aquinas might have approved of Hume's observation: "Probability is founded on the presumption of a resemblance between those objects of which we have had experience and those of which we had none; and therefore it is impossible that this presumption can arise from probability"[8].

The philosophical instructiveness of Hume's works is inexhaustible and so is its scientific applicability. A remark, though very brief, of our colleague, Geraci, deserves recalling. The genes, he noted, are the same in most different organisms although, I would add, one should expect the very opposite on the basis of a consistently universal theory of evolution. Another example would be the absolute identity of gluons, of quarks, of nuclei, of electrons, wherever we find them. No physicist would readily speak about a statistical distribution of electron spin or of proton mass. Were he to do so, he would have to look for a non-statistical layer that supports the one where interactions can be cast into a statistical framework.

In other words, the simple is the basis of the complex, the certain of the probable, both in the logical and in the real order. Such a point should have received appropriate attention during this Plenary Session.

Of course, men of science need not be constantly preoccupied with

---

[8] D. HUME, *A Treatise of Human Nature*, London: J.M. Dent, 1911, vol. II, p. 97 (Bk.T, Pt. III, Sec. vi).

philosophical foundations. These may simply be assumed by at least not implying that foundations need not precede the suprastructure. A splendid example of what can be achieved scientifically, with this restriction in mind, was provided by Prof. Rao on supercomplex molecules. The only point to which I would take exception relates to his references to Nature not liking disorder and avoiding it as much possible. I am certain that such references on Prof. Rao's part were merely a customary shorthand perfectly legitimate in itself.

The use of such and similar shorthand expressions has, however, its perils. The shorthand in question has time and again become a sleight of hand whereby Nature was invested with personal powers. The heavy reliance on such shorthand expression may create the illusion that the superstructure of complexity can float in mid-air without resting on solid foundations. This remark, and the last in my reflections, has more than a purely logical objective. In so far as this Academy is the "senatus scientificus" of the Holy See, its role is not an information retrieval system in the sense of listing data. This can now be done by any high class computer. The very etymology of "in-formation" points far beyond mere data to where science and philosophy inform one another by acting together, without compromising their respective methods and objectives. Then the complex will be safeguarded by the simple and the simple will be secured as the complex is investigated.

RESUMÉ

J. LEJEUNE*

Institut de progenèse, Paris

Monsieur le Président, je suis fort déçu qu'il n'y ait pas une discussion, car cela auralt peut-etre eclairé ma lanterne avant d'essayer de faire un résumé, mais puisque vous me donnez la parole je commencerais par une réflexion de Pascal:

"C'est un héritier qui trouve les titres de sa maison. Dira-t-il peut-etre qu'ils sont faux, et négligera-t-il de les examiner?"

J'ai eu l'impression pendant toute cette semaine que nous étions cet héritier. Nous avons les titres de notre maison, chacun la nõtre, chacun notre discipline, et j'ai remarqué avec le plus grand intéret que chacun des orateurs réellement se demandait: "Mais les titres de ma maison sont-ils faux?" et très honnètement il les a examinés devant nous. C'est à la lumière de ce phénomène, observé pendant cette semaine, que je voudrais non pas faire un résumé, il serait tout-à-fait déplacé d'essayer de résumer la "Weltanschauung" de chacun des orateurs qui se sont succédés, mais rappeler simplement quelques faits saillants à propos de l'exposé de chacun.

L'ouverture par Monsieur Berti a été vraiment triomphale. Il a commencé: "Cosmos veut dire beau", et la cosmologie c'est simplement la science que les hommes ont inventée en s'apercevant que le ciel etait beau. Les Grecs étaient vraiment des poètes mais ils étaient en même temps de profonds réa-

---

* Professor J. Lejeune died on 3 April 1994. His extraordinary personality and scientific achievement have been the object of a discourse delivered by Academician B. Pullman at the 1994 Plenary Session of the Pontifical Academy of Sciences (G.D.R.).

listes. C'est vrai que l'homme est le seul être actuellement vivant sur cette planète qui regarde le ciel étoilé pour y découvrir la beauté. C'est une étrange connivence qui est inscrite en nous mais qui est caractéristique des hommes. D'ailleurs Aristote ne s'était pas trompé, *toute science commence par l'admiration*, et c'est peut-être la meilleure définition d'un travail académique: admirer l'objet dont chacun des dépositaires d'une partie de la science nous a fait miroiter quelques facettes.

Monsieur Thirring nous a présenté des formules pour nous montrer comment se faisait ce travail étrange de l'esprit humain qui ose écrire quelques symboles supposés representer un extraordinaire - quel mot emploirai-je - un extraordinaire panorama, pour ne pas parler justement de choses trop compliquées. Cela peut sembler, pour les non-physiciens, un orgueil démesuré de notre espèce. Et pourtant, au-fur et à mesure que se sont developpées les différentes communications, nous avons progressivement compris, spécialement en écoutant l'heureuse formulation de Monsieur Thom, que la mathématique est en quelque sorte le résultat d'une succession d'engagements déontologiques qui édifient une hiérarchie de structures de plus en plus complexe. De ce mode émergent des structures et c'est une sorte de paradigme pour les structures concrètes, celles des outils, celles des organismes, grâce à une veritable implication locale de ces mécanismes d'émergence dans leur substrat. J'ai dit que c'etait une très heureuse formule résumant Monsieur Thom. (Ce était pas pour me vanter, Monsieur Thom a bien voulu me la fournir). La formule est veritablement heureuse.

Ensuite nous avons eu un duo. Pas un vrai duo comme au theâtre, mais un duo l'un après l'autre, de Monsieur Lovasz et de Monsieur Shamir.

Monsieur Lovasz nous a parlé de quelque chose qu'en tant que français j'appellerais fortuit, ce qui vient de l'exterieur, ce qui n'est pas voulu par celui qui l'observe. Nous avons appris grâce à Monsieur Thom que random venait d'une façon de chasser, où, comme autrefois on chassait à la bille-baude, c'est-à-dire à la bille qui saute, là on chassait en randonnée, au hasard, au random, à random, randomness. Ce qui m'a frappé c'est que la définition du fortuit du random est ce qui permet la plus grande efficacité pour crypter un message. La révélation faite par Monsieur Shahir a été que plus la clef est "random", mieux le message est caché. Je n'ai pas entendu nos collègues philosophes se poser des questions sur ce point. Or, ce point est fort étrange, tout au moins pour le nonspécialiste que je suis, puisqu'il soulève deux questions, une évidence sur la géométrie et l'arithmétique, et une autre encore plus évidente tout au moins pour le spécialiste à propos de l'A.D.N. J'ai entendu dire - peut-etre me suis-je trompé -, que si l'on observait les décimales du chiffre π, elles représentaient, prises en n'importe quel endroit de ce nombre immense qu'on a découpé par ordinateur, elles représentaient des séries de nombres qui sont apparemment au hasard. Or, il est certain que le rapport entre le diamètre et la circonférence ne doit rien au hasard mais tout à la précision géométrique et arithmétique. Eh bien, c'est la meme chose dans l'A.D.N. Regardez une longue molécule d'A.D.N. - et ici je fais référence à un autre exposé qui est venu plus tard mais qui s'insère curieusement dans celui de Monsieur Shamir - regardez une longue molécule d'A.D.N. Vous allez voir des endroits associés aux "box" dont on vous a parlé. Vous allez les reconnaître, elles ne sont pas random, c'est des T.A.T.A. box, qu'on reconnaît très bien. Prenez ensuite un segment qui ne ressemble à rien du tout, impossible à déchiffrer par un ordinateur. Eh bien, c'est justement celui-là qui vous donne la formule d'une protéine particulière qui sera régie par la T.A.T.A. box. Autrement dit, si vous regardez l'information de la vie, plus elle vous paraîtra ne rien signifier, ou plus exactement moins elle évoquera des notions déjà connues chez vous et plus elle aura de signification! Peut-être y a-t-il là matière à réflexion pour déchiffrer une partie du message de la nature qui justement est d'autant plus signifiant qu'il paraît mieux crypté.

Monsieur Richter nous a emmenés dans d'autres domaines, sur les solides de Kepler qui ont toujours pour l'amoureux de la géométrie, - on ne peut pas être généticien sans etre amoureux de la géométrie, j'essaierai de vous expliquer pourquoi dans un instant - un pouvoir de fascination tout-à-fait extraordinaire. J'ai été tres étonné de voir qu'ils ne sont pas si démodés qu'on le disait dans ma jeunesse. Ce qui m'a frappé c'est la solution du problème des trois corps par de très complexes machines qui arrivent progressivement à s'apercevoir que dans certaines régions on est encore tout-à-fait dans le chaos, alors qu'on est tout près cependant d'une solution compréhensible. Ce que je n'ai pas saisi du point de vue mathématique c'est ce qui fait l'impossibilité de comprendre quand on est tout proche de la zone compréhensible, pourquoi est-ce qu'il n'y a pas une zone grise entre les taches qu'ont révélé les ordinateurs? La façon dont l'ordinateur est réglé n'est peut-être pas aussi fine que la façon dont la nature a institué ses lois fondamentales.

Pour M. Arecchi, il nous a montré remarquablement les limitations du langage formel, ou plus exactement du langage formalisé. D'après lui, si j'ai compris le résumé qu'il a bien voulu me donner, aucun point de la science n'est completement décrit par une seule discipline, ce qui fait que chacune a son role mais qu'il faut bien les unir par quelque chose. Ce quelque chose peut aussi bien être les analogies de Thomas ou encore un mot qui n'a plus cours en ce moment dans le milieu scientifique, la métaphysique. Et pourtant c'était un mot si beau. Il était déjà au-dessus de la physique.

Grâce aux remarquables images de M. Rao, j'ai été aussi séduit par la notion que la nature — puisqu'il a employé ce mot nature —, ou tout au moins les formes pseudo-cristallines ou véritablement cristallines ont tendance à mettre en ordre meme les erreurs. C'est assez curieux que ce ne soit pas les atomes qui se déplacent pour compenser les difficultés, mais que ce soit l'erreur elle-même qui essaye en quelque sorte de devenir une sorte de sous-ordre qui s'insère dans l'ordre supérieur. J'avoue que ce qui m'a ravi c'est la tendance des petites particules en s'associant de refaire des icosaèdres. Tout francais sait bien que c'est la seule façon de disposer les miroirs qui permettaient a Cyrano de Bergerac d'aller dans la lune! Et pour nous tous c'est une figure admirable.

Reste la belle démonstration de M. Moshinsky qui nous a appris que dans le fond il suffisait d'injecter un tout petit changement algébrique pour obtenir un changement modéré des courbes si bien que ce qui était auparavant une parabole devenait une hyperbole a deux branches, si j'ai bien compris, l'une étant négative. Cela m'a paru un petit peu plus qu'un résultat mathématique fort important, puisque la branche négative sert ensuite pour les antiprotons, enfin les antimatières, si j'ai un peu compris. Finalement c'était, qu'on me pardonne ce pseudo-jeu de mots, c'était une véritable parabole qu'en lançant trop loin cette hyperbole on découvre brusquement une chose négative qui ne devrait pas exister et qui lorsqu'on la cherche fait comprendre un peu mieux les possibilités cachées dans la nature.

M. Heller nous a parlé des phénomenes créatifs en cosmologie et a essayé de nous faire sentir que pour que l'émergence puisse se manifester il ne faut pas que tout soit trop linéaire et trop strictement prévisible. Il faudrait laisser une sorte de petit chaos, pas trop éloigné de l'ordre, quand même pour qu'il y ait assez de souplesse pour la nouveauté et assez d'ordre pas trop loin pour récupérer un peu d'intelligibilité. J'avoue que j'ai eu un moment de malaise au début de son texte à propos d'un jeu avec une épingle lorsqu'il nous a parlé de Dieu, en soulignant "s'il joue honnêtement". Le mot "honnêtement" m'a un peu choqué, car il sous-entendait, si j'ai compris, se soumettant volontairement à des lois du hasard ou de la randomness. J'avoue que plutôt que de croire l'Etre Suprême "honnête" en ce sens-là, je préfère tout simplement savoir qu'Il soit "bon", même si je suis incapable de Le comprendre.

Finalement nous avons-eu une évocation des galaxies et de leur début. On n'a pas beaucoup parlé de leur fin. M. Rees nous a présenté un diorama extraordinaire. Je n'ai pas besoin de vous rappeler, vous avez tous encore dans l'esprit ou dans les yeux ce qu'il nous a fait voir par les images, et pressentir par sa parole. Mais j'ai retenu un seul mot, qui aurait été je crois pour les anciens Romains une merveilleuse consolation, j'entends ceux qui se faisaient brûler sur un bûcher et dont on conservait précieusement les quelques poussières dans une petite urne funéraire. Il nous a dit qu'il fallait qu'il y ait eu des étoiles dans lesquelles les corps chimiques se sont composés et furent libérés ensuite par leur explosion. Finalement nous sommes faits d'une poudre céleste. Il est profondément vrai, il est consolant, il est éclairant, que cet argile dont nous sommes pétris, ce limon de la terre soit en réalité de la cendre d'étoiles!

Malheureusement nous n'avons pas eu M. Lions et nous ne savons pas encore pourquoi l'environnement de la terre est si stable. Je le regrette beaucoup car en bon gaulois, j'ai toujours eu un peu peur que "le ciel nous tombe sur la tete". Apparemment il tient bon et j'aurais bien aimé savoir pourquoi, mais ça sera pour une autre séance de l'Académie. M. Raven nous a parlé de toute la terre. Oh, il l'a pris de très loin, il nous a raconté toute son histoire, toute l'histoire de la vie survenant sur elle, et nous en a tracé un tableau a la fois émouvant et impressionnant. Et qui nous a rendu bien perplexes sur le fait que nous gaspillons les richesses de notre planète. (Je ne voudrais pas relancer une discussion, il paraît qu'il n'y a plus de discussion ensuite, je ne sais pas si vous en déciderez ainsi, Monsieur le Président). J'ai été très inquiété par la notion qu'il a émis juste en passant que même si l'on ne se souciait pas de trois ou quatre milliards d'hommes qui n'ont pas le train de vie que possèdent les nations dites développées le problème de la pollution serait aussi catastrophique. Je pense qu'il a voulu parler de phénomènes statistiques, enfin de contamination générale du milieu, et non pas d'un phénomène beaucoup plus grave, celui qu'une partie des hommes soit suffisamment aveugle pour ne pas etre sensible à la misère d'une grande partie de leur espèce!

Et nous avons eu après un essai qui par moments - pardonnez-moi le mot - a été pathétique de la part de M. Del Re. Il a essayé d'utiliser une logique solide pour défendre des théories portant sur des entités réagissant entre elles selon des modes incomplètement définis: les origines de la vie sont difficiles a déceler. Pour le généticien que je suis ce fut un secours que nous ayons ensuite M. Geraci nous apportant la notion de signification de séquence. J'ai déja dit tout à l'heure comment ces qualités de l'A.D.N. se rattachent à l'exposé sur la cryptographie que vous avait présenté M. Shamir.

De M. Chauvin, je voudrais dire que j'accueille, hélas, le faire-part qu'il nous a apporté: le Darwinisme a vécu! J'avoue très simplement que c'est ce que j'enseigne depuis longtemps aux étudiants en leur disant que Ptolémée avait construit un remarquable système astronomique avec les épicycles mais qu'on ne l'enseigne plus que dans l'histoire des sciences. Il devrait en être de meme pour le Darwinisme, comme M. Chauvin nous l'a montré par de nombreux exemples d'histoire naturelle. Malheureusement nous n'avons pas encore le Copernic de l'évolution, et comme on n'a pas d'autres modèles à proposer, le modèle mutation-sélection reste pour l'instant la seule simplification qu'on peut offrir aux étudiants. Mais il est au moins sage de leur faire sentir que nous savons parfaitement que ce modèle n'est pas suffisant pour expliquer le peu que nous sachions.

M. Seifert nous a parlé de l'irréductibilité de la vie à des systèmes chaotiques ou non-chaotiques, et j'avoue qu'en l'écoutant me revenait en mémoire la vieille aporie de Démocrite "tout dans la nature est le fruit du hasard et de la nécessité", et en l'écoutant je me disais: "Mais si nous reprenions la vieille aporie, pourquoi ne pas lui donner, maintenant que nous en savons beaucoup plus, un but, une sortie, le contraire d'une aporie en disant: *"la vie c'est l'art de conformer le hasard à nos propres nécessités"*. Nous voyons que ce que fait un être vivant c'est très exactement d'utiliser comme énergie de base la plus dénaturée des énergies, la chaleur. Il faut l'agitation des molécules pour que tous les systèmes puissent éventuellement trier des particules. Nous sommes des machines qui carburent avec l'agitation des molécules, c'est-à-dire que nous sommes de petits moulins dont les ailes sont mues par le temps. Cela est vrai de tous les êtres vivants. C'est même tellement vrai que si l'on arrête le temps apparemment la vie s'arrête, mais cependant ce qui la rendait possible persiste.

Il suffit que la chaleur soit revenue, pour que le temps retrouvé fasse remarquer la machine. Nous le faisons tous les jours en conservant des cellules dans nos laboratoires. Abaisser la température c'est ralentir le temps. Le temps est un concept bien difficile puisque les physiciens s'en servent dans des sens assez divers.

En tant que simple biologiste, j'avoue que je suis toujours très impressionné par les discussions sur des temps très, très petits et très, très lointains, dix puissances moins, je ne sais pas, trente-deux, trente-six ou quarante-quatre, je ne sais pas exactement le chiffre, juste après le premier événement bigbangesque. Qu'est-que ça veut dire? Que signifie une minuscule fraction de seconde à un moment où il n'y avait pas de planètes tournant autour du soleil, donc pas d'années, où il n'y avait même pas d'atomes, donc pas de vibrations qui eussent permis de se référer à des horloges atomiques! Est-que ce concept ne devient pas particulièrement flou lorsqu'il remonte vers l'origine? Mais la façon purement empirique de la vie de mesurer le temps, si je puis dire, par la température (et ce n'est pas une faiblesse de langage, c'est je crois une extraordinaire découverte de la sagesse des hommes que temps et température aient la meme racine latine) est peut-être difficilement extrapolable au-delà de l'univers habituel de la biologie.

Reste ce joyau du logique, cette interdiction d'être à la fois telle chose, et de ne l'etre pas. Dans sa nouvelle quête de la Nature Mr. Nicolescu nous a proposé, entre autres, d'utiliser, avec modération certes, mais impavidement l'éventualité réciproque: *le tiers inclus.*

Il s'agit là d'une inclusion à laquelle le biologiste n'entend guère se risquer. Mais si le rappel des communications faites à notre session n'avait pas mentionné cette intervention remarquable, qu'il me soit permis de réparer sur épreuves cette erreur bien involontaire.

Quant à la complexité, Sir John nous en a parlé et nous en a montré l'image. Notre cerveau est, comme il l'a fort bien dit, certainement, et de beaucoup, la chose la plus complexe de tout l'univers. Ce qui est d'ailleurs tout-à-fait troublant, car lorsqu'on accuse très souvent tous les scientifiques et surtout les biologistes de faire de l'anthropomorphisme, il faut bien reconnaître que nous sommes juchés, si je puis dire, sur l'édifice le plus complexe qui existe véritablement dans l'univers, à moins qu'il y en ait d'autres beaucoup plus loin et que personnellement je ne connais pas. A ce propos je n'ai pas entendu, et j'en aurais été pourtant très friand, de discussion sur les petits martiens verts ou sur les anges. Et pourtant une longue expérience des discussions non pas académiques mais avec des collègues académiciens, m'a amené à observer qu'il n'y avait que deux catégories de scientifiques dans le monde: ceux qui croient aux petits martiens verts ou ceux qui croient aux anges. Je n'ai pratiquement rencontré que ces deux catégories, qui se superposent légèrement dans certains cas. Je n'ai guère jusqu'ici pas entendu l'affirmation absolue: "nous sommes la seule manifestation de la pensée dans l'univers". C'est une observation que je me permets de soumettre à votre jugement. Il est remarquable que pratiquement la notion de pensée soit tellement associée avec la notion d'existence que nous imaginons difficilement qu'il n'y ait pas d'autre pensée dans le monde que la pensée strictement humaine, celle que nous connaissons. Je ne m'étendrai pas sur ce que nous a dit Sir John, qui est un nouveau monument pour tenter de comprendre ce qui est notre intuition immédiate, l'esprit. Comme le disait si bien le philosophe, il n'y a qu'une chose que nous connaissions directement, c'est l'esprit: la matière ne se connait que par notre esprit.

Mais, Monsieur le Président, je risque d'etre trop long et je m'aperçois que l'heure tourne. Essayer de résumer une telle mosaïque m'a forcément pris un certain temps en tentant de ne pas trop déformer la pensée de chacun.

Je me demande à la fin si toute cette semaine n'est pas en quelque sorte un témoignage du fait que la science est trop découpée en disciplines et que chaque scientifique oublie un tout petit peu trop qu'il est un homme. Je m'explique; il me semble que nous devrions, contrairement à tout ce qu'on nous enseigne, être plus antropomorphes que nous ne le sommes, tout au moins quand nous faisons de la science. Je m'explique, et pour m'expliquer je vous demanderai la liberté de prendre trois, quatre minutes pour vous raconter des histoires. Toutes mes histoires sont vraies, comme le disait Marett, mais certaines sont plus vraies que d'autres. Il me semble que nous avons tort d'oublier les circonstances des grandes découvertes, car nous verrions que peut-être les plus savants ne font que redécouvrir sur un point tout-à-fait particulier, la façon dont nous sommes faits.

On raconte, et je l'ai lu dans de très bons livres, que Galilée pour découvrir la chute des corps n'a pas regardé une pendule, car aucune pendule ne pouvait battre assez vite pour mesurer le temps que mettait une bille à rouler sur une glissière. Il avait une craie à la main et quand revenait le temps fort d'une chanson qu'il fredonnait en suivant la bille, il marquait un point sur la glissière. Il est pour un médecin absolument, je dirais, familier, et en meme temps délicieux, de voir que c'est l'organe qui nous apporte la musique, et avec elle le temps - je parle de l'organe de Corti - qui permit à Galilée de mesurer la durée par le rythme d'une chanson. Que la gravité soit découverte à l'aide de la musique la plus simple, celle qu'on fredonne dans un laboratoi-re, fait comprendre peut-être que ce qui nous paraît extrêmement abstrait - et qui l'est réellement - ne peut être perçu que quand nous utilisons toutes les facultés de notre être.

Autre histoire: Newton est sous son pommier. Il voit la lune dans le ciel, une pomme tombe, et voilà qu'il invente les lois de la gravitation. Mais avez-vous réfléchi qu'en été, quand vous êtes sous un pommier et qu'une pomme tombe, vous ne voyez pas tomber la pomme: vous l'entendez! Vous l'entendez qui froisse les feuilles en tombant, et qu'est-ce que vous faites? Bon, vous relevez la tete pour voir d'où vient ce danger imminent, et à ce moment-là votre organe de l'équilibre - dont les trois canaux semi-circulaires ferment l'espace euclidien - vous apprend à la fois la vitesse, l'accélération et la masse, tous ingrédients indispensables pour inventer la formule de la gravi-tation. Newton ne nous a pas menti, il était bien sous un pommier. C'est l'histoire qui nous a été mal racontée, *il entendait* tomber la pomme, tandis qu'il voyait la lune. A côté du nerf cochléaire, qui nous amène la musique, (c'est celui de Galilée), il y a le nerf vestibulaire, celui qui nous indique l'accélération, l'espace (c'est celui de Newton). Il suffisait de les rassembler l'un et l'autre pour concevoir une sorte d'espace lié au temps. C'est l'opéra-tion anatomique à laquelle s'est fié Einstein. Effectivement le nerf acoustique et le nerf vestibulaire sont réunis l'un à l'autre pour constituer le nerf auditif chez chacun d'entre nous. Il suffisait de s'en aviser!

Quand Descartes découvrit la geometrie analytique il eut une vision admirable, et chacun de nous peut s'en donner le spectacle simplement le soir en appuyant sur ses paupières closes pour voir apparaître un échiquier exactement orienté sur la verticale de votre corps et l'horizontale des deux yeux. Cela suffit pour s'apercevoir que la position d'un point se trouve définie par les colonnes et les rangées de l'échiquier. Pour ce faire il suffisait de se frotter les yeux, geste tout-à-fait approprié pour un philosophe qui doute et veut voir s'il a bien vu!

La physique particulaire manque-t-elle au rendez-vous? Pas du tout. Au niveau des neurones, un canal ionique se trouve excité lorsque la vesicule s'est rompue dans l'intervalle synaptique et que le médiateur chimique est arrivé sur la membrane réceptrice. Soudain ce canal ionique engouffre un par un les ions positifs. C'est un compteur de particules d'une extrême vélocité! Nous retrouvons dans ce qui est la pièce opératoire de notre système à mettre de l'ordre dans l'univers un trieur de particules qui dépasse en efficacité le petit démon de Maxwell!

Ces réflexions sont purement physiologiques. Elles ne prétendent absolument pas que la physiologie vous permet d'inventer la science expérimentale, mais elles nous obligent peut-être à ne pas oublier qu'il y a dans la façon dont nous sommes faits une sorte de résumé très succinct de la science.

Si ce résumé ne nous était pas donné par notre nature humaine, la science serait impossible.

J'avais dit pour commencer — et je terminerai bien sûr par le commencement — j'avais dit comme Aristote "toute science commence par l'admiration".

A la lumière de nos travaux au cours de cette semaine, qu'il me soit permis de poursuivre cette pensée:

Toute science commence par l'admiration,

— s'affine par la disputation,

— et s'épanouit par la contemplation.

DAY FIVE
31 OCTOBER 1992

SOLEMN PONTIFICAL AUDIENCE

On 31 October, at the conclusion of the 1992 Plenary Session, Pope John Paul II received the Pontifical Academicians, members of the Curia and of the Diplomatic Corps at the solemn audience held in the Sala Regia in the Vatican Palace. The Holy Father's allocution, which was pronounced in French, is reprinted here both in French and in English, preceded by the addresses of Academician G.V. Coyne SJ, acting president with N. Dallaporta, and of Paul Cardinal Poupard, President of the Commission for the study of Ptolemaic-Copernican controversies established by Pope John Paul II in 1981. In the course of the solemn audience, Professor Adi Shamir received the Pius XI Gold Medal from the Holy Father.

ADDRESS OF THE ACTING PRESIDENT FATHER
G.V. COYNE SJ

Dear Holy Father,

It is always a great honor and joy for the Academicians to be received by you on these annual occasions when we meet for our Plenary Sessions. On this occasion, however, our joy is mixed with some sadness in that we must report to you that our President, Professor Giovanni Battista Marini-Bettòlo, is seriously ill and cannot be here to greet you. He has requested that I bring to you his fondest regards and asks that you would please invoke the Lord's blessing upon him and his family at this difficult time.

In our continuing search to revitalize the Pontifical Academy of Sciences we come before you, when the circumstances require it, with the presentation of new candidates for membership to the Academy. On 18 September last, Holy Father, you graciously nominated four new Academicians and we are honored that today during this Solemn Audience you will bestow upon them the insignia of their membership. They are: Bernardo Maria Colombo of Italy, Minoru Oda of Japan, Wolf Joachim Singer of Germany, and Richard Southwood of Great Britain.

In a further effort to revitalize the Academy we have held a special session to examine anew and in a critical manner the role and functioning of the Academy in modern times, fifty-six years after its refoundation. This is the first such inquiry since the last one carried out in 1937 and will certainly help us to direct our energies into the future.

It is always a great privilege for our Academy to be able to promote research among young astronomers by awarding the Gold Medal established by your predecessor of happy memory Pius XI. This year we are happy that Professor Adi Shamir, a mathematician and specialist in computer sciences of The Weizmann Institute of Science of Israel, will accept this honor from you.

I would like to report to you, Holy Father, that during this past week of our annual Plenary Meetings we have discussed "The Emergence of Complexity in Mathematics, Physics, Chemistry and Biology». This is surely one of the most important developments in modern scientific thought and has many repercussions for philosophy and theology. In dedicating our Plenary Session to this theme we have sought to make significant contributions to the interdisciplinary dialogue. In September the Academy, under the leadership of Academician Sir Martin J. Rees, sponsored a research meeting on the "Epoch of Galaxy Formation", one of the most intriguing and important topics in modern cosmology.

As you well know, Your Holiness has been very closely involved in the work of the Commission which you established in 1981 for the Study of Ptolemaic-Copernican controversies of the 16th and 17th centuries. It is with great interest, therefore, that we await your address in which you wish to bring to a conclusion the work of that Commission. His Eminence Cardinal Paul Poupard, whom you have nominated to coordinate that conclusion, will now speak of the Commission's work.

DISCOURS DE M. LE CARDINAL PAUL POUPARD

Très Saint-Père,

Voici treize ans déjà, en recevant l'Académie Pontificale des Sciences, dans cette même Salle Royale, pour le 1er Centenaire d'Albert Einstein, vous rameniez l'attention du monde de la culture et de la science sur un autre savant, Galileo Galilei[1].

1. *Vous souhaitiez qu'une recherche interdisciplinaire soit entreprise sur les rapports difficiles de Galilée avec l'Église.* Et vous avez institué, le 3 juillet 1981, une Commission Pontificale pour l'étude de la controverse ptoléméo-copernicienne aux XVIe et XVIIe siècles, dans laquelle s'insère le cas Galilée[2], dont vous aviez confié au Cardinal Garrone le soin de coordonner les recherches. Vous m'avez demandé de vous en rendre compte.

Cette Commission était constituée en quatre groupes de travail, avec pour responsables: Son Em. le Cardinal Carlo Martini, pour la section exégétique; moi-même pour la section culturelle; le Professeur Carlos Chagas et le R.P. George Coyne pour la section scientifique et épistemologique; Mgr Michele Maccarrone pour les questions historiques et juridiques; le R.P. Enrico di Rovasenda, secrétaire. *Le but de ces groupes* devait être de répondre aux attentes du monde de la science et de la culture au sujet de la question Galilée, de repenser toute cette question, en pleine fidélité aux faits histori-

---

[1] Discours du Pape Jean-Paul II à l'Académie Pontificale des Sciences le 10 novembre 1979, dans *AAS* t. LXXI, 1979, p. 1464-1465.
[2] Cf. *Edizione Nazionale delle Opere di Galileo Galilei*, dir. Antonio Favaro, Florence, Giunti Barbera, 1890-1909; réimpression, 1929-1939. 20 vol. Cf. Mons. Pio Paschini, *Vita e Opere di Galileo Galilei*, 2 vol., Città del Vaticano, 1964, cité dans *Gaudium et Spes*, 1ère Partie, Ch. III, n. 36, *Juste autonomie des réalités terrestres*, note 7.

quement établis et en conformité aux doctrines et à la culture du temps, et de reconnaître loyalement, dans l'esprit du Concile Œcuménique Vatican II, les torts et les raisons, de quelque côté qu'ils proviennent. Il ne s'agissait pas de réviser un procès, mais d'entreprendre une réflexion sereine et objective, en tenant compte de la conjoncture historico-culturelle. L'enquête fut large, exhaustive, et conduite dans tous les domaines intéressés. Et l'ensemble des études, mémoires et publications de la Commission ont suscité par ailleurs de nombreux travaux en divers milieux.

2. *La Commission s'est posée trois questions: Que s'est-il passé? Comment cela s'est-il passé? Pourquoi les faits se sont-ils passés ainsi?* À ces trois questions, les réponses fondées sur l'examen critique des textes mettent plusieurs points importants en lumière.

L'édition critique des documents et en particulier des pièces émanant de l'*Archivio Segreto Vaticano*, permet de consulter facilement et avec toutes les garanties souhaitables le dossier complet des deux procès et en particulier les comptes-rendus détaillés des interrogatoires auxquels Galilée fut soumis[3]. La publication de la déclaration du Cardinal Bellarmin à Galilée, jointe à celle d'autres documents, éclaire l'horizon intellectuel de ce personnage-clé de toute l'affaire[4]. La rédaction et la publication d'une série d'études ont mis en lumière le contexte culturel, philosophique et théologique du XVIIe siècle[5], et une meilleure compréhension des prises de position de Galilée par rapport aux décrets du Concile de Trente[6], et aux orientations exégétiques de son temps[7], rendant possible une appréciation mesurée de l'immense littéra-

---

[3] *I Documenti del Processo di Galileo Galilei*, a cura di S.M. PAGANO, Pontificiae Academiae Scientiarum Scripta Varia 53, Città del Vaticano 1984. Cfr. M. D'ADDIO, *Considerazioni sui processi a Galileo.* Quaderni della Rivista di Storia della Chiesa in Italia, n. 8, Herder Editrice et Libreria, Roma, 1985.

[4] *The Louvain Lectures (Lectiones Lovanienses) of Bellarmine and the Autograph Copy of his 1616 Declaration to Galileo*, U. BALDINI and P. George V. COYNE, ed., Texts. Commentary and Notes, Studi Galileiani, vol. I, n. 2, Specola Vaticana, 1984.

[5] *Galileo Galilei, 350 ans d'histoire, 1633-1983*, sous la dir. du Cardinal Paul POUPARD, Coll. *Cultures et Dialogue*, n. 1, Desclée International, Paris, 1983; *Galileo Galilei, 350 anni di storia (1633-1983), Studi e Ricerche, Coll. Culture e Dialogo, n. 1*, Piemme, Casale Monferrato (AL), 1984; *Galileo Galilei. Toward a Resolution of 350 years of Debate, 1633-1983*, Dusquesne University Press, Pittsburgh (PA), 1986; *Sprawa Galileusza*, Wybór i redakcja J. ZYCINSKI, Znak, Kraków 1991.

[6] O. PEDERSEN, *Galileo and the Council of Trent*, Studi Galileiani, vol. I, n. 1, Specola Vaticana, 1983.

[7] R. FABRIS, *Galileo Galilei e gli orientamenti esegetici del suo tempo*, Pontificiae Academiae Scientiarum Scripta Varia 62, Città del Vaticano, 1986.

ture consacrée à Galileé, du siècle des lumières à nos jours[8].

Le Cardinal Robert Bellarmin avait déjà exposé dans une lettre du 12 avril 1615 adressée au Carme Foscarini les deux vraies questions soulevées par le système de Copernic: l'astronomie copernicienne est-elle *vraie*, dans le sens qu'elle est *appuyée par des preuves réelles et vérifiables*, ou repose-t-elle seulement sur des conjectures ou des vraisemblances? Les thèses coperniciennes *sont-elles compatibles* avec les énoncés de la Sainte Ecriture? Selon Robert Bellarmin, aussi longtemps qu'il n'y avait pas de preuve de l'orbitation de la Terre autour du Soleil, il fallait *interpréter avec une grande circonspection* les passages bibliques déclarant la Terre immobile. Si jamais l'orbitation terrestre venait à être démontrée comme certaine, alors les théologiens devraient, selon lui, *revoir leurs interprétations* des passages bibliques apparemment opposés aux nouvelles théories coperniciennes, de façon à ne pas traiter de fausses des opinions dont la vérité aurait été prouvée: «Je dis que, s'il était vraiment démontré que le soleil est au centre du monde et la terre au 3e ciel, et que ce n'est pas le soleil qui tourne autour de la terre, mais la terre autour du soleil, il faudrait alors procéder avec beaucoup de circonspection dans l'explication des Ecritures qui paraissent contraires à cette assertion, et plutôt dire que nous ne les comprenons pas, que de dire que ce qui est démontré est faux»[9].

3. *En fait, Galilée n'avait pas réussi à prouver de façon irréfutable* la double mobilité de la Terre, son orbitation annuelle autour du soleil et sa rotation journalière autoir de l'axe des pôles, alors qu'il avait la conviction d'en avoir trouvé la preuve dans les marées océaniques, dont Newton seulement devait démontrer la véritable origine. Galilée proposa une autre esquisse de preuve dans l'existence des vents alizés, mais personne ne possédait alors les connaissances indispensables pour en tirer les éclaircissements nécessaires.

---

[8] *The Galileo Galilei Affair. A Meeting of Faith and Science. Proceedings of the Cracow Conference 1984*, G. COYNE, M. HELLER, J. ZYCINSKI ed., Vatican Observatory Publications, vol. 1, n. 3, 1985; J. ZYCINSKI, *The idea of unification in Galileo's Epistemology, ibid.*, vol. 1, n. 4, 1988; R.S. WESTFALL, *Essays on the trial of Galileo, ibid.*, vol. 1, n. 5, 1989; W. BRANDMÜLLER, *Galilei und die Kirche oder Das Recht auf Irrtum*, Pustet, Regensburg 1982; *Galileo y la Iglesia*, Rialp, Madrid 1987; *Galilei e la Chiesa ossia il diritto ad errare*, LEV, Città del Vaticano 1992.

[9] Lettre du Card. Bellarmin au P. Carme Foscarini, le 12 avril 1615: «... Dico che quando ci fusse vera demostratione che il sole stia nel centro del mondo e la terra nel 3° cielo, e che il sole non circonda la terra, ma la terra circonda il sole, allhora bisogneria andar con molta consideratione in esplicare le Scritture che paiono contrarie, e più tosto dire che non l'intendiamo, che dire che sia falso quello che si dimostra», *Opere di Galileo Galilei, op. cit.*, vol. XII, p. 172.

Il fallut plus de 150 ans encore pour trouver les preuves optiques et mécaniques de la mobilité de la Terre. De leur côté, les adversaires de Galilée n'ont, ni avant lui ni après lui, rien découvert qui pût constituer une réfutation convaincante de l'astronomie copernicienne. Les faits s'imposèrent et firent bientôt apparaître le caractère relatif de la sentence donnée en 1633. Celle-ci n'avait pas un caractère irréformable. En 1741, devant la preuve optique de l'orbitation de la Terre autour du Soleil, Benoît XIV fit donner par le Saint-Office l'«*imprimatur*» à la première édition des *Oeuvres complètes de Galilée*.

4. *Cette réforme implicite de la sentence de 1633 s'explicita* dans le décret de la Sacrée Congrégation de l'Index qui retirait de l'édition de 1757 du *Catalogue des Livres Interdits* les ouvrages en faveur de la théorie héliocentrique. En fait, malgré ce décret, nombreux furent ceux qui demeurèrent réticents à admettre l'interprétation nouvelle. En 1820, le chanoine Settele, professeur à l'Université de Rome «*La Sapienza*», s'apprêtait à publier ses *Eléments d'optique et d'astronomie*. Il se heurta au refus du Père Anfossi, Maître du Sacré Palais, de lui concéder l'«*Imprimatur*». Cet incident donna l'impression que la sentence de 1633 était bien restée irréformée parce qu'irréformable. L'auteur injustement censuré interjeta appel auprès du Pape Pie VII, dont il reçut en 1822 une sentence favorable. Fait décisif, le Père Olivieri, ancien Maître Général des Frères Prêcheurs et Commissaire du Saint-Office, rédigea un rapport favorable à la concession de l'«*Imprimatur*» aux ouvrages qui exposaient l'astronomie copernicienne comme une *thèse*, et non plus seulement comme une hypothèse[10].

La décision pontificale devait trouver son actuation pratique en 1846, lors de la publication d'un nouvel Index mis à jour des livres prohibés[11].

5. *En conclusion, la relecture des documents d'archives* le montre encore une fois: tous les acteurs d'un procès, sans exception, ont droit au bénéfice de la bonne foi, en l'absence de documents extra-processuels contraires. Les qualifications philosophiques et théologiques abusivement données aux théories alors nouvelles sur la centralité du soleil et la mobilité de la terre furent la conséquence d'une *situation de transition* dans le domaine des connaissances astronomiques, et d'une *confusion* exégétique concernant la cosmologie. Héritiers de la conception unitaire du monde, qui s'imposa uni-

---

[10] P. M.B. OLIVIERI, o.p., *Di Copernico e di Galileo, scritto postumo*, Bologna 1872.
[11] Cfr. Pont. Acad. Scientiarum, *Copernico, Galilei et la Chiesa. Fine della controversia (1820). Gli atti del Sant'Uffizio*, di W. Brandmüller e E.J. Greipl, Leo Olschki ed., Firenze 1992.

versellement jusqu'à l'aube du XVIIe siècle, certains théologiens contemporains de Galilée n'ont pas su interpréter la signification profonde, non littérale, des Ecritures, lorsqu'elles décrivent la structure physique de l'univers créé, ce qui les conduisit à transposer indûment une question d'observation factuelle dans le domaine de la foi.

C'est dans cette conjoncture historico-culturelle, bien éloignée de notre temps, que les juges de Galilée, incapables de dissocier la foi d'une cosmologie millénaire, crurent, bien à tort, que l'adoption de la révolution copernicienne, par ailleurs non encore définitivement prouvée, était de nature à ébranler la tradition catholique, et qu'il était de leur devoir d'en prohiber l'enseignement. Cette erreur subjective de jugement, si claire pour nous aujourd'hui, les conduisit à une mesure disciplinaire dont Galileé «eut beaucoup à souffrir». Il faut loyalement reconnaître ces torts, comme vous l'avez demandé, Très Saint-Père.

Tels sont les fruits de l'enquête interdisciplinaire que vous avez demandé à la Commission d'entreprendre. Tous ses membres, par mon intermédiaire, vous remercient de l'honneur et de la confiance que vous leur avez témoignés, en leur laissant toute latitude d'explorer, de rechercher et de publier, dans la totale liberté qu'exigent les études scientifiques.

Daigne Votre Sainteté en agréer le fervent et filial hommage.

ALLOCUTION DE SA SAINTETÉ JEAN PAUL II

Messieurs les Cardinaux,
Excellences,
Mesdames, Messieurs,

1. La conclusion de la session plénière de l'Académie pontificale des Sciences me donne l'heureuse occasion de rencontrer ses illustres membres, en présence de mes principaux collaborateurs et des Chefs des Missions diplomatiques accréditées auprès du Saint-Siège. À tous, j'adresse un salut chaleureux.

Ma pensée se tourne en ce moment vers Monsieur le Professeur Marini-Bettòlo que la maladie empêche de se trouver parmi nous; je forme des voeux fervents pour sa santé et je l'assure de ma prière.

J'aimerais aussi saluer les quatre personnalités qui siègent pour la première fois dans votre Académie; je les remercie d'apporter à vos travaux la contribution de leurs hautes qualifications.

D'autre part, il m'est agréable de saluer la présence de Monsieur le Professeur Adi Shamir, professeur au «Weizmann Institute of Science» de Rehovot (Israël), lauréat de la médaille d'or de Pie XI, décernée par l'Académie, et de lui offrir mes cordiales félicitations.

Deux sujets retiennent aujourd'hui notre attention. Ils viennent d'être présentés avec compétence et je voudrais dire ma gratitude à Monsieur le Cardinal Paul Poupard et au Révérend Père George Coyne pour leurs exposés.

– I –

2. En premier lieu, je désire féliciter l'Académie pontificale des Sciences d'avoir choisi, pour sa session plénière, de traiter un problème de grande

importance et de grande actualité: celui de *l'émergence de la complexité en mathématique, en physique, en chimie et en biologie.*

L'émergence du thème de la complexité marque probablement, dans l'histoire des sciences de la nature, une étape aussi importante que le fut l'étape à laquelle a été attaché le nom de Galilée, alors qu'un modèle univoque de l'ordre semblait devoir s'imposer. La complexité indique précisément que, pour rendre compte de la richesse du réel, il est nécessaire de recourir à une pluralité de modèles.

Ce constat pose une question qui intéresse scientifiques, philosophes et théologiens: comment concilier l'explication du monde — et ceci dès le niveau des entités et des phénomènes élémentaires — avec la reconnaissance de cette donnée que «le tout est plus que la somme des parties»?

Dans son effort de description rigoureuse et de formalisation des données de l'expérience, le scientifique est conduit à recourir à *des concepts métascientifiques* dont l'usage est comme exigé par la logique de sa démarche. Il convient de préciser avec exactitude la nature de tels concepts, pour éviter que l'on ne procède à des extrapolations indues qui lient les découvertes strictement scientifiques à une vision du monde ou à des affirmations idéologiques ou philosophiques qui n'en sont nullement des corollaires. On saisit ici l'importance de la philosophie qui considère les phénomènes aussi bien que leur interprétation.

3. Pensons, à titre d'exemple, à l'élaboration de théories nouvelles au niveau scientifique pour rendre compte de *l'émergence du vivant.* En bonne méthode, on ne saurait les interpréter immédiatement et dans le cadre homogène de la science. Notamment, quand il s'agit de ce vivant qu'est l'homme et de son cerveau, on ne peut pas dire que ce théories constituent par elles-mêmes une affirmation ou une négation de l'âme spirituelle, ou encore qu'elle fournissent une preuve de la doctrine de la création, ou au contraire qu'elles la rendent inutile.

Un travail d'interprétation ultérieure est nécessaire: *c'est précisément l'objet de la philosophie,* laquelle est recherche du sens global des données de l'expérience, et donc également des phénomènes recueillis et analysés par les sciences.

La culture contemporaine exige *un effort constant de synthèse des connaissances et d'intégration des savoirs.* Certes, c'est à la spécialisation des recherches que sont dus les succès que nous constatons. Mais si elle n'est pas équilibrée par une réflexion soucieuse de marquer l'articulation des savoirs, le risque est grand d'aboutir à une «culture éclatée», qui serait en fait la négation de la vraie culture. Car celle-ci ne se conçoit pas sans humanisme et sagesse.

– II –

4. J'étais animé par des préoccupations similaires, le 10 novembre 1979, lors de la célébration du premier centenaire de la naissance d'Albert Einstein, quand j'exprimai devant cette même Académie le souhait que «des théologiens, des savants et des historiens, animés par un esprit de sincère collaboration, approfondissent *l'examen du cas Galilée* et, dans une reconnaissance loyale des torts de quelque côté qu'ils viennent, fassent disparaître la défiance que cette affaire oppose encore, dans beaucoup d'esprits, à une concorde fructueuse entre science et foi»[1]. *Une commission d'étude* a été constituée dans ce but le 3 juillet 1981. L'année même où l'on célèbre le trois cent cinquantième anniversaire de la mort de Galilée, la commission présente aujourd'hui, en conclusion de ses travaux, un ensemble de publications que j'apprécie vivement. Je désire exprimer ma sincère reconnaissance au Cardinal Poupard, chargé de coordonner les recherches de la commission en sa phase conclusive. À tous les experts qui ont participé de quelque manière aux travaux des quatre groupes qui ont mené cette étude pluridisciplinaire, je dis ma profonde satisfaction et ma vive gratitude. Le travail effectué depuis plus de dix ans répond à une orientation suggérée par le Concile Vatican II et permet de mieux mettre en lumière plusieurs points importants de la question. À l'avenir, on ne pourra pas ne pas tenir compte des conclusions de la commission.

On s'étonnera peut-être qu'au terme d'une semaine d'études de l'Académie sur le thème de l'émergence de la complexité dans les diverses sciences, je revienne sur le cas Galilée. Ce cas n'est-il pas depuis longtemps classé et les erreurs commises n'ont-elles pas été reconnues?

Certes, cela est vrai. Cependant, *les problèmes sous-jacents à ce cas touchent à la nature de la science comme à celle du message de la foi*. Il n'est donc pas à exclure que l'on se trouve un jour devant une situation analogue, qui demandera aux uns et aux autres une conscience avertie du champ et des limites de ses propres compétences. L'approche du thème de la complexité pourrait en fournir une illustration.

---

[1] *AAS* 71 (1979), pp. 1464-1465.

5. *Une double question* est au coeur du débat dont Galilée fut le centre.

La première est d'ordre épistémologique et concerne *l'herméneutique biblique*. A ce propos, deux points sont à relever. D'abord, comme la plupart de ses adversaires, Galilée ne fait pas de distinction entre ce qu'est l'approche scientifique des phénomènes naturels et la réflexion sur la nature, l'ordre philosophique, qu'elle appelle généralement. C'est pourquoi il a refusé la suggestion qui lui était faite de présenter comme une hypothèse le système de Copernic, tant qu'il n'était pas confirmé par des preuves irréfutables. C'était pourtant là *une exigence de la méthode expérimentale* dont il fut le génial initiateur.

Ensuite, la représentation géocentrique du monde était communément admise dans la culture du temps comme pleinement concordante avec l'enseignement de la Bible dont certaines expressions, prises à la lettre, semblaient constituer des affirmations de géocentrisme. Le problème que se posèrent donc les théologiens de l'époque est celui de la compatibilité de l'héliocentrisme et de l'Écriture.

Ainsi la science nouvelle, avec ses méthodes et la liberté de recherche qu'elles supposent, obligeait les théologiens à s'interroger sur leurs propres critères d'interprétation de l'Écriture. La plupart n'ont pas su le faire.

Paradoxalement, Galilée, croyant sincère, s'est montré plus perspicace sur ce point que ses adversaires théologiens. «Si l'Écriture ne peut errer, écrit-il à Benedetto Castelli, certains de ses interprètes et commentateurs le peuvent et de plusieurs façons»[2]. On connaît aussi sa lettre à Christine de Lorraine (1615) qui est comme un petit traité d'herméneutique biblique[3].

6. Nous pouvons déjà ici émettre une première conclusion. L'irruption d'une manière nouvelle d'affronter l'étude des phénomènes naturels impose une *clarification de l'ensemble des disciplines du savoir*. Elle les oblige à mieux délimiter leur champ propre, leur angle d'approche, leurs méthodes, ainsi que la portée exacte de leurs conclusions. En d'autres termes, cette apparition oblige chacune des disciplines à prendre une conscience plus rigoureuse de sa propre nature.

Le bouleversement provoqué par le système de Copernic a ainsi exigé un effort de réflexion épistémologique sur les sciences bibliques, effort qui

---

[2] Lettre du 21 décembre 1613, in *Edizione nazionale delle Opere di Galileo Galilei*, dir. A. FAVARO, réédition de 1968, vol. V, p. 282.

[3] Lettre à Christine de Lorraine, 1615, in *Edizione nazionale delle Opere di Galileo Galilei*, dir. A. FAVARO, réédition de 1968, vol. V, pp. 307-348.

devait porter plus tard des fruits abondants dans les travaux exégétiques modernes et qui a trouvé dans la Constitution conciliaire *Dei Verbum* une consécration et une nouvelle impulsion.

7. La crise que je viens d'évoquer n'est pas le seul facteur à avoir eu des répercussions sur l'interprétation de la Bible. Nous touchons ici au *deuxième aspect du problème, l'aspect pastoral.*

En vertu de sa mission propre, l'Église a le devoir d'être attentive aux incidences pastorales de sa parole. Qu'il soit clair, avant tout, que cette parole doit correspondre à la vérité. Mais il s'agit de savoir comment prendre en considération une donnée scientifique nouvelle quand elle semble contredire des vérités de foi. Le jugement pastoral que demandait la théorie copernicienne était difficile à porter dans la mesure où le géocentrisme semblait faire partie de l'enseignement lui-même de l'Écriture. Il aurait fallu tout ensemble vaincre des habitudes de pensée et inventer une pédagogie capable d'éclairer le peuple de Dieu. Disons, d'une manière générale, que le pasteur doit se montrer prêt à une authentique audace, évitant le double écueil de l'attitude timorée et du jugement précipité, qui l'un et l'autre peuvent faire beaucoup de mal.

8. *Une crise analogue* à celle dont nous parlons peut être ici évoquée. Au siecle passé et au début du nôtre, le progrès des sciences historiques a permis d'acquérir *de nouvelles connaissances sur la Bible et le milieu biblique.* Le contexte rationaliste dans lequel, le plus souvent, les acquis étaient présentés, a pu sembler les rendre ruineux pour la foi chrétienne. Certains, dans le souci de défendre la foi, ont pensé qu'il fallait rejeter des conclusions historiques sérieusement établies. Ce fut là une décision précipitée et malheureuse. L'oeuvre d'un pionnier comme le Père Lagrange aura été de savoir opérer des discernements nécessaires sur la base de critères sûrs.

Il faut répéter ici se que j'ai dit plus haut. C'est un devoir pour les théologiens de se tenir régulièrement informés des acquisitions scientifiques pour examiner, le cas échéant, s'il y a lieu ou non de les prendre compte dans leur réflexion ou d'opérer des révisions dans leur enseignement.

9. Si la culture contemporaine est marquée par une tendance au scientisme, l'horizon culturel de l'époque de Galilée était unitaire et portait l'empreinte d'une formation philosophique particulière. Ce caractère unitaire de la culture, qui est en soi positif et souhaitable aujourd'hui encore, fut une

des causes de la condamnation de Galilée. La majorité des théologiens ne percevaient pas *la distinction formelle entre l'Écriture Sainte et son interprétation*, ce qui les conduisit à transposer indûment dans le domaine de la doctrine de la foi une question de fait relevant de l'investigation scientifique.

En réalité, comme l'a rappelé le Cardinal Poupard, Robert Bellarmin, qui avait perçu le véritable enjeu du débat, estimait pour sa part que, devant d'éventuelles preuves scientifiques de l'orbitation de la terre autour du soleil, on devait «interpréter avec une grande circonspection» tout passage de la Bible qui semble affirmer que la terre est immobile et «dire que nous ne comprenons pas, plutôt que d'affirmer que ce qui est démontré est faux»[4]. Avant lui, c'était déjà la même sagesse et le même respect de la Parole divine qui inspiraient saint Augustin lorsqu'il écrivait: «S'il arrive que l'autorité des Saintes Écritures soit mise en opposition avec une raison manifeste et certaine, cela veut dire que celui qui [interprète l'Écriture] ne la comprend pas correctement. Ce n'est pas le sens de l'Écriture qui s'oppose à la vérité, mais le sens qu'il a voulu lui donner. Ce qui s'oppose à l'Écriture ce n'est pas ce qui est en elle, mais ce qu'il y a mis lui-même, croyant que cela constituait son sens»[5]. Il y a un siècle, le Pape Léon XIII faisait écho à ce conseil dans son encyclique *Providentissimus Deus*: «Puisque le vrai ne peut en aucune façon contredire le vrai, on peut être certain qu'une erreur s'est glissée soit dans l'interpretation des paroles sacrées, soit dans une autre partie de la discussion»[6].

Le Cardinal Poupard nous a également rappelé comment la sentence de 1633 n'était pas irréformable et comment le débat, qui n'avait cessé d'évoluer, fut clos en 1820 avec l'*imprimatur* accordé à l'ouvrage de chanoine Settele[7].

10. A partir du siècle des Lumières et jusqu'à nos jours, *le cas Galilée* a constitué une sorte de mythe, dans lequel l'image que l'on s'était forgée des événements était passablement éloignée de la réalité. Dans cette perspective, le cas Galilée était le symbole du prétendu refus par l'Église du progrès scientifique, ou bien de l'obscurantisme «dogmatique» opposé à la libre

---

[4] Lettre au Père A. Foscarini, 12 avril 1615, in *Edizione nazionale delle Opere di Galileo Galilei*, dir. A. FAVARO, vol. XII, pp. 172.

[5] S. Augustin, *Epistula 143*, n. 7; *PL* 33, col. 588.

[6] *Leonis XIII Pont. Max. Acta*, vol. XIII (1894), p. 361.

[7] Cfr. Pontificia Academia Scientiarum, *Copernico, Galilei e la Chiesa. Fine della controversia (1820). Gli atti del Sant'Ufficio*, a cura di W. BRANDMÜLLER e E.J. GREIPL, Firenze, Olschki 1992.

recherche de la vérité. Ce mythe a joué un rôle culturel considérable; il a contribué à ancrer de nombreux scientifiques de bonne foi dans l'idee qu'il y avait incompatibilité entre, d'un côté, l'esprit de la science et son éthique de recherche et, de l'autre, la foi chrétienne. *Une tragique incompréhension réciproque* a été interprétée comme le reflet d'une opposition constitutive entre science et foi. Les élucidations apportées par les récentes études historiques nous permettent d'affirmer que ce douloureux malentendu appartient désormais au passé.

11. On peut tirer de l'affaire Galilée *un enseignement qui reste d'actualité* par rapport à des situations analogues qui se présentent aujourd'hui et peuvent se présenter demain.

Au temps de Galilée, il était inconcevable de se représenter un monde qui fût dépourvu d'un point de référence physique absolu. Et comme le cosmos alors connu était pour ainsi dire contenu dans le seul système solaire, on ne pouvait situer ce point de référence que sur la terre ou sur le soleil. Aujourd'hui, après Einstein et dans la perspective de la cosmologie contemporaine, aucun de ces deux points de référence n'a plus l'importance qu'ils présentaient alors. Cette remarque ne vise pas, cela va de soi, la validité de la position de Galilée dans le débat; elle entend indiquer que souvent, au-delà de deux visions partielles et contrastées, *il existe une vision plus large qui les inclut et les dépasse l'une et l'autre.*

12. Un autre enseignement qui se dégage est le fait que *les diverses disciplines du savoir appellent une diversité de méthodes.* Galilée, qui a pratiquement inventé la méthode expérimentale, avait compris, grâce à son intuition de physicien de génie et en s'appuyant sur divers arguments, pourquoi seul le soleil pouvait avoir fonction de centre du monde, tel qu'il était alors connu, c'est-à-dire comme système planétaire. L'erreur des théologiens d'alors, quand ils soutenaient la centralité de la terre, fut de penser que notre connaissance de la structure du monde physique était, d'une certaine manière, imposée par le sens littéral de l'Écriture Sainte. Rappelons-nous le mot célèbre attribué à Baronius: «*Spiritui Sancto mentem fuisse nos docere quomodo ad coelum eatur, non quomodo coelum gradiatur*». En réalité, l'Écriture ne s'occupe pas des détails du monde physique, dont la connaissance est confiée à l'expérience et au raisonnement humains. Il existe deux domaines du savoir, celui qui a sa source dans la Révélation et celui que la raison peut découvrir par ses seules forces. A ce dernier appartiennent notamment les

sciences expérimentales et la philosophie. La distinction entre les deux domaines du savoir ne doit pas être comprise comme une opposition. Les deux domaines ne sont pas purement extérieurs l'un à l'autre, ils ont des points de rencontre. Les méthodologies propres à chacun permettent de mettre en évidence des aspects différents de la réalité.

– III –

13. Votre Académie conduit ses travaux dans cet état d'esprit. Sa tâche principale est de promouvoir le développement des connaissances, selon la légitime autonomie de la science[8] que le Siège apostolique reconnaît expressément dans les statuts de votre institution.

Ce qui importe, dans une théorie scientifique ou philosophique, c'est avant tout qu'elle soit vraie ou, du moins, sérieusement et solidement établie. Et le but de votre Académie est précisément de discerner et de faire connaître, dans l'état actuel de la science et pour le domaine qui est le sien, ce qui peut être regardé comme une vérité acquise ou du moins comme jouissant d'une telle probabilité qu'il serait imprudent et déraisonnable de le rejeter. Ainsi pourront être évités des conflits inutiles.

Le sérieux de l'information scientifique sera ainsi la meilleure contribution que l'Académie pourra apporter à l'énoncé exact et à la solution des problèmes angoissants auxquels l'Église, en vertu de sa mission propre, a le devoir de porter attention – problèmes qui ne concernent plus seulement l'astronomie, la physique et la mathématique, mais également des disciplines relativement nouvelles comme *la biologie* et *la biogénétique*. Bien des découvertes scientifiques récentes et leurs applications possibles *ont une incidence plus directe que jamais sur l'homme lui-même,* sur sa pensée et son action, au point de sembler menacer les fondements mêmes de l'humain.

14. Il y a, pour l'humanité, *un double mode de développement.* Le premier comprend la culture, la recherche scientifique et technique, c'est-à-dire *tout ce qui appartient à l'horizontalité de l'homme* et de la création, et qui s'accroît à un rythme impressionnant. Pour que ce développement ne demeure pas totalement extérieur à l'homme, il suppose un approfondissement concomitant de la conscience ainsi que son actuation. Le second mode de développement concerne ce qu'il y a de plus profond dans l'être humain quand, trans-

---

[8] Cf. Concile Vatican II, const. past. *Gaudium et Spes*, n. 36, § 2.

cendant le monde et se transcendant lui-même, l'homme se tourne vers Celui qui est le Créateur de toute chose. Cette *démarche verticale* peut seule, en définitive, donner tout son sens à l'être et à l'agir de l'homme, car elle le situe entre son origine et sa fin. Dans cette double démarche horizontale et verticale, l'homme se réalise pleinement comme être spirituel et comme *homo sapiens*. Mais on observe que le développement n'est pas uniforme et rectiligne, et que la progression n'est pas toujours harmonieuse. Cela rend manifeste le désordre qui affecte la condition humaine. Le scientifique, qui prend conscience de ce double développement et en tient compte, contribue à la restauration de l'harmonie.

Celui qui s'engage dans la recherche scientifique et technique admet comme présupposé à sa démarche que le monde n'est pas un chaos, mais un «cosmos» , c'est-à-dire qu'il y a un ordre et des lois naturelles, qui se laissent appréhender et penser, et qui ont par là une certaine affinité avec l'esprit. Einstein disait volontiers: «Ce qu'il y a, dans le monde, d'éternellement incompréhensible, c'est qu'il soit compréhensible»[9]. Cette intelligibilité, attestée par les prodigieuses découvertes des sciences et des techniques, renvoie en définitive à la Pensée transcendante et originelle dont toute chose porte l'empreinte.

Mesdames, Messieurs, en concluant cet entretien, je forme les meilleurs voeux afin que vos recherches et vos réflexions contribuent à offrir à nos contemporains des orientations utiles pour bâtir une société harmonieuse dans un monde plus respectueux de l'humain. Je vous remercie pour les services que vous rendez au Saint-Siège, et je demande à Dieu de vous combler de ses dons.

---

[9] In «The Journal of the Franklin Institute», vol. 221, n. 3, mars 1936.

ALLOCUTION OF THE HOLY FATHER JOHN PAUL II

Your Eminences,
Your Excellencies,
Ladies and Gentlemen,

1. The conclusion of the plenary session of the Pontifical Academy of Sciences gives me the pleasant opportunity to meet its illustrious members, in the presence of my principal collaborators and the Heads of the Diplomatic Missions accredited to the Holy See. To all of you I offer a warm welcome.

My thoughts go at this moment to Professor Marini-Bettòlo, who is prevented by illness from being among us, and, assuring him of my prayers, I express fervent good wishes for his restoration to health.

I would also like to greet the members taking their seats for the first time in this Academy; I thank them for having brought to your work the contribution of their lofty qualifications.

In addition, it is a pleasure for me to note the presence of Professor Adi Shamir, of the Weizmann Institute of Science at Rehovot, Israel, holder of the Gold Medal of Pius XI, awarded by the Academy, and to offer him my cordial congratulations.

Two subjects in particular occupy our attention today. They have just been ably presented to us, and I would like to express my gratitude to Cardinal Paul Poupard and Fr George Coyne for having done so.

I

2. In the first place, I wish to congratulate the Pontifical Academy of Sciences for having chosen to deal, in its plenary session, with a problem of great importance and great relevance today: the problem of *the emergence of complexity in mathematics, physics, chemistry and biology.*

The emergence of the subject of complexity probably marks in the history of the natural sciences a stage as important as the stage which bears relation to the name of Galileo, when a univocal model or order seemed to be obvious. Complexity indicates precisely that, in order to account for the rich variety or reality, we must have recourse to a number of different models.

This realization poses a question which concerns scientists, philosophers and theologians: how are we to reconcile the explanation of the world — beginning with the level of elementary entities and phenomena — with the recognition of the fact that "the whole is more than the sum of its parts"?

In his effort to establish a rigorous description and formalization of the data of experience, the scientist is led to have recourse to *metascientific concepts* the use of which is, as it were, demanded by the logic of his procedure. It is useful to state exactly the nature of these concepts in order to avoid proceeding to undue extrapolations which link strictly scientific discoveries to a vision of the world, or to ideological or philosophical affirmations, which are in no way corollaries of it. Here one sees the importance of philosophy which considers phenomena just as much as their interpretation.

3. Let us think, for example, of the working out of new theories at the scientific level in order to take account of the *emergence of living beings*. In a correct method, one could not interpret them immediately and in the exclusive framework of science. In particular, when it is a question of the living being which is man, and of his brain, it cannot be said that these theories of themselves constitute an affirmation or a denial of the spiritual soul, or that they provide a proof of the doctrine of creation, or that, on the contrary, they render it useless.

A further work of interpretation is needed. *This is precisely the object of philosophy* which is the study of the global meaning of the data of experience, and therefore also of the phenomena gathered and analysed by the sciences.

Contemporary culture demands *a constant effort to synthesize knowledge and to integrate learning.* Of course, the successes which we see are due to the specialization of research. But unless this is balanced by a reflection concerned with articulating the various branches of knowledge, there is a great risk that we shall have a "shattered culture", which would in fact be the negation of true culture. A true culture cannot be conceived of without humanism and wisdom.

II

4. I was moved by similar concerns on 10 November 1979, at the time of the first centenary of the birth of Albert Einstein, when I expressed the hope

before this same Academy that "theologians, scholars and historians, animated by a spirit of sincere collaboration, will study the *Galileo case* more deeply and, in frank recognition of wrongs from whatever side they come, dispel the mistrust that still opposes, in many minds, a fruitful concord between science and faith".[1] A *Study Commission* was constituted for this purpose on 3 July 1981. The very year when we are celebrating the 350th anniversary of Galileo's death, the Commission is presenting today, at the conclusion of its work, a number of publications which I value highly. I would like to express my sincere gratitude to Cardinal Poupard, who was entrusted with coordinating the Commission's research in its concluding phase. To all the experts who in any way took part in the proceedings of the four groups that guided this multidisciplinary study, I express my profound satisfaction and my deep gratitude. The work that has been carried out for more than 10 years responds to a guideline suggested by the Second Vatican Council and enables us to shed more light on several important aspects of the question. In the future, it will be impossible to ignore the Commission's conclusions.

One might perhaps he surprised that at the end of the Academy's study week on the theme of the emergence of complexity in the various sciences, I am returning to the Galileo case. Has not this case long been shelved and have not the errors committed been recognized?

That is certainly true. However, *the underlying problems of this case concern both the nature of science and the message of faith.* It is therefore not to be excluded that one day we shall find ourselves in a similar situation, one which will require both sides to have an informed awareness of the field and of the limits of their own competencies. The approach provided by the theme of complexity could provide an illustration of this.

5. *A twofold question* is at the heart of the debate of which Galileo was the centre.

The first is of the epistemological order and concerns *biblical hermeneutics.* In this regard, two points must again be raised. In the first place, like most of his adversaries, Galileo made no distinction between the scientific approach to natural phenomena and a reflection on nature, of the philosophical order, which that approach generally calls for. That is why he rejected the suggestion made to him to present the Copernican system as a hypothesis, inasmuch as it had not been confirmed by irrefutable proof. Such, however, was *an exigency of the experimental method* of which he was the inspired founder.

---

[1] *AAS* 71 (1979), pp. 1464-1465.

Secondly, the geocentric representation of the world was commonly admitted in the culture of the time as fully agreeing with the teaching of the Bible, of which certain expressions taken literally, seemed to affirm geocéntrism. The problem posed by the theologians of that age was, therefore, that of the compatibility between heliocentrism and Scripture.

Thus the new science with its methods and the freedom of research which they implied, obliged the theologians to examine their own criteria of scriptural interpretation. Most of them did not know how to do so.

Paradoxically, Galileo, a sincere believer, showed himself to be more perceptive in this regard than the theologians who opposed him. "If Scripture cannot err", he wrote to Benedetto Castelli, "certain of its interpreters and commentators can and do so in many ways".[2] We also know of his letter to Christine de Lorraine (1615) which is like a short treatise on biblical hermeneutics.[3]

6. From this we can now draw our first conclusion. The birth of a new way of approaching the study of natural phenomena demands *a clarification on the part of all disciplines of knowledge*. It obliges them to define more clearly their own field, their approach, their methods, as well as the precise import of their conclusions. In other words, this new way requires each discipline to become more rigorously aware of its own nature.

The upset caused by the Copernican system thus demanded epistemological reflection on the biblical sciences, an effort which later would produce abundant fruit in modern exegetical works and which has found sanction and a new stimulus in the Dogmatic Constitution *Dei Verbum* of the Second Vatican Council.

7. The crisis that I have just recalled is not the only factor to have had repercussions on biblical interpretation. Here we are concerned with *the second aspect of the problem, its pastoral dimension*.

By virtue of her own mission, the Church has the duty to be attentive to the pastoral consequences of her teaching. Before all else, let it be clear that this teaching must correspond to the truth. But it is a question of knowing how to judge a new scientific datum when it seems to contradict the truths of faith. The pastoral judgement which the Copernican theory required was difficult to make, in so far as geocentrism seemed to be a part of scriptural

---

[2] Letter of 21 November 1613, in *Edizione nazionale delle Opere di Galileo Galilei*, dir. A. Favaro, edition of 1968, vol. V, p. 282

[3] Letter to Christine de Lorraine 1615 in *Edizione nazionale delle Opere di Galileo Galilei*, dir. A. Favaro, edition of 1968, vol. V, pp. 307-348

teaching itself. It would have been necessary all at once to overcome habits of *thought* and to devise a way of teaching capable of enlightening the people of God. Let us say, in a general way, that the pastor ought to show a genuine boldness, avoiding the double trap of a hesitant attitude and of hasty judgement, both of which can cause considerable harm.

8. *Another crisis* similar to the one we are speaking of, can be mentioned here. In the last century and at the beginning of our own, advances in the historical sciences made it possible to acquire *a new understanding of the Bible and of the biblical world*. The rationalist context in which these data were most often presented seemed to make them dangerous to the Christian faith. Certain people, in their concern to defend the faith, thought it necessary to reject firmly-based historical conclusions. That was a hasty and unhappy decision. The work of a pioneer like Fr Lagrange was able to make the necessary discernment on the basis of dependable criteria.

It is necessary to repeat here what I said above. It is a duty for theologians to keep themselves regularly informed of scientific advances in order to examine if such be necessary, whether or not there are reasons for taking them into account in their reflection or for introducing changes in their teaching.

9. If contemporary culture is marked by a tendency to scientism, the cultural horizon of Galileo's age was uniform and carried the imprint of a particular philosophical formation. This unitary character of culture, which in itself is positive and desirable even in our own day, was one of the reasons for Galileo's condemnation. The majority of theologians did not recognize *the formal distinction between Sacred Scripture and its interpretation*, and this led them unduly to transpose into the realm of the doctrine of the faith a question which in fact pertained to scientific investigation.

In fact as Cardinal Poupard has recalled, Robert Bellarmine, who had seen what was truly at stake in the debate personally felt that, in the face of possible scientific proofs that the earth orbited round the sun, one should "interpret with great circumspection" every biblical passage which seems to affirm that the earth is immobile and "say that we do not understand, rather than affirm that what has been demonstrated is false".[4] Before Bellarmine, this same wisdom and same respect for the divine Word guided St Augustine when he wrote: "If it happens that the authority of Sacred Scripture is set in opposition to clear and certain reasoning this must mean that the person

---

[4] Letter to Fr A. Foscarini, 12 April 1615, cf. *Edizione nazionale delle Opere di Galileo Galilei,* dir. A. Favaro vol. XII, p. 172.

who *interprets Scripture* does not understand it correctly. It is not the meaning of Scripture which is opposed to the truth, but the meaning which he has wanted to give to it. That which is opposed to Scripture is not what is in Scripture but what he has placed there himself, believing that this is what Scripture meant".[5] A century ago, Pope Leo XIII echoed this advice in his Encyclical *Providentissimus Deus:* "Truth cannot contradict truth, and we may be sure that some mistake has been made either in the interpretation of the sacred words, or in the polemical discussion itself".[6]

Cardinal Poupard has also reminded us that the sentence of 1633 was not irreformable, and that the debate, which had not ceased to evolve thereafter was closed in 1820 with the *imprimatur* given to the work of Canon Settele.[7]

10. From the beginning of the Age of Enlightenment down to our own day, *the Galileo case* has been a sort of "myth", in which the image fabricated out of the events was quite far removed from reality. In this perspective, the Galileo case was the symbol of the Church's supposed rejection of scientific progress, or of "dogmatic" obscurantism opposed to the free search for truth. This myth has played a considerable cultural role. It has helped to anchor a number of scientists of good faith in the idea that there was an incompatibility between the spirit of science and its rules of research on the one hand and the Christian faith on the other. *A tragic mutual incomprehension* has been interpreted as the reflection of a fundamental opposition between science and faith. The clarifications furnished by recent historical studies enable us to state that this sad misunderstanding now belongs to the past.

11. From the Galileo affair we can learn a *lesson which remains valid* in relation to similar situations which occur today and which may occur in the future.

In Galileo's time, to depict the world as lacking an absolute physical reference point was, so to speak, inconceivable. And since the cosmos, as it was then known, was contained within the solar system alone, this reference point could only be situated in the Earth or in the Sun. Today, after Einstein and within the perspective of contemporary cosmology, neither of these two reference points has the importance they once had. This observation, it goes without saying, is not directed against the validity of Galileo's position in the

---

[5] Saint Augustine, *Epistula 143*, n. 7; *PL* 33, col. 588.
[6] *Leonis XIII Pont. Max. Acta*, vol. XIII (1894), p. 361.
[7] Cf. Pontificia Academia Scientiarum *Copernico, Galilei e la Chiesa. Fine della controversia (1820). Gli atti del Sant'Ufficio*, a cura di W. Brandmüller e E. J. Griepl, Firenze, Olschki, 1992.

debate; it is only meant to show that often, beyond two partial and contrasting perceptions, *there exists a wider perception which includes them and goes beyond both of them.*

12. Another lesson which we can draw is that *the different branches of knowledge call for different methods.* Thanks to his intuition as a brilliant physicist and by relying on different arguments, Galileo, who practically invented the experimental method, understood why only the Sun could function as the centre of the world, as it was then known, that is to say as a planetary system. The error of the theologians of the time, when they maintained the centrality of the Earth, was to think that our understanding of the physical world's structure was, in some way, imposed by the literal sense of the Sacred Scripture. Let us recall the celebrated saying attributed to Baronius: *"Spiritui Sancto mentem fuisse nos docere quomodo ad coelum eatur non quomodo coelum gradiatur."* In fact the Bible does not concern itself with the details of the physical world, the understanding of which is the competence of human experience and reasoning. There exist two realms of knowledge, one which has its source in Revelation and one which reason can discover by its own power. To the latter belong especially the experimental sciences and philosophy. The distinction between the two realms of knowledge ought not to be understood as opposition. The two realms are not altogether foreign to each other; they have points of contact. The methodologies proper to each make it possible to bring out different aspects of reality.

III

13. Your Academy conducts its work with this outlook. Its principal task is to promote the advancement of knowledge, with respect for the legitimate freedom of science[8] which the Apostolic See expressly acknowledges in the statutes of your institution.

What is important in a scientific or philosophic theory is above all that it should be true or, at least, seriously and solidly grounded. And *the purpose of your Academy* is precisely *to discern and to make known* in the present state of science and within its proper limits, *what can be regarded as an acquired truth* or at least as enjoying such a degree of probability that it would be imprudent and unreasonable to reject it. In this way unnecessary conflicts can be avoided.

---

[8] Cf. Second Vatican Ecumenical Council, Pastoral Constitution *Gaudium et spes* n. 36 par. 2.

The seriousness of scientific knowledge will thus be the best contribution that the Academy can make to the exact formulation and solution of the serious problems to which the Church, by virtue of her specific mission, is obliged to pay close attention, problems no longer related merely to astronomy, physics and mathematics, but also to relatively new disciplines such as *biology* and *biogenetics*. Many recent scientific discoveries and their possible applications *affect man more directly than ever before* his thought and action, to the point of seeming to threaten the very basis of what is human.

14. Humanity has before it *two modes of development*. The first involves culture, scientific research and technology, that is to say *whatever falls within the horizontal aspect of man* and creation, which is growing at an impressive rate. In order that this progress should not remain completely external to man, it presupposes a simultaneous raising of conscience, as well as its actuation. The second mode of development involves what is deepest in the human being, when, transcending the world and transcending himself, man turns to the One who is the Creator of all. It is only this *vertical direction* which can give full meaning to man's being and action, because it situates him in relation to his origin and his end. In this twofold direction, horizontal and vertical, man realizes himself fully as a spiritual being and as *homo sapiens*. But we see that development is not uniform and linear, and that progress is not always well ordered. This reveals the disorder which arrests the human condition. The scientist who is conscious of this twofold development and takes it into account contributes to the restoration of harmony.

Those who engage in scientific and technological research admit, as the premise of its progress, that the world is not a chaos but a "cosmos"; that is to say, that there exist order and natural laws which can be grasped and examined, and which, for this reason, have a certain affinity with the spirit. Einstein used to say: "What is eternally incomprehensible in the world is that it is comprehensible".[9] This intelligibility, attested to by the marvellous discoveries of science and technology, leads us, in the last analysis, to that transcendent and primordial Thought imprinted on all things.

Ladies and gentlemen, in concluding these remarks, I express my best wishes that your research and reflection will help to give our contemporaries useful directions for building a harmonious society in a world more respectful of what is human. I thank you for the service you render to the Holy See, and I ask God to fill you with his gifts.

---

[9] In *The Journal of the Franklin Institute* vol. 221, n. 3, March 1936.

TIPOGRAFIA EDITRICE M. PISANI S.A.S. - ISOLA DEL LIRI (FR) - ROMA